AF670267

Springer Proceedings in Physics **431**

Indexed by Scopus

The series Springer Proceedings in Physics, founded in 1984, is devoted to timely reports of state-of-the-art developments in physics and related sciences. Typically based on material presented at conferences, workshops and similar scientific meetings, volumes published in this series will constitute a comprehensive up to date source of reference on a field or subfield of relevance in contemporary physics. Proposals must include the following:

- Name, place and date of the scientific meeting
- A link to the committees (local organization, international advisors etc.)
- Scientific description of the meeting
- List of invited/plenary speakers
- An estimate of the planned proceedings book parameters (number of pages/articles, requested number of bulk copies, submission deadline).

Please contact:

For Americas and Europe: Dr. Zachary Evenson; zachary.evenson@springer.com
For Asia, Australia and New Zealand: Dr. Loyola DSilva; loyola.dsilva@springer.com

Nat Gopalswamy · Daniela Banyś ·
Sharafat Gadimova
Editors

Proceedings of the United Nations/Germany Workshop on the International Space Weather Initiative

ISWI 2024, 10–14 June, Neustrelitz, Germany

Springer

Editors
Nat Gopalswamy
Heliophysics Code 671
NASA Goddard Space Flight Center
Greenbelt, MD, USA

Daniela Banyś
Institute for Solar-Terrestrial Physics
German Aerospace Center (DLR)
Neustrelitz, Mecklenburg-Vorpommern, Germany

Sharafat Gadimova
United Nations Office for Outer Space Affairs (UNOOSA)
Vienna, Austria

ISSN 0930-8989 ISSN 1867-4941 (electronic)
Springer Proceedings in Physics
ISBN 978-981-95-1120-4 ISBN 978-981-95-1121-1 (eBook)
https://doi.org/10.1007/978-981-95-1121-1

© The Editor(s) (if applicable) and The Author(s) 2026. This book is an open access publication.

Open Access This book is licensed under the terms of the Creative Commons Attribution 4.0 International License (http://creativecommons.org/licenses/by/4.0/), which permits use, sharing, adaptation, distribution and reproduction in any medium or format, as long as you give appropriate credit to the original author(s) and the source, provide a link to the Creative Commons license and indicate if changes were made.
The images or other third party material in this book are included in the book's Creative Commons license, unless indicated otherwise in a credit line to the material. If material is not included in the book's Creative Commons license and your intended use is not permitted by statutory regulation or exceeds the permitted use, you will need to obtain permission directly from the copyright holder.
The use of general descriptive names, registered names, trademarks, service marks, etc. in this publication does not imply, even in the absence of a specific statement, that such names are exempt from the relevant protective laws and regulations and therefore free for general use.
The publisher, the authors and the editors are safe to assume that the advice and information in this book are believed to be true and accurate at the date of publication. Neither the publisher nor the authors or the editors give a warranty, expressed or implied, with respect to the material contained herein or for any errors or omissions that may have been made. The publisher remains neutral with regard to jurisdictional claims in published maps and institutional affiliations.

This Springer imprint is published by the registered company Springer Nature Singapore Pte Ltd.
The registered company address is: 152 Beach Road, #21-01/04 Gateway East, Singapore 189721, Singapore

If disposing of this product, please recycle the paper.

Dedicated to the Memory of Patricia Doherty

Former ISWI Workshop Coordinator

Former director of the ISR at Boston College

Foreword

The International Space Weather Initiative (ISWI), launched in 2009, has enabled scientists to use global navigation satellite system (GNSS) data to study space weather. This data has brought together scientists from various disciplines (such as seismology and the study of the ionosphere and the atmosphere) to work in the field of space weather and have made it possible to apply the fundamental physics of Sun-Earth relations to applications in everyday life, which is of great importance to policymakers.

The *United Nations/Germany workshop on ISWI: Preparing for the Solar Maximum* provided a framework for collaboration between scientific teams that serves as an example of remarkable international work in instrument operation, data collection, analysis, and the publication of scientific results. This work is fundamentally important from the United Nations perspective. It contributes to promoting global, sustainable, and high-quality space weather services for all and ensures that all nations, including those who may not have the national capacity to carry out such work, also benefit.

The ISWI workshop considered advanced technologies and applications in space weather. It highlighted the status of global space weather research, and strategies to improve the collection, exchange, and delivery of space weather data, modelling and forecasting methods, and how to improve accuracy, reliability and interoperability, and how to promote synergies between different communities addressing space weather impacts.

I do look forward to the time when we can share interoperable data that will further our understanding and management of space weather phenomena for the sustainable development of all humankind.

Aarti Holla-Maini

Preface

Space weather is becoming a central topic that requires improved and sustained international coordination to respond to severe space weather events, including improved international data-sharing. The International Space Weather Initiative (ISWI), launched in 2009 under the umbrella of the United Nations, has brought together scientists from various disciplines to work in the field of space weather and have made it possible to apply the fundamental physics of Sun-Earth relations to applications in everyday life, which is of great importance to policymakers. The ISWI instrument network data in conjunction with spaceborne observations has enabled scientists to address wide-ranging problems in space weather from the origin at the Sun to the interplanetary medium and geospace manifestation. ISWI has enabled extensive use of global navigation satellite system (GNSS) data in studies of space weather in the ionosphere and the atmosphere.

These are the proceedings of the United Nations Workshop on the International Space Weather Initiative: *Preparing for the Solar Maximum* that was organized by the Office for Outer Space Affairs, supported by the German Aerospace Center (DLR) and co-sponsored by the International Committee on Global Navigation Satellite Systems (ICG). It was held in Neustrelitz, Germany, from 10 to 14 June 2024, a place with a long tradition in radio wave propagation studies for more than 100 years. DLR has been involved in studying ionospheric processes and modelling for mitigating space weather impact on modern technological infrastructures.

The workshop programme, developed by the Office for Outer Space Affairs, DLR and the National Aeronautics and Space Administration (NASA) in cooperation with an international scientific organizing committee, consisted of technical sessions and discussions of observations and conclusions, followed by closing remarks by the co-organizers and participants. With 85 participants from over 30 countries, the workshop encompassed a total of 18 posters and 52 presentations given at eleven technical sessions, here summarized in seven parts:

1. Solar eruptions, flares, CMEs and stream interaction regions – their sources at the Sun and impact on geospace (magnetosphere, ionosphere, atmosphere, ground);
2. Space weather extreme events and geomagnetic storms;
3. Solar energetic particles and the associated phenomena such as coronal/interplanetary radio bursts;
4. Spacecraft anomaly, impact on GNSS, ionospheric irregularities,
5. Space weather prediction using various techniques including machine learning;
6. Space weather instrumentation; and
7. Operational space weather.

The workshop participants made the following recommendations:

(a) capacity-building and technical guidance should continue to be provided to countries that wish to be engaged in space weather science and education;

(b) opportunities for continued partnerships with capacity-building entities and activities within the United Nations should be further developed;
(c) increased international cooperation should be promoted in order to meet current and future needs for space weather services, and a coordination mechanism should be established, with participation on a voluntary basis, which would be supported by the Office for Outer Space Affairs as appropriate.

All materials, including the Workshop programme and book of abstracts can be found on the Office for Outer Space Affairs web page at: https://www.unoosa.org/oosa/en/ourwork/psa/schedule/2024/2024-iswi-workshop.html

We thank the workshop scientific organizing committee, the Conference Organization Team for their enormous efforts, and participants for their valuable contributions to a successful workshop. We appreciate the time and effort of the reviewers in bringing out these proceedings.

We are grateful to NASA's Living With a Star program for covering the publication cost of these Proceedings. We also thank the Scientific Committee on Solar-Terrestrial Physics (SCOSTEP) and its Scientific Programme on Predictability of the variable Solar-Terrestrial Coupling (PRESTO) for their financial support.

January 2025

Nat Gopalswamy
Daniela Banyś
Sharafat Gadimova

International Space Weather Initiative
ISWI
United Nations / Germany Workshop
Neustrelitz 2024

Organization

Workshop Conveners

Gopalswamy Nat	ISWI/NASA, USA
Banyś Daniela	DLR, Germany
Gadimova Sharafat	UNOOSA, Austria

Scientific Organizing Committee

Fung Shing (Chair)	NASA/GSFC, USA
Cho Kyung-suk	KASI, Republic of Korea
Foullon Claire	U. Exeter, UK
Hurlburt Neal	Lockheed-Martin, USA
Georgieva Katya	BAS, Bulgaria
Kliem Bernard	U. Potsdam, Germany
Moldwin Mark	U. Michigan, USA
Molina Graciela	U. Nacional de Tucumán, Argentina
Nava Bruno	ICTP, Italy
Ostgaard Nikolai	U. Bergen, Norway
Rabiu Babatunde	NASRDA, Nigeria
Rajaguru S. P.	IIA, India
Raulin Jean-Pierre	U. Mackenzie, Brazil
Shiokawa Kazuo	Nagoya U., Japan
Sinnhuber Miriam	KIT, Germany
Skone Susan	U. Calgary, Canada
Sterling Alphonse	NASA/MSFC, USA
Stolle Claudia	Leibniz Institue, Germany
Warmuth Alexander	U. Potsdam. Germany
Yan Yihua	NSSC/CAS, China

Conference Organization Team

Banyś Daniela (Co-chair)
Wenzel David (Co-chair)
Amaral Ferreira Arthur
Böhme Andrea
Bröse Malte
Cahuasquí J. Andrés
Dühnen Hanna
Geib Reik

Heymann Frank
Gadimova Sharafat
Gindler Patrick
Lubyk Kateryna
Mohanty Shradha
Sato Hiroatsu
Scherdin Antje
Schulmeister Robert
Tandler Frank
Vasylyev Dmytro

Reviewers

Adolfs Marjorlijn
Amory Christine
Berdermann Jens
Bröse Malte
Buzulukova Natalia
Coster Anthea
Dasso Sergio
Dresing Nina Dresing
Fung Shing
Gonzales Gilda
Heymann Frank
Hoque Mainul
Gopalswamy Nat
Ishii Mamoru Kliem Bernhard
Kuznetsova Masha
Molina Graciela
Muglach Karin
Muhamed Johan
Munakata Kazuoki
Nava Bruno
Shiokawa Kazuo
Skone Susan
Temmer Manuela
Wimmer-Schweingruber Robert
Xu Long
Yan Yihua

Participants

Akerele Aderonke	Nigeria
Aksonova Kateryna	Czech Republic
Amaral Ferreira Arthur	Germany
Amory Christine	France
Ameen Muhammad Ayyaz	Pakistan
Banyś Daniela	Germany
Beckwith-Chandler William	UK
Berdermann Jens	Germany
Bernhardt Sebastian	Germany
Bothmer Volker	Germany
Bröse Malte	Germany
Buzulukova Natalia	USA
Chakrabarty Dibyendu	India
Chi Yutian	China
Coster Anthea	USA
Del Moro Dario	Italy
Dröge Henrik	Germany

Dusabe Mary	Kenya
Eiselbrecher Florian	Germany
Foullon Claire	UK
Fung Shing	USA
Gadimova Sharafat	United Nations
Galkin Ivan	USA
Gindler Patrick	United Nations
Glover Alexi	France
Gonzalez Gilda	USA
Grodji Oswald Didier Franck	Côte d'Ivoire
Groves Keith	USA
Gopalswamy Nat	USA
Hoque Mainul	Germany
Holla-Maini Aarti	United Nations
Jakowski Norbert	Germany
Kilcik Ali	Türkiye
Kliem Bernhard	Germany
Kraemer Kathleen	USA
Kriegel Martin	Germany
Kuznetsova Maria	USA
Li Ting	China
Lomotey Solomon Otoo	Ghana
Lubyk Kateryna	Germany
Malandraki Olga	Greece
Malatji Mmatjie Sarah	South Africa
Matzka Jürgen	Germany
Miloch Wojciech Jacek	Norway
Molina Maria Graciela	Argentina
Mohanty Shradha	Germany
Muglach Karin	USA
Muhamad Johan	Indonesia
Nah Hyon Jun	Republic of Korea
Ngwira Chigomezyo	USA
Nndanganeni Rendani	South Africa
Pandit Drabindra	Nepal
Picanço Giorgio Arlan da Silva	Brazil
Ramos Yvette	France
Ranasinghe Manjula	Sri Lanka
Rabiu Babatunde	Nigeria
Reuter Wolf J.	Germany
Savic Mihailo	Serbia
Schlüter Stefan	Germany

Schmidt Andreas	Germany
Shimeis Amira	Egypt
Shprits Yuri	Germany
Silva Picanço Giorgio Arlan da	Brazil
Skone Susan	Canada
Sori Takuya	Japan
Stolle Claudia	Germany
Supnithi Pornchai	Thailand
Temmer Manuela	Austria
Tsugawa Takuya	Japan
Török Tibor	USA
Drew Turner	USA
Urbar Jaroslav	Czech Republic
Uwamahoro Jean	Rwanda
Vasylyev Dmytro	Germany
Verronen Pekka T.	Finland
Veselinovic Nikola	Serbia
Viola Massimo	Italy
Wang Wei	China
Wenzel David	Germany
Westrich Lukas	Germany
Wissing Jan Maik	Germany
Xu Long	China
Yan Yihua	China
Yoon Ki Chang	Republic of Korea
Zhu Xiaoshuai	China

Contents

About the Editors

Dr. Natchimuthukonar "Nat" Gopalswamy, Fellow of the American Geophysical Union, the International Science Council, and SCOSTEP, is an astrophysicist at NASA's Goddard Space Flight Center, Solar Physics Laboratory, Heliophysics Science Division. His research is in the area of solar and solar terrestrial physics including coronal mass ejections, solar flares, solar energetic particles, solar radio bursts, interplanetary shocks, and space weather. He is involved in space missions such as SOHO, STEREO, Solar Orbiter, and CLPS. He is the executive director of the International Space Weather Initiative (ISWI), past president of the Scientific Committee on Solar Terrestrial Physics (SCOSTEP), and vice chair of COSPAR's panel on Space Weather. He was conferred with a Doctor Honoris Causa by the Bulgarian Academy of Sciences (2019). He has published more than 475 articles that have been cited more than 27,000 times. Dr. Gopalswamy received his BSc and MSc degrees from the University of Madras. His PhD degree is from Indian Institute of Science Bangalore (1982) and postdoctoral training at the University of Maryland, College Park (1985).

Dr. Daniela Banyś is a physicist at the German Aerospace Center DLR, Institute for Solar-Terrestrial Physics. During her studies in Geophysics, she gained insight into seismic risk analysis and radio wave propagation. Her research area now focusses on ground-based space weather observations via very low frequency up to very high frequency measurements. She is European and National Coordinator of the International Space Weather Initiative (ISWI) and Vice President of the Commitee "Near-Earth Space" of the working group on extraterrestrial research (Arbeitsgemeinschaft Extraterrestrische Forschung, AEF). She develops and operates the Global Ionospheric Flare Detection System (GIFDS), and is its ISWI instrument lead.

Sharafat Gadimova is a scientific programme officer at the United Nations Office for Outer Space Affairs. She leads the United Nations Programme on Space Applications' activities on global navigation satellite systems (GNSS) and space weather. She coordinates the work of an International Committee on Global Navigation Satellite Systems (ICG) and its Providers' Forum. She is a chair of the ICG Working Group on Capacity Building and Information Dissemination. She spearheaded the development of education curriculum in GNSS for the United Nations-affiliated Regional Centres for Space Science and Technology Education. She is Vice Chair of the Committee on Space Research (COSPAR) panel on Space Weather.

Solar Eruptions, Flares, CMEs and Stream Interaction Regions – Their Sources at the Sun and Impact on Geospace (Magnetosphere, Ionosphere, Atmosphere, Ground)

CME Observations from Sun to Impact on Geospace

Manuela Temmer(✉)

Institute of Physics, University of Graz, Graz, Austria
manuela.temmer@uni-graz.at
https://swe.uni-graz.at

Abstract. Our Sun is an active star expelling dynamic phenomena known as coronal mass ejections (CMEs). The magnetic field configuration on the Sun and related solar wind structures affect the propagation behavior of CMEs, dominate its transit time and embedded magnetic field properties when impacting Earth. Since the conditions on the Sun constantly change, the impact of CMEs on the different regimes of geospace is quite variable and may differ significantly from event to event. This short review summarizes the different manifestations of CMEs on the Sun, their appearance in interplanetary space, and how CMEs trigger a cascade of reactions as they interact with Earth.

Keywords: Coronal mass ejections · Solar Wind · Interplanetary Space · Earth · Space Weather research

1 Introduction

Coronal mass ejections (CMEs) are the most energetic events from the Sun, appearing as rapidly moving and expanding magnetic plasma structures. Frequently CMEs are associated with flares (Yashiro et al., 2006), which actively drive CMEs, as well as solar energetic particles (SEPs) which are accelerated by the huge energy release in the wake of the magnetic reconnection process and by the shock that is driven by the CME (Reames, 2013). CMEs, flares, and SEPs are recognized as key factors in investigating Space Weather (e.g., Luhmann et al., 2020; Temmer, 2021; Gopalswamy, 2022). Out of that CMEs are known to cause the most severe Space Weather effects, such as geomagnetic storms that can induce electric currents in power lines, potentially leading to widespread electrical grid failures and damage to infrastructure (Pulkkinen et al., 2017). Intense research for better understanding the physical processes in the complex and interdisciplinary area of Space Weather is therefore essential for safeguarding technology, infrastructure, and human activities both in space and on Earth (see Fig. 1).

There are many open questions, e.g., about the interaction processes between CMEs and the ambient solar wind flow. Especially interaction of CMEs with other large-scale solar wind structures lying in the path of the CME, produces

© The Author(s) 2026
N. Gopalswamy et al. (Eds.): ISWI 2024, SPPHY 431, pp. 3–13, 2026.
https://doi.org/10.1007/978-981-95-1121-1_1

complex interplanetary magnetic field structures significantly enhancing Space Weather effects (e.g., Dumbović et al., 2015). Most of all, we face the fact that the "plain" background solar wind and its embedded structures, such as stream interaction regions and high-speed solar wind streams, are not well represented by state-of-the-art models (see e.g., COSPAR Space Weather Roadmap update by Temmer et al., 2023). This lowers the reliability of Space Weather forecasts, especially during high solar activity phases when interactions might happen frequently (Lugaz et al., 2017).

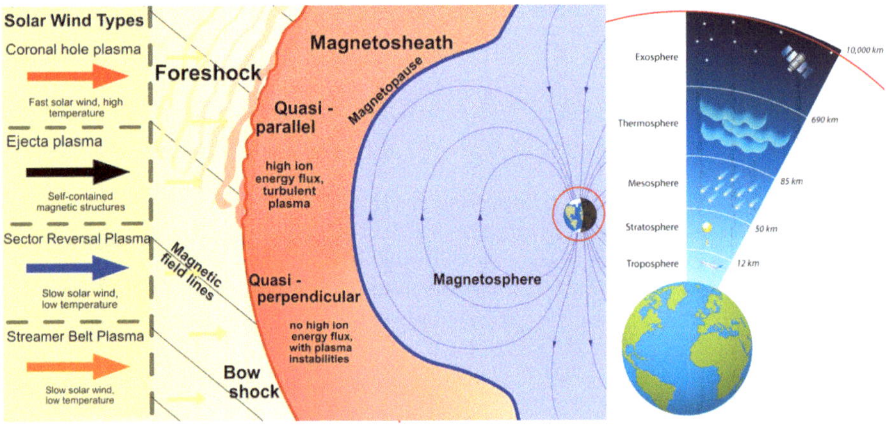

Fig. 1. Left: Graphic representation of the boundary region between the solar wind and the Earth's magnetic field (adapted from Koller et al., 2024). Right: The atmospheric layers of Earth - a zoom in version of the red circle marked in the left panel (source: European Centre for Medium-Range Weather Forecasts).

2 Towards More Reliable Background Solar Wind Models

It is acknowledged that the open magnetic flux (OMF) "shapes" interplanetary space, magnetically connects the Sun to the planets, and dominates the motion of SEPs (e.g., Owens and Forsyth, 2013). Coronal holes (CHs) are assumed to mainly represent the OMF (e.g., Hofmeister et al., 2018). CHs are also the sources of fast solar wind streams which generate, when interacting with the ambient slow solar wind, so-called stream interaction regions (SIRs). A CME encountering a SIR and related high-speed stream may experience acceleration, changes in the magnetic field configuration and direction of motion, resulting in altered transit times and geomagnetic effects at Earth.

Investigating the dynamic interplay between open (fast wind) and closed (slow wind) magnetic field on the Sun is key in deriving answers on how interplanetary space is structured. However, there is a strong discrepancy found when calculating the in-situ OMF and the OMF coming from the solar surface with an underestimation by a factor of two (e.g., Linker et al., 2017). If we assume the accuracy of the in-situ OMF calculation, any unaccounted-for OMF must be sought at the Sun. There might be outflows from active regions that could contribute to the OMF (e.g., van Driel-Gesztelyi et al., 2012). Recent studies derived that the OMF evolution correlates with CH open flux, but not with CH area evolution (Heinemann et al., 2024). Hence, the uncertainty in detecting boundaries of CHs might not be the crucial point in the missing OMF and, hence, might not be the main parameter in improving the performance of background solar wind models (Linker et al., 2021). Rather, the currently missing polar magnetic field information might play a key role. ESA's mission Solar Orbiter (SolO; Müller et al., 2017) will move out of the ecliptic plane starting in 2026. It will deliver high-resolution magnetic field information from the Sun's polar region, which might enable us to get one step closer towards solving the OMF problem and more reliable background solar wind models.

3 Solar Signatures Related to CMEs

The regions on the Sun from which CMEs originate give limitations in the driving force of a CME (Bein et al., 2011). Active regions with strong magnetic fields and complex sunspot configurations are often associated with greater solar eruptive potential (e.g., Sammis et al., 2000). The common ground for the occurrence of CMEs and flares is magnetic reconnection and magnetohydrodynamic instability, however, recent studies highlight that the trigger mechanisms for CMEs can vary (see review by Green et al., 2018).

3.1 Signatures of Eruptive Events

Besides the high association rate between flares and CMEs (Yashiro et al., 2006), there are also well-known active regions producing confined flares of high energy release (e.g., Thalmann et al., 2015). While confined flares can still produce intense bursts of electromagnetic radiation across various wavelengths, they do not lead to CMEs. It is therefore important to be able to distinguish confined from eruptive flares.

For that radio signatures are favorable. Especially the detection of type II and type III radio bursts are important indicators of solar eruptive events. Type II radio bursts are related to fast CMEs, and are indicative of the passage of shock waves through the solar corona and interplanetary space (see e.g., Gopalswamy et al., 2008). Type III bursts are characterized by rapidly drifting frequency signatures as the electron beams propagate away from the Sun along magnetic field lines and are indicative of the opening of the magnetic field. Hence, the observation of type III followed by type II bursts can be used to infer the eruptive and energetic nature of a flare. E.g., for continuously monitoring the Sun's radio emissions, the e-CALLISTO (Enhanced Coordinated Low-cost Low-frequency Instrument for Spectroscopy and Transportable Observatory; Benz et al., 2009) network consists of multiple stations distributed worldwide, each equipped with a radio spectrometer ranging typically from a few tens of megahertz to a few gigahertz.

Before entering the field-of-view of the coronagraph, typically solar surface signatures collectively provide evidence of the occurrence and characteristics of eruptive flare events and CME characteristics. It is therefore important to monitor the Sun especially in EUV or X-rays from which, together with magnetic field information, we can assess the energy storage and release in the corona. The close connection between the energy release of flares and CMEs in various wavelengths gives us an estimate of the driving forces of a CME (Zhang et al., 2001; Temmer et al., 2008). EUV dimming regions and their dynamics tell us about the actual mass release, the CME speed as well as the CME's direction of motion (Dissauer et al., 2018; Chikunova et al., 2023). Propagating surface waves show us the expansion behavior of the CME (e.g., Patsourakos et al., 2020). Post-eruptive arcades can be used as proxy for the magnetic structure of the flux rope and the amount of reconnected flux, frequently used in models (e.g., Scolini et al., 2020). The coronal hole influence parameter (CHIP) determines the influence of nearby CHs on the CME trajectory (Gopalswamy et al., 2009).

4 CME Structures and Their Differences

Focusing on the predictability of Space Weather, we first need to consider the different structures of a CME. In rough terms, we may differentiate between the shock-sheath and the magnetic ejecta of a CME (e.g., Vourlidas et al., 2013). Based on statistical results, the difference in the arrival time of the shock-sheath and magnetic ejecta is on average about 10 h (Russell and Mulligan, 2002). Geomagnetic effects associated with the shock-sheath include sudden increases in solar wind speed, pressure, and magnetic field intensity, leading to enhanced geomagnetic activity, which are often short-lived but can be intense. The CME magnetic structure plays a more crucial role in enhanced geomagnetic effects as a southward directed magnetic field component (negative B_z, which is opposite to Earth's magnetic field), can efficiently couple with Earth's magnetosphere and trigger magnetic reconnection (e.g. review by Tsurutani et al., 2023). In that respect we may ask ourselves if we are consistent in our understanding of different CME structures and how well we are able to identify them in remote-sensing image data and to track them from Sun to Earth (see e.g., Verbeke et al., 2023, and related ISSI team).

White-light remote sensing image data give us a global view on the CME, whereas in-situ measurements reveal local variations. The closeness of Parker Solar Probe (PSP; Fox et al., 2016) and SolO spacecraft to the Sun, together with well established and newly developed triangulation methods, enable us to investigate for the first time small-scale structures at different regions of the CME (see Fig. 2). Results from Cappello et al. (2024) revealed that blob-like density structures distributed in the rear-part of the CME, might reveal the interaction process with the ambient solar wind. This helps us in better understanding the propagation behavior of CMEs in interplanetary space as well as to connect globally and locally derived measurements.

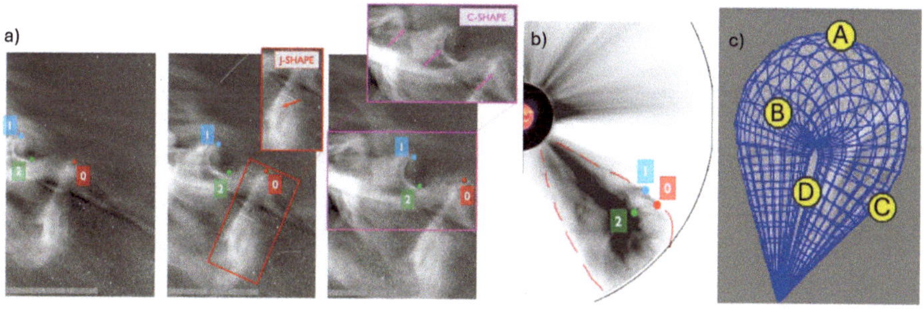

Fig. 2. a): WISPR-I white-light data from 2022-12-08 showing in detail some identified small-scale internal CME structures (0, 1, 2). b): STEREO-A COR2 coronagraph image, showing the tracked features in the global view. Adapted from Cappello et al. (2024). c): Cartoon of a GCS reconstructed CME with regions of interest highlighted: front (A), internal structure (B), flank (C), and current sheet (D).

5 Imprint of CMEs as They Propagate Through the Heliosphere

As the evolving CME compresses the solar wind ahead of it, it forms turbulent high-pressure regions known as sheath, and if fast enough, these steepen up into shocks. The sheath is a high plasma-beta region, hence, it holds implications for the draping of the interplanetary magnetic field and accumulation of plasma (Siscoe and Odstrcil, 2008). How the sheath region might affect the expansion behavior of the CME's magnetic structure and how it is related to interplanetary magnetic field draping is another open issue. A recent comprehensive study on CMEs with and without clear sheath regions over the inner and outer heliosphere is presented by Larrodera and Temmer (2024). The dataset covers more than 2000 CMEs over the time range 1975–2022. As can be seen in Fig. 3, it is found that both types (CMEs with and without sheath) increase in size from the inner heliosphere to 1 AU by ca. 47%, with the strongest size increase around 0.75 AU. There were no CMEs found with sheath structures ahead beyond 1.5 AU. CMEs without sheath seemed to almost double their size beyond 1.5 AU compared to 1 AU, however, the data sample is rather low for the outer heliosphere. A review on CME sheaths is given by Kilpua et al. (2017) and there are interesting aspects found in CME sheath regions as shown in other studies (e.g., Salman et al., 2021).

During high solar activity we frequently observe multiple CMEs (see e.g., episodes of enhanced geomagnetic effects due to multiple CMEs in March, April, and November 2023, or most recent in May 2024). In the late phase of solar activity, coronal hole structures at low latitudes are frequently observed, and SIRs dominate in interplanetary space. These multiple eruptions can lead to the formation of very complex magnetic structures in the heliosphere significantly complicating the prediction of their impact on Earth (Gopalswamy et al., 2001; Burlaga et al., 2002; Harrison et al., 2012) and causing to the most intense

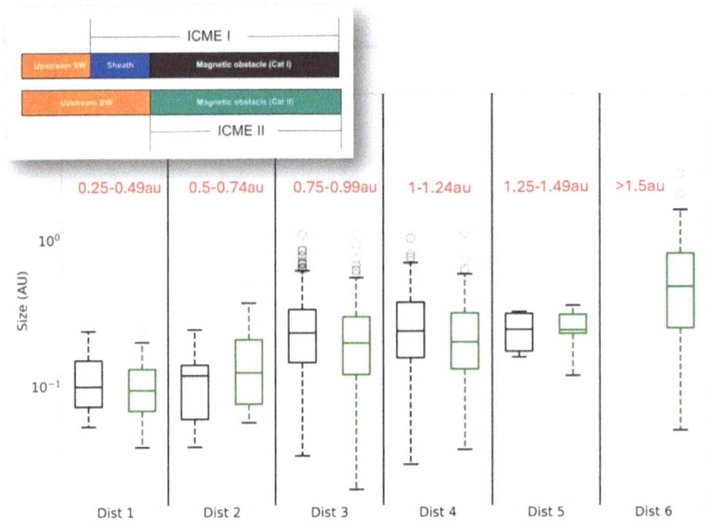

Fig. 3. Size of CME magnetic obstacles driving a clear sheath (black) and those w/o sheath (green). The size (mean speed over structure × duration) is calculated for measurements derived at different distances from the Sun covering the inner heliosphere to 1 AU. Adapted from Larrodera & Temmer, 2024.

geomagnetic storms (Farrugia et al., 2006; Xie et al., 2004; Dumbović et al., 2015). Besides the mutual interaction between the various structures, each CME (also SIR) generates perturbations in the smooth outflow of the slow solar wind and changes the magnetic field configuration. With that interplanetary space and near-Earth conditions experience a preconditioning effect that might last for several days (Temmer et al., 2017; Janvier et al., 2019). The occurrence of strong eruptions can also lead to transient open field conditions making subsequent CME propagation super-fast (Liu et al., 2014).

6 Geospace Impact and Interdisciplinary Research

Due their complex nature, the detailed impact of CMEs on geospace is highly variable. When CMEs collide with Earth, they trigger a cascade of reactions which are started at the magnetosphere. These reactions include the opening of the magnetic field, compression, and the occurrence of substorms in the magnetotail (see e.g., Russell and McPherron, 1973; Gosling et al., 1991). The strongly varying solar and magnetospheric energy input leads to significant fluctuations in the Earth's ionosphere (see e.g., Buonsanto, 1999; Tsurutani et al., 2004). The thermosphere gets heated and expands due to the absorption of EUV and X-ray radiation and, as the CME reconnects with the magnetosphere, due to injection of additional particles into that atmospheric region. This leads to an increase in the neutral density in larger heights of the atmosphere, leading to stronger drag for low Earth orbiting (LEO) satellites (see e.g., Knipp et al., 2004; Krauss et al., 2015; Bruinsma et al., 2023).

Recent studies found that density enhancements in the magnetosheath, so-called magnetosheath jets (MJ), represent a significant coupling effect between the solar wind and the Earth's magnetosphere (see review by Plaschke et al., 2018). A significant variation in the number of MJs is derived with CMEs/SIRs actually lowering/enhancing the MJ production rate (Koller et al., 2022). More interdisciplinary research is needed to better understand the coupling processes.

7 Conclusion

CMEs are pivotal in shaping Space Weather, influencing both the heliosphere and geospace. The driving forces of CMEs are related to the conditions on the Sun that lead to CME formation. The factors influencing their propagation through the heliosphere are likewise related to slow and fast solar wind regions in terms of CHs. Knowledge on the surface structures and how these further evolve into interplanetary space is therefore crucial for accurate forecasting. Small scale structures embedded in the CME or signatures of solar wind interaction, need yet to be studied to find out more about their effects on Space Weather.

Forecasting the impact of CMEs, however, remains challenging due to their complex and variable nature, the not-well understood interaction processes with the solar wind, as well as lack in accurate knowledge of the background solar wind itself. Advanced coronal and heliospheric models incorporating real-time solar wind data and detailed magnetic field measurements are essential for improving predictions. The unprecedented data gathered by PSP and SolO, currently provide and will provide detailed information of the near-Sun environment, enhancing our understanding of CME initiation and interaction processes with the solar wind.

At the impact of the various solar wind structures on Earth, the cascade of processes in the magnetosphere and lower atmospheric layers is not fully understood. More interdisciplinary research is therefore needed to identify universal processes and to tackle problems from different aspects.

References

Bein, B.M., Berkebile-Stoiser, S., Veronig, A.M., et al.: Impulsive acceleration of coronal mass ejections. I. Stat. Coronal Mass Eject. Source Reg. Characteristics. Astrophys. J. **738**(2), 191 (2011). https://doi.org/10.1088/0004-637X/738/2/191, 1108.0561

Benz, A.O., Monstein, C., Meyer, H., et al.: A world-wide net of solar radio spectrometers: e-CALLISTO. Earth Moon Planet. **104**(1–4), 277–285 (2009). https://doi.org/10.1007/s11038-008-9267-6

Bruinsma, S.: Cluster g: G2a summary and ti2 cluster paper - on atmospheric variability. Adv. Space Res. (2023)

Buonsanto, M.J.: Ionospheric storms — a review. Space Sci. Rev. **88**, 563–601 (1999). https://doi.org/10.1023/A:1005107532631

Burlaga, L.F., Plunkett, S.P., St Cyr, O.C.: Successive CMEs and complex ejecta. J. Geoph. Res. (Space Phys.) **107**(A10), 1266 (2002). https://doi.org/10.1029/2001JA000255

Cappello, G.M., Temmer, M., Vourlidas, A., et al.: Internal magnetic field structures observed by PSP/WISPR in a filament related coronal mass ejection. arXiv e-prints (2024). arXiv:2402.14682, https://doi.org/10.48550/arXiv.2402.14682, 2402.14682

Chikunova, G., et al.: Three-dimensional relation between coronal dimming, filament eruption, and CME. A case study of the 28 October 2021 X1.0 event. Astron. Astrophys. **678**, A166 (2023), https://doi.org/10.1051/0004-6361/202347011, 2308.09815

Dissauer, K., Veronig, A.M., Temmer, M., Podladchikova, T., Vanninathan, K.: statistics of coronal dimmings associated with coronal mass ejections. I. characteristic dimming properties and flare association. Astrophys. J. **863**(2), 169 (2018). https://doi.org/10.3847/1538-4357/aad3c6, 1807.05056

Dumbović, M., Devos, A., Vršnak, B., et al.: Geoeffectiveness of coronal mass ejections in the SOHO Era. Solar Phys. **290**(2), 579–612 (2015). https://doi.org/10.1007/s11207-014-0613-8, 1410.3303

Farrugia, C.J., Jordanova, V.K., Thomsen, M.F., Lu, G., Cowley, S.W.H., Ogilvie, K.W.: A two-ejecta event associated with a two-step geomagnetic storm. J. Geoph. Res. (Space Phys.) **111**(A11), A11104 (2006). https://doi.org/10.1029/2006JA011893

Fox, N.J., Velli, M.C., Bale, S.D., et al.: The solar probe plus mission: humanity's first visit to our star. Space Sci. Rev. **204**(1-4), 7–48 (2016). https://doi.org/10.1007/s11214-015-0211-6

Gopalswamy, N.: The sun and space weather. Atmosphere **13**(11), 1781 (2022). https://doi.org/10.3390/atmos13111781, 2211.06775

Gopalswamy, N., Yashiro, S., Kaiser, M.L., Howard, R.A., Bougeret, J.L.: Radio signatures of coronal mass ejection interaction: coronal mass ejection cannibalism? Astrophys. J. Lett. **548**(1), L91–L94 (2001). https://doi.org/10.1086/318939

Gopalswamy, N., Yashiro, S., Akiyama, S., et al.: Coronal mass ejections, type II radio bursts, and solar energetic particle events in the SOHO era. Ann. Geophys. **26**(10), 3033–3047 (2008). https://doi.org/10.5194/angeo-26-3033-2008

Gopalswamy, N., Mäkelä, P., Xie, H., Akiyama, S., Yashiro, S.: CME interactions with coronal holes and their interplanetary consequences. J. Geoph. Res. (Space Phys.) **114**(A3), A00A22 (2009), https://doi.org/10.1029/2008JA013686

Gosling, J.T., McComas, D.J., Phillips, J.L., Bame, S.J.: Geomagnetic activity associated with earth passage of interplanetary shock disturbances and coronal mass ejections. J. Geophys. Res. **96**(A5), 7831–7839 (1991). https://doi.org/10.1029/91JA00316

Green, L.M., Török, T., Vršnak, B., Manchester, W., Veronig, A.: The origin, early evolution and predictability of solar eruptions. Space Sci. Rev. **214**(1), 1–52 (2018). https://doi.org/10.1007/s11214-017-0462-5

Harrison, R.A., Davies, J.A., Möstl, C., et al.: An analysis of the origin and propagation of the multiple coronal mass ejections of 2010 august 1. Astrophys. J. **750**(1), 45 (2012). https://doi.org/10.1088/0004-637X/750/1/45

Heinemann, S.G., Owens, M.J., Temmer, M., et al.: On the origin of the sudden heliospheric open magnetic flux enhancement during the 2014 pole reversal. Astrophys. J. **965**(2), 151 (2024), https://doi.org/10.3847/1538-4357/ad2b69, 2402.12805

Hofmeister, S.J., Veronig, A., Temmer, M., Vennerstrom, S., Heber, B., Vršnak, B.: The dependence of the peak velocity of high-speed solar wind streams as measured in the ecliptic by ACE and the STEREO satellites on the area and co-latitude of their solar source coronal holes. J. Geophys. Res. Atm. **123**(3), 1738–1753 (2018). https://doi.org/10.1002/2017JA024586, 1804.09579

Janvier, M., Winslow, R.M., Good, S., et al.: Generic magnetic field intensity profiles of interplanetary coronal mass ejections at mercury, venus, and earth from superposed epoch analyses. J. Geoph. Res. (Space Phys.) **124**(2), 812–836 (2019), https://doi.org/10.1029/2018JA025949, 1901.09921

Kilpua, E., Koskinen, H.E.J., Pulkkinen, T.I.: Coronal mass ejections and their sheath regions in interplanetary space. Living Rev. Sol. Phys. **14**(1), 1–83 (2017). https://doi.org/10.1007/s41116-017-0009-6

Knipp, D.J., Tobiska, W.K., Emery, B.A.: Direct and indirect Thermospheric heating sources for solar cycles 21–23. Solar Phys. **224**(1–2), 495 (2004). https://doi.org/10.1007/s11207-005-6393-4

Koller, F., Temmer, M., Preisser, L., et al.: Magnetosheath jet occurrence rate in relation to CMEs and SIRs. J.Geoph. Res. (Space Phy.) **127**(4), e30124 (2022). https://doi.org/10.1029/2021JA03012410.1002/essoar.10508761.2

Koller, F., Raptis, S., Temmer, M., Karlsson, T.: The effect of fast solar wind on ion distribution downstream of earth's bow shock. Astrophys. J. Lett. **964**(1), L5 (2024). https://doi.org/10.3847/2041-8213/ad2ddf

Krauss, S., Temmer, M., Veronig, A., Baur, O., Lammer, H.: Thermospheric and geomagnetic responses to interplanetary coronal mass ejections observed by ACE and GRACE: Statistical results. J. Geoph. Res. (Space Phys.) **120**(10), 8848–8860 (2015), https://doi.org/10.1002/2015JA021702, 1510.03549

Larrodera, C., Temmer, M.: Evolution of coronal mass ejections with and without sheaths from the inner to the outer heliosphere: statistical investigation for 1975 to 2022. Astron. Astrophys. **685**, A89 (2024), https://doi.org/10.1051/0004-6361/202348641, 2402.16653

Linker, J.A., et al.: The open flux problem. Astrophys. J. —848(1), 70 (2017), https://doi.org/10.3847/1538-4357/aa8a70, 1708.02342

Linker, J.A., Heinemann, S.G., Temmer, M., et al.: Coronal hole detection and open magnetic flux. Astrophys. J. **918**(1), 21 (2021), https://doi.org/10.3847/1538-4357/ac090a, 2103.05837

Liu, Y.D., Luhmann, J.G., Kajdič, P., et al.: Observations of an extreme storm in interplanetary space caused by successive coronal mass ejections. Nat. Commun. **5**, 3481 (2014), https://doi.org/10.1038/ncomms4481, 1405.6088

Lugaz, N., Temmer, M., Wang, Y., Farrugia, C.J.: The interaction of successive coronal mass ejections: a review. Solar Phys. **292**(4) (2017), https://doi.org/10.1007/s11207-017-1091-6, http://dx.doi.org/10.1007/s11207-017-1091-6, 1612.02398

Luhmann, J.G., Gopalswamy, N., Jian, L.K., Lugaz, N.: ICME evolution in the inner heliosphere. Sol. Phys. **295**(4), 1–32 (2020). https://doi.org/10.1007/s11207-020-01624-0

Müller, D., Nicula, B., Felix, S., et al.: JHelioviewer. Time-dependent 3D visualisation of solar and heliospheric data. Astron. Astrophys. **606**, A10 (2017), https://doi.org/10.1051/0004-6361/201730893, 1705.07628

Owens, M.J., Forsyth, R.J.: The Heliospheric magnetic field. Living Rev. Sol. Phys. **10**(1), 1–52 (2013). https://doi.org/10.12942/lrsp-2013-5

Patsourakos, S., et al.: Decoding the pre-eruptive magnetic field configurations of coronal mass ejections. Space Sci. Rev. **216**(8), 1–63 (2020). https://doi.org/10.1007/s11214-020-00757-9

Plaschke, F., et al.: Jets downstream of Collisionless shocks. Space Sci. Rev. **214**(5), 1–77 (2018). https://doi.org/10.1007/s11214-018-0516-3

Pulkkinen, A., Bernabeu, E., Thomson, A., et al.: Geomagnetically induced currents: Science, engineering, and applications readiness. Space Weather **15**(7), 828–856 (2017). https://doi.org/10.1002/2016SW001501

Reames, D.V.: The two sources of solar energetic particles. Space Sci. Rev. **175**(1-4), 53–92 (2013), https://doi.org/10.1007/s11214-013-9958-9, 1306.3608

Russell, C.T., McPherron, R.L.: The magnetotail and Substorms. Space Sci. Rev. **15**(2–3), 205–266 (1973). https://doi.org/10.1007/BF00169321

Russell, C.T., Mulligan, T.: On the magnetosheath thicknesses of interplanetary coronal mass ejections. Planet. Space Sci. **50**(5–6), 527–534 (2002). https://doi.org/10.1016/S0032-0633(02)00031-4

Salman, T.M., Lugaz, N., Winslow, R.M., Farrugia, C.J., Jian, L.K., Galvin, A.B.: Categorization of coronal mass ejection-driven sheath regions: characteristics of STEREO events. Astrophys. J. **921**(1), 57 (2021), https://doi.org/10.3847/1538-4357/ac11f3, 2106.12076

Sammis, I., Tang, F., Zirin, H.: The dependence of large flare occurrence on the magnetic structure of sunspots. Astrophys. J. **540**(1), 583–587 (2000). https://doi.org/10.1086/309303

Scolini, C., Chané, E., Temmer, M., et al.: CME–CME interactions as sources of CME Geoeffectiveness: the formation of the complex ejecta and intense geomagnetic storm in 2017 early September. Astrophys. J. Suppl. Ser. **247**(1), 21 (2020), https://doi.org/10.3847/1538-4365/ab6216, 1911.10817

Siscoe, G., Odstrcil, D.: Ways in which ICME sheaths differ from magnetosheaths. J. Geophys. Res. **113**(A9), A00B07 (2008). https://doi.org/10.1029/2008JA013142

Temmer, M.: Space weather: the solar perspective. Living Rev. Sol. Phys. **18**(1), 1–80 (2021). https://doi.org/10.1007/s41116-021-00030-3

Temmer, M., et al.: Acceleration in fast halo CMEs and synchronized flare HXR bursts. Astrophys. J. Lett. **673**(1), L95 (2008). https://doi.org/10.1086/527414

Temmer, M., Reiss, M.A., Nikolic, L., Hofmeister, S.J., Veronig, A.M.: Preconditioning of interplanetary space due to transient CME disturbances. Astrophys. J. **835**(2), 1–6 (2017), https://doi.org/10.3847/1538-4357/835/2/141, http://dx.doi.org/10.3847/1538-4357/835/2/141

Temmer, M., Scolini, C., Richardson, I.G., et al.: CME Propagation through the heliosphere: status and future of observations and model development (2023). arXiv e-prints arXiv:2308.04851, https://doi.org/10.48550/arXiv.2308.04851, 2308.04851

Thalmann, J.K., Su, Y., Temmer, M., Veronig, A.M.: The confined x-class flares of solar active region 2192. Astrophys. J. Lett. **801**(2), L23 (2015), https://doi.org/10.1088/2041-8205/801/2/L23, 1502.05157

Tsurutani, B., Mannucci, A., Iijima, B., et al.: Global dayside ionospheric uplift and enhancement associated with interplanetary electric fields. J. Geophys. Res. (Space Phys.) **109**(A8), A08302 (2004). https://doi.org/10.1029/2003JA010342

Tsurutani, B.T., Zank, G.P., Sterken, V.J., et al.: Space plasma physics: a review. IEEE Trans. Plasma Sci. **51**(7), 1595–1655 (2023), https://doi.org/10.1109/TPS.2022.3208906, 2209.14545

van Driel-Gesztelyi, L., Culhane, J.L., Baker, D., et al.: Magnetic topology of active regions and coronal holes: implications for coronal outflows and the solar wind. Solar Phys. **281**(1), 237–262 (2012). https://doi.org/10.1007/s11207-012-0076-8

Verbeke, C., Mays, M.L., Kay, K., et al.: Quantifying errors in 3D CME parameters derived from synthetic data using white-light reconstruction techniques. Adv. Space Res. (2023). https://doi.org/10.1016/j.asr.2022.08.056

Vourlidas, A., Lynch, B.J., Howard, R.A., Li, Y.: How many CMEs have flux ropes? Deciphering the signatures of shocks, flux ropes, and prominences in coronagraph observations of CMEs. Solar Phys. **284**(1), 179–201 (2013), https://doi.org/10.1007/s11207-012-0084-8, 1207.1599

Xie, H., Ofman, L., Lawrence, G.: Cone model for halo CMEs: application to space weather forecasting. J. Geophys. Res. Space Phys. **109** (2004), https://doi.org/10.1029/2003JA010226

Yashiro, S., Akiyama, S., Gopalswamy, N., Howard, R.A.: Different power-law indices in the frequency distributions of flares with and without coronal mass ejections. Astrophys. J. Lett. **650**(2), L143–L146 (2006), https://doi.org/10.1086/508876, astro-ph/0609197

Zhang, J., Dere, K.P., Howard, R.A., Kundu, M.R., White, S.M.: On the temporal relationship between coronal mass ejections and flares. Astrophys. J. **559**(1), 452–462 (2001). https://doi.org/10.1086/322405

Open Access This chapter is licensed under the terms of the Creative Commons Attribution 4.0 International License (http://creativecommons.org/licenses/by/4.0/), which permits use, sharing, adaptation, distribution and reproduction in any medium or format, as long as you give appropriate credit to the original author(s) and the source, provide a link to the Creative Commons license and indicate if changes were made.

The images or other third party material in this chapter are included in the chapter's Creative Commons license, unless indicated otherwise in a credit line to the material. If material is not included in the chapter's Creative Commons license and your intended use is not permitted by statutory regulation or exceeds the permitted use, you will need to obtain permission directly from the copyright holder.

On the Non-Force-Free Magnetic Fields and Solar Eruptions

Xiaoshuai Zhu(✉)

Key Laboratory of Solar Activity and Space Weather, National Space Science Center, Chinese Academy of Sciences, Beijing 100190, People's Republic of China
zhuxiaoshuai@nssc.ac.cn

Abstract. Modeling the three-dimensional (3D) magnetic fields of the solar active region across multiple layers is crucial for understanding the nature of solar eruptions. The main approach is to extrapolate the magnetic field from magnetograms measured in the photosphere into chromosphere and corona. In the past few decades, a prevailing extrapolation technique, known as force-free field (FFF) extrapolation, assumed a complete neglect of all plasma effects. While the force-free assumption is well justified in the inner solar corona, it is not the case in the photosphere and chromosphere. To better describe the lower solar atmosphere, a non-force-free state called magnetohydrostatic (MHS) equilibrium, which takes into account plasma forces, has gained prominence and witnessed rapid development in recent years. Several tests have demonstrated that the MHS extrapolation yields a more accurate reconstruction of the magnetic field compared to FFF extrapolation. Therefore, it helps to enhance our understanding of the initiation of solar eruptions. In this paper, I will provide an overview of various MHS extrapolation methods, including their most common model assumptions, tests, and most importantly, applications to studies of solar activities.

Keywords: magnetic field · magneto-hydro-statics · solar eruptions

1 Introduction

The electromagnetic and particle radiation emitted by the Sun, as the nearest star to Earth, exerts direct influence on the spatial environment around our planet. Large solar activities, including solar flares, coronal mass ejections, and high-energy particle radiation, pose potential risks to aerospace operations, communication navigation systems, as well as long-range transmission networks and critical infrastructure like oil pipelines.

These solar activities originate from active regions on the Sun, and the magnetic field plays a key role in determining when and where the dynamic processes are taking place, how the energy is released and transformed. However, routine and reliable measurements of the magnetic field are only available in the photosphere. While there have been significant advancements in measuring the chromospheric and coronal magnetic field in recent years [1, 2], there still remains substantial error and uncertainty. Therefore, the primary approach for obtaining the three-dimensional (3D) magnetic field configuration is by extrapolating the magnetic field from the photospheric measurements.

© The Author(s) 2026
N. Gopalswamy et al. (Eds.): ISWI 2024, SPPHY 431, pp. 14–23, 2026.
https://doi.org/10.1007/978-981-95-1121-1_2

The overall magnetic field structure in a solar active region is generally stable for majority of the time. Therefore, it is reasonable to assume that the active region is in a static state without any plasma and magnetic field motion. Based on this assumption, magnetic field extrapolation methods have been developed by solving the following magnetohydrostatic (MHS) equations:

$$\mathbf{J} \times \mathbf{B} - \nabla \mathrm{p} + \rho \mathbf{g} = 0, \tag{1}$$

$$\nabla \times \mathbf{B} = 4\pi \mathbf{J}, \tag{2}$$

$$\nabla \cdot \mathbf{B} = 0, \tag{3}$$

where $\mathbf{J}$, $\mathbf{B}$, p, ρ, and $\mathbf{g}$ are electric current, magnetic field, plasma pressure, plasma density, and gravitational acceleration, respectively. The above equations take plasma forces into account, capturing the strong interaction between the magnetic field and plasma in the lower atmosphere. In contrast, the force-free field (FFF, zero Lorentz force) neglects the plasma forces and is only suitable for coronal magnetic field extrapolation. For comprehensive understanding of the nonlinear FFF (NLFFF) extrapolation, we refer interested readers to reviews [3–5]. In this paper, MHS and non-force-free are used interchangeably.

A review focuses on the methods of the MHS extrapolation can be found in [6]. However, in this particular review, the emphasis in on the applications rather than methods. The present review starts with a brief introduction to both analytical and numerical approaches for the MHS extrapolation (Sect. 2). Their applications to the study of solar eruptions are presented in Sect. 3. Discussion and conclusion are provided in Sect. 4.

2 The MHS Extrapolation Methods

By using the photospheric magnetoram as the bottom boundary input, Eqs. (1–3) can be solved either analytically or numerically. The analytical method, which relies on additional assumptions regarding the MHS equilibrium, is very fast but fails to recover magnetic configurations with strong electric current concentration in part of the domain. In contrast, several numerical methods have been proposed that require more computational resources but are capable of dealing with the general MHS case.

2.1 The Analytical Method

An analytical and complete treatment of the MHS Eqs. (1–3) in 3D is not tractable. Many efforts have been made to solve the equations by ignoring a coordinate which renders the problem tractable [7, 8]. However, active regions rarely exhibit a high degree of geometric symmetry, which means these solutions are not able to accurately model the magnetic field on the Sun.

The first analytical MHS solution applicable to model the real solar atmosphere was found by Low in 1985 [9]. The key is to assume the electric current to be everywhere perpendicular to the gravitational force. The assumption enables the reduction of the nonlinear Eq. (3) to a single, scalar PDE. In 1990, Low [10] extended this solution by incorporating a field-aligned electric current component. The two current systems can be expressed as:

$$\nabla \times \mathbf{B} = \alpha \mathbf{B} + a e^{-\kappa z} \nabla B_z \times \hat{z}, \tag{4}$$

where α, a, and κ are free parameters. κ is the reciprocal of characteristic length of Lorentz force, and typically we set $1/\kappa = 2$ Mm. α can be determined using vector magnetogram following the same approach used in the linear FFF case [11]. Wiegelmann et al. [12] proposed a relationship between a and net Lorentz force computed using magnetogram to determin a. Equation (4) is linear and can be easily solved. Based on assumption made in [9], other analytical solutions have also been discovered [13, 14].

2.2 The Numerical Method

The analytical method has limitations in modeling the nonlinear magnetic field on the Sun. Therefore, numerical methods have been developed in recent years. It is worth noting that each numerical method is a generalization of the method used in the NLFFF extrapolation.

Optimization Method For MHS extrapolation was proposed in [15, 16] and further developed in [17]. This method aims to find MHS solutions by minimizing functional

$$\mathrm{L}(\mathbf{B}, \rho, \mathrm{p}) = \int_V [\omega_a B^{-2} (\mathbf{J} \times \mathbf{B} - \nabla \mathrm{p} + \rho \mathbf{g})^2 + \omega_b (\nabla \cdot \boldsymbol{B})^2] dV, \tag{5}$$

where ω_a and ω_b are weighting functions. The monotonic decrease of L can be achieved by iteration formula designed based on functional derivative of Eq. (5). The method has been tested by using an analytical MHS solution as a reference model, as well as a more realistic case which is a snapshot of a radiative magnetohydrodynamic simulation [18]. The method has found applications in solar activity studies.

MHD Relaxation Method For MHS extrapolation has been developed in [19, 20]. The idea is to relax either full or simplified MHD system in which a realistic solar atmosphere is included. The MHS equilibrium can be achieved with the help of a friction term. This method has been tested both by an analytical solution and an MHD simulation snapshot. It has also been applied to extrapolate the magnetogram.

Grad-Rubin Method For MHS extrapolation has been developed in [21, 22]. The method decomposes the MHS equations into two hyperbolic parts for evolving plasma pressure and σ (counterpart of the force-free factor α in the force-free field) along the magnetic lines and an elliptic equation to update the magnetic field. Basically, it replaces the nonlinear MHS equations by a system of linear equations, and then solve the linear equations iteratively until a fixed point is reached. The method has been tested by using a known analytical solution.

It is worth noting that iterations of aforementioned methods are initiated with the potential field.

3 Applications to Studies of Solar Activities

Both analytical and numerical MHS extrapolations have been used to study various solar activities.

3.1 Major Eruptive Events in the Corona

The MHS extrapolation provides with detailed 3D magnetic field structure which helps us to understand the nature of major eruptive events.

Song et al. [23] studied a M5.7 white-light flare event associated with a filament eruption. By using the MHS extrapolation, the authors found the magnetic field shows a fan-spine feature and a flux rope lies under the dome-like structure (see Fig. 1 left). They suggest either reconnection between the magnetic field of the filament and the overlying magnetic field or reconnection within the flux rope leads to the white-light flare. In another C2.3 white-light flare event, Song et al. [24] reconstructed a U-shaped magnetic field configuration by using the MHS extrapolation (see Fig. 1 right). The authors suggest, by combining the extrapolation result and observations in other wavelengths, that the white-light flare is powered by the magnetic reconnection in the lower part of the solar atmosphere rather than in the corona.

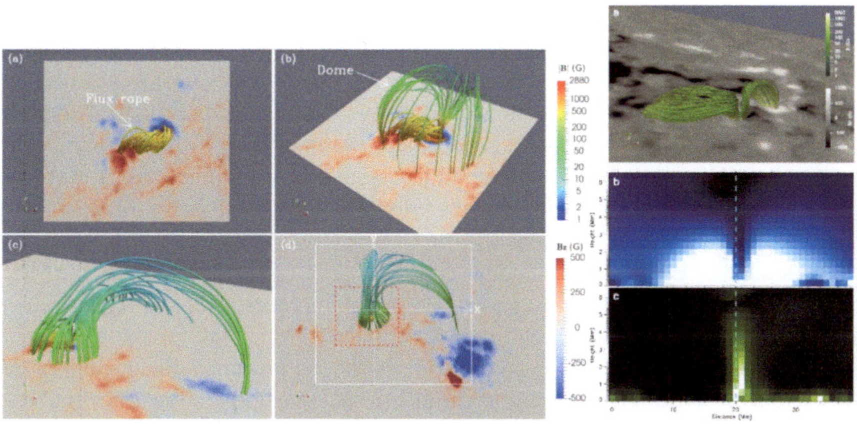

Fig. 1. Left four panels: Magnetic flux rope structure obtained at the source region of the M5.7 white-light flare on 2012 May 10. Right three panels: U-shaped magnetic structure associated with the C2.3 white-light flare.

Huang et al. [25] studied the slipping brightening and damped quasi-periodic pulsations in a circular ribbon flare. The MHS extrapolation shows a fan-spine structure at the source region and the dome displays a twisting feature (see Fig. 2). The authors propose, based on the magnetic field configuration, that the slipping reconnection occurs in the null point of the fan-spine structure, and the twisting of the dome continuously promotes new reconnection.

The MHS extrapolation has also been applied to investigate other solar activities in the corona, such as an atypical confined flare [26], coronal loop formation [27], source

Fig. 2. Left: The circular flare observed by AIA at 94 Å. Right: Fan-spine structure of the magnetic field associated with the flare.

of the material within coronal mass ejection [28], and magnetic flux rope in AR 12158 [29].

3.2 Small-Scale Events in the Lower Atmosphere

Some applications focus on activities in the lower solar atmosphere where the magnetic field is not force-free.

By using the MHS extrapolation successively before and after a blowout jet, Zhu et al. [30] found that a magnetic flux rope existed at source region before the jet blows out and the rope disappeared after the jet (see Fig. 3). This provides strong evidence that the blowout jet is essentially the ejection of the magnetic flux rope. In a filament eruption event which causes one of the largest geomagnetic storm of solar cycle 24, Wang et al. [31] used the MHS extrapolation and found that a small-scale filament has a magnetic flux rope structure. They confirmed that the eruption of the small-scale filament is the trigger of the large filament eruption.

In the studies of UV burst events, Zhao et al. [32] and Chen et al. [33] found the MHS extrapolation shows consistent results between bright arcades in the emerging region observed by IRIS and magnetic field lines passing through the bald patches (magnetic dips) (see Fig. 4). Tian et al. [34] also investigated the 3D magnetic field configuration of the UV bursts in the emerging region through the MHS extrapolation. They find that almost all bursts are located in regions of large squashing factor at the height of 1 Mm, which reinforces their conclusion that UV bursts are produced by reconnection in the lower atmosphere.

The MHS extrapolation results have also been compared with magnetic field in the chromosphere obtained by inversion [35] and fibril structures observed at lower solar atmosphere as well. By using the IMaX high-resolutional (100 km) magnetogram [36], Zhu et al. [37] found that the magnetic field lines trace the fibril structures observed at 3968 Å in the chromosphere nicely. Through quantitative comparison, the authors further found that the magnetic field lines are more aligned with the fibrils than the NLFFF lines are. Similar results were also found in [38] by comparison with H α images obtained by New Vacuum Solar Telescope [39].

The MHS extrapolation has also been applied to investigate temperature oscillation in the chromosphere with ALMA observation [40] and chromospheric heating [41].

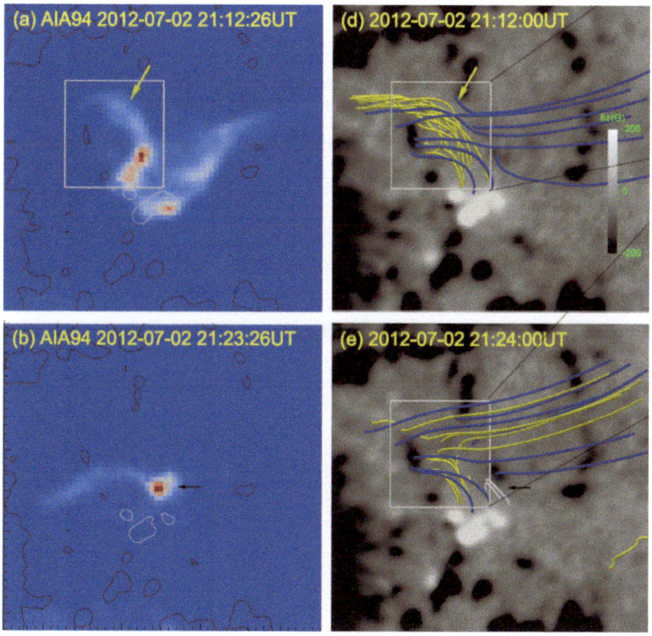

Fig. 3. AIA 94 Å. Image and magnetic field structure at the jet's source region before and after it blows out.

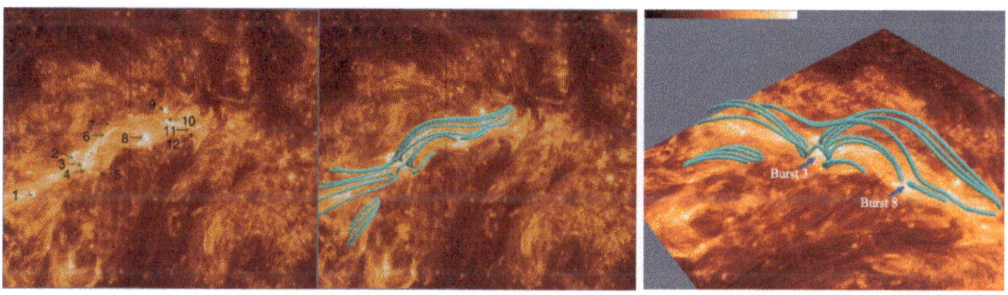

Fig. 4. Comparison between IRIS imaging observation and magnetic field lines from the extrapolation.

3.3 Global Magnetic Structure of the Corona

In the global scale, the analytical MHS solution has been applied to model the corona for nearly 40 years.

Zhao et al. [42, 43] developed a global model to depict the magnetic field in corona and interplanetary space. The authors used Bogdan & Low's analytical MHS solution [13] to build the model in the inner corona (below the cusp surface). The Bogdan & Low's solution has also been applied to reconstruct the coronal structure observed during total solar eclipse [44, 45] (see Fig. 5).

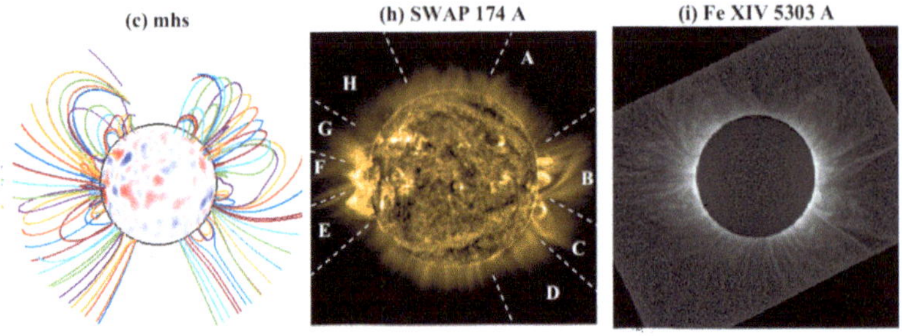

Fig. 5. Left: Magnetic field structure in the plane-of-sky, as viewed from the Earth at the eclipse time. Middle: EUV 174 Å image of the real corona from PROBA2/SWAP. Right: Corona in the Fe XIV 5405 Å line.

4 Discussion and Conclusion

The MHS extrapolation has been developing fast over the past decade, it still faces following challenges.

1. The basic assumption for the MHS extrapolation is that the active region is in an MHS equilibrium. However, we found some magnetograms do not fulfill the criteria (3 integrals, see Eqs. (7, 8, 12) in [46]) for MHS equilibrium, which has a negative impact in the extrapolation. Zhu et al. [46] developed a preprocessing method by using the MHS criteria to derive a more suitable boundary for the MHS extrapolation. It is worth noting that the method is an extension of the preprocessing method for the NLFFF extrapolation [47].
2. The MHS extrapolation needs plasma pressure or density distribution on the photosphere which is difficult to be obtained from observation. The optimization method assumes a balance of the total pressure in the photosphere to derive the plasma pressure: $p+\frac{B_z^2}{2}=p_{quiet}$. The MHS extrapolation still needs 3D temperature or pressure scale height to close the MHS equations. We may use a fixed 1D temperature model instead of 3D temperature model. In the current optimization method, a 1D temperature model is used to create the initial atmosphere for iteration. However, the temperature is free to change during the optimization which results in large deviation of the plasma reconstruction higher than 1 Mm compared with the reference model [18].
3. The MHS extrapolation is more computationally expensive than the NLFFF extrapolation. In the optimization method, the MHS code is 50 times slower than the corresponding NLFFF code. The main reason is that the plasma pressure has a big difference in magnitude through out the computational box. Recently, Zhu et al. [48] restricted the MHS extrapolation in the non-force-free layer, derived the magnetogram by cutting the MHS result at 1 Mm, and then perform the NLFFF extrapolation on the new magnetogram. By doing this, the computational efficiency has been improved for 7 times (still as accurate as the pure MHS extrapolation).

To further develop the MHS extrapolation method, an ISSI/ISSI-BJ team consists of 12 experts in the magnetic field modeling, application, and inversion was convened in 2022 (https://teams.issibern.ch/magnetohydrostaticsolaratmosphere/). The team aims to improve the already available MHS extrapolation methods by testing them with the unified reference model and applying them to observations from DKIST, GST, SolO, and SDO, etc. Currently, results from different methods applied to the test are evaluating.

References

1. Yang, Z., Bethge, C., Tian, H., et al.: Global maps of the magnetic field in the solar corona. Science **369**, 6504 (2020)
2. Li, W., Grumer, J., Yang, Y., et al.: A novel method to determine magnetic fields in low-density plasma facilitated through accidental degeneracy of quantum states in Fe9+. ApJ **807**, 69 (2015)
3. Régnier, S.: Magnetic field extrapolations into the corona: Succe. Solar Phys. **288**(2), 481–505 (2013). https://doi.org/10.1007/s11207-013-0367-8
4. Guo, Y., Cheng, X., Ding, M.: Origin and structures of solar eruptions II: magnetic modeling. Sci. China Earth Sci. **60**, 1408 (2017)
5. Wiegelmann, T., Sakurai, T.: Solar force-free magnetic fields. Living Rev. Sol. Phys. **18**, 1 (2021)
6. Zhu, X., Neukirch, T., Wiegelmann, T.: Magnetohydrostatic modeling of the solar atmosphere. Sci. China Technol. Sci. **65**, 1710 (2022)
7. Dungey, J.: A family of solutions of the magneto-hydrostatic problem in a conducting atmosphere in a gravitational field. MN, 113, 180 (1953)
8. Low, B.: Nonisothermal magnetostatic equilibria in a uniform gravity field. I. Mathematical formulation. ApJ **197**, 251 (1975)
9. Low, B.: Three-dimensional structures of magnetostatic atmospheres I. Theory. ApJ **293**, 31 (1985)
10. Low, B.: Three-dimensional structures of magnetostatic atmospheres III. A general formulation. ApJ **370**, 427 (1991)
11. Hagino, M., Sakurai, T.: Latitude variation of helicity in solar active regions. Publ. Astron. Soc. Jpn. **56**, 831 (2004)
12. Wiegelmann, T., Neukirch, T., Nickeler, D., et al.: Magneto-static modeling from sunrise/imax: application to an active region observed with sunrise II. ApJS **229**, 18 (2017)
13. Bogdan, T., Low, B.: The three-dimensional structure of magnetostatic atmospheres. II. Modeling the large-scale corona. ApJ **306**, 271 (1986)
14. Neukirch, T.: On self-consistent three-dimensional analytic solutions of the magnetohydrostatic equations. A&A **301**, 628 (1995)
15. Wiegelmann, T., Neukirch, T.: An optimization principle for the computation of MHD equilibria in the solar corona. A&A **457**, 1053 (2006)
16. Wiegelmann, T., Neukirch, T., Ruan, P., et al.: Optimization approach for the computation of magnetohydrostatic coronal equilibria in spherical geometry. A&A **475**, 701 (2007)
17. Zhu, X., Wiegelmann, T.: On the extrapolation of magnetohydrostatic equilibria on the Sun. ApJ **866**, 130 (2018)
18. Zhu, X., Wiegelmann, T.: Testing magnetohydrostatic extrapolation with radiative MHD simulation of a solar flare. A&A **631**, A162 (2019)
19. Zhu, X., Wang, H., Du, Z., et al.: Forced field extrapolation: testing a magnetohydrodynamic (MHD) relaxation method with a flux-rope emergence model. ApJ **768**, 119 (2013)
20. Miyoshi, T., Kusano, K., Inoue, S.: A magnetohydrodynamic relaxation method for non-force-free magnetic field in magnetohydrostatic equilibrium. ApJS **247**, 6 (2020)

21. Gilchrist, S., Wheatland, M.: A magnetostatic grad-rubin code for coronal magnetic field extrapolations. Sol. Phys. **282**, 283 (2013)
22. Gilchrist, S., Braun, D., Barnes, G.: A fixed-point scheme for the numerical construction of magnetohydrostatic atmospheres in three dimensions. Sol. Phys. **291**, 3583 (2016)
23. Song, Y., Guo, Y., Tian, H., et al.: Observations of a white-light flare associated with a filament eruption. ApJ **854**, 64 (2018)
24. Song, Y., Tian, H., Zhu, X., et al.: A White-light flare powered by magnetic reconnection in the lower solar atmosphere. ApJ **893**, 13 (2020)
25. Huang, J., Tan, B., Zhang, Y., et al.: The slipping magnetic reconnection and damped quasiperiodic pulsations in a circular ribbon flare. ApJ **965**, 137 (2024)
26. Joshi, N., Zhu, X., Schmieder, B., et al.: Generalization of the magnetic field configuration of typical and atypical confined flares. ApJ **871**, 165 (2019)
27. Hou, Z., Tian, H., Chen, H., et al.: Formation of solar quiescent coronal loops through magnetic reconnection in an emerging active region. ApJ **915**, 39 (2021)
28. Fu, H., Harrison, R., Davies, J., et al.: The high helium abundance and charge states of the interplanetary CME and its material source on the sun. ApJL **900**, 18 (2020)
29. Yu, F., Zhao, J., Su, Y., et al.: Magnetic field extrapolation in active region well comparable to observations in multiple layers. ApJ **951**, 54 (2023)
30. Zhu, X., Wang, H., Cheng, X., et al.: A solar blowout jet caused by the eruption of a magnetic flux rope. ApJL **844**, 20 (2017)
31. Wang, R., Liu, Y., Zimovets, I., et al.: Sympathetic solar filament eruptions. ApJL **827**, 12 (2016)
32. Zhao, J., Schmieder, B., Li, H., et al.: Observational evidence of magnetic reconnection for brightenings and transition region arcades in IRIS observations. ApJ **836**, 52 (2017)
33. Chen, Y., Tian, H., Zhu, X., et al.: Solar ultraviolet bursts in a coordinated observation of IRIS, Hinode and SDO. Sci. China Technol. Sci. **62**, 1555 (2019)
34. Tian, H., Zhu, X., Peter, H., et al.: Magnetic reconnection at the earliest stage of solar flux emergence. ApJ **854**, 174 (2018)
35. Visser, G., Danilovic, S., Zhu, X., et al.: Active region chromospheric magnetic fields: Observational inference versus magnetohydrostatic modelling. A&A **662**, 88 (2022)
36. Solanki, S., Riethmüller, T., Barthol, P., et al.: The second flight of the sunrise balloon-borne solar observatory: overview of instrument updates, the flight, the data, and first results. ApJS **229**(2) (2017)
37. Zhu, X., Wiegelmann, T., Solanki, S.: Magnetohydrostatic modeling of AR11768 based on a SUNRISE/IMaX vector magnetogram. A&A **640**, A103 (2020)
38. Zhu, X., Wang, H., Du, Z., et al.: Forced field extrapolation of the magnetic structure of the H fibrils in the solar chromosphere. ApJ **826**, 51 (2016)
39. Liu, Z., Xu, J., Gu, B., et al.: New vacuum solar telescope and observations with high resolution. RAA **14**, 705 (2014)
40. Jafarzadeh, S., Wedemeyer, S., Fleck, B., et al.: An overall view of temperature oscillations in the solar chromosphere with ALMA. Phil. Trans. R. Soc. A **379**, 20200174 (2021)
41. da Silva Santos, J., Danilovic, S. Leenaarts, J., et al.: Heating of the solar chromosphere through current dissipation. A&A **661**, 59 (2022)
42. Zhao, X., Hoeksema, T.: Unique determination of model coronal magnetic fields using photospheric observations. Sol. Phys. **143**, 41 (1993)
43. Zhao, X., Hoeksema, T.: A coronal magnetic field model with horizontal volume and sheet currents. Sol. Phys. **151**, 91 (1994)
44. Ambrož, P., Druckmüller, M., Galal, A., et al.: 3D coronal structures and magnetic field during the total solar eclipse of 29 March 2006. Sol. Phys. **258**, 243 (2009)
45. Yeates, A., Amari, T., Contopoulos, I., et al.: Global non-potential magnetic models of the solar corona during the March 2015 eclipse. Space Sci. Rev. **214**, 99 (2018)

46. Zhu, X., Wiegelmann, T., Inhester, B.: Preprocessing of vector magnetograms for magnetohydrostatic Extrapolations. A&A **644**, 57 (2020)
47. Wiegelmann, T., Inhester, B., Sakurai, T.: Preprocessing of vector magnetograph data for a nonlinear force-free magnetic field reconstruction. Sol. Phys. **233**, 215 (2006)
48. Zhu, X., Wiegelmann, T.: Toward a fast and consistent approach to modeling solar magnetic fields in multiple layers. A&A **658**, 37 (2022)

Open Access This chapter is licensed under the terms of the Creative Commons Attribution 4.0 International License (http://creativecommons.org/licenses/by/4.0/), which permits use, sharing, adaptation, distribution and reproduction in any medium or format, as long as you give appropriate credit to the original author(s) and the source, provide a link to the Creative Commons license and indicate if changes were made.

The images or other third party material in this chapter are included in the chapter's Creative Commons license, unless indicated otherwise in a credit line to the material. If material is not included in the chapter's Creative Commons license and your intended use is not permitted by statutory regulation or exceeds the permitted use, you will need to obtain permission directly from the copyright holder.

Impact of Interplanetary Conditions on the Magnetosphere/ionosphere Produced by an ICME/SIR Double Structure

Sergio Dasso[1,2(✉)], María Graciela Molina[3,4], Pascal Démoulin[5], Noelia Ayelén Santos[1], Christian Gutierrez[1,2], and Adriana María Gulisano[2,6]

[1] Universidad de Buenos Aires, Facultad de Ciencias Exactas y Naturales, Departamento de Ciencias de la Atmósfera y los Océanos, LAMP, Buenos Aires, Argentina
sdasso@iafe.uba.ar

[2] CONICET - Universidad de Buenos Aires, Instituto de Astronomía y Física del Espacio, LAMP, Buenos Aires, Argentina

[3] Tucumán Space Weather Center (TSWC), Facultad de Ciencias Exactas y Tecnología, Universidad Nacional de Tucumán (FACET-UNT), Tucumán & Consejo Nacional de Investigaciones Científicas y Tecnológicas (CONICET), Tucumán, Argentina

[4] Instituto Nazionale Di Geofísica E Vulcanología (INGV) Roma, Rome, Italy

[5] LESIA, Observatoire de Paris, Université PSL, CNRS, Sorbonne Université, Université Paris Cité, Meudon, France

[6] Instituto Antártico Argentino, Dirección Nacional del Antártico, Buenos Aires, Argentina

Abstract. Stream interaction regions (SIRs) are created from the interaction between fast and slow interplanetary plasma. SIRs at 1 au can cause significant perturbations in geospace, including geomagnetic and ionospheric storms. In this paper we present a case study involving an additional structure in the solar wind before the typical SIR signatures. We find that this first preceding structure corresponds to the flank of an interplanetary coronal mass ejection while the second one is associated to an SIR. Both structures cause significant perturbations in the magnetosphere and ionosphere.

Keywords: Solar Wind · Space Weather · Stream Interaction Regions · Interplanetary Coronal Mass Ejections

1 Introduction

Plasma coming from different regions of the Sun can access to the inter-planetary medium creating solar wind with different bulk velocities. While it is known that solar coronal holes (CHs) are the source regions for fast solar wind, source regions for slow solar wind are still controversial, where the vicinity of coronal regions with closed magnetic loops, typically active regions, is a strong candidate.

Streams interaction regions (SIRs) arise when fast solar wind flow reaches slow solar wind flow. Since the solar wind velocity is radial from the Sun, this interaction is only

© The Author(s) 2026
N. Gopalswamy et al. (Eds.): ISWI 2024, SPPHY 431, pp. 24–32, 2026.
https://doi.org/10.1007/978-981-95-1121-1_3

possible due to solar rotation. The interaction between fast and slow solar wind generates a stream interface, a region where interplanetary plasma and magnetic field are compressed and heated. Plasma flows on both sides are deflected in opposite directions, and interplanetary shocks can form (e.g., [1] and references therein). SIRs develop progressively during their travel from the Sun to the outer heliosphere, and their properties intensify with solar distance. The observation of interplanetary shocks associated with SIRs is more frequent beyond 1 au than closer to the Sun, with a trend to shocks form beyond 2 au (e.g., [2–4]).

The key interplanetary properties affecting geo-space include the interplanetary magnetic field (IMF), in particular the modulus and southern (B_s) component of the IMF, the Sun-Earth solar wind bulk velocity ($V_{Sun\text{-}Earth}$), the solar wind mass density (associated to the proton density, n_p), the level/kind of fluctuations, and the Alfvén Mach number. Two major quantities affecting the geo-space are the dawn-dusk electric field (which can be estimated by combining B_s and $V_{Sun\text{-}Earth}$) and the ram pressure (which can be estimated by combining n_p and $V_{Sun\text{-}Earth}$). All these quantities can significantly change with respect to the typical solar wind conditions when an SIR passes around Earth, and thus produce geomagnetic and ionospheric storms.

Interplanetary coronal mass ejections (ICMEs) are formed by plasma and magnetic field ejected during a solar eruption, typically coming from an active region or a chromospheric filament. Its front part is typically faster than the encountered solar wind, so ICMEs are generally preceded by a sheath region containing plasma and magnetic field of larger values than the ambient solar wind (for more details about ICMEs, see e.g., [5]). ICMEs can be strongly influenced by fast flows coming from behind, with serious consequences on their expected expansion [6].

For some cases, both an ICME and an SIR are present nearby 1 au [7]. This configuration requires specific solar conditions, it depends on the relative position of a coronal hole (CH) and the ICME source region. The presence of near Earth ICMEs and SIRs can strongly affect the interplanetary conditions and thus their impact on the space near Earth.

In the Sect. 2 we present a case study where a non-typical geomagnetic and ionospheric storm were produced in relation with an ICME/SIR configuration. In Sect. 2.1 we present the data and methodology. In Sect. 2.2 we describe the analysis of the interplanetary forcing and the geomagnetic and ionospheric answers. Then, we provide a revised interpretation for the solar origin for these interplanetary conditions. Finally, in Sect. 3 we present our conclusions.

2 An Ionospheric Storm Related with an SIR

In this section we revise a case study of an SIR event that triggered a double geomagnetic storm that was study by [8]. It is also associated with two ionospheric storms. We present observations and describe the geomagnetic and ionospheric storms which occurred around 7 October 2015, and analyze the possible origin of this double structure.

2.1 Data and Methodology Used

We analyze here the IMF and interplanetary bulk velocity, observed by MAG [9] and SWEPAM [10] instruments aboard the Advanced Composition Explorer (ACE) spacecraft, located at the L1 point (i.e., the solar wind conditions before arriving Earth). We analyzed the modulus (B_t) and the three components of the interplanetary magnetic field vector, measured in geocentric solar magnetospheric (GSM) coordinate system.

We analyze the dst index provided by the World Data Center for Geomagnetism, Kyoto [11], as the proxy to study the magnetospheric conditions.To study the state of the ionosphere we present data from instruments of the Tucumán Space Weather Center (TSWC), at Tucumán Argentina (26 ° 51S, 65 ° 12W). We analyze the ionosphere from the study of two quantities:

(a) The critical frequency foF2 corresponds to the maximum frequency which is reflected by the ionosphere when sounded by a Earth-based radio emitter (ionospheric sounder). This provides the temporal evolution of the maximum electron density present in the ionosphere. For more details of the instrument, its calibration and others details about how foF2 was computed, see [8].
(b) The vertically integrated total electron content (vTEC), derived from measurements from a GPS receiver system located at Tucumán, TUCU station (26°50S, 65°13W). This station is part of the Red Argentina de Monitoreo Satelital Continuo (RAMSAC). Details of the calibration and others about how vTEC was computed are also in [8].

For foF2 and vTEC a reference curve is computed, performing an hourly averaged curve using five quiet days before the beginning of the event (see more details in [8]).

2.2 Analysis of the Geomagnetic and Ionospheric Storm of October 2015

SIRs are typically not producing the most intense geomagnetic storms (e.g. [12]). While geomagnetic storms associated with a pure SIR are typically weak (-50 nT $<$ dst ≤ -30 nT) to moderate (-100 nT $<$ dst ≤ -50 nT), [13] and [14] showed that only about 33% of SIRs are associated with moderate to intense storms (dst ≤ -100 nT). The most intense geomagnetic storms are mostly associated with an SIR in interaction with a ICME [7].

The panels (a) and (b) of Fig. 1 show interplanetary conditions for 6–10 October 2015. They show the arrival of the interplanetary driver for the geomagnetic and ionospheric perturbations observed during 7–8 October. Panel (a) shows that the IMF is not a typical magnetic field for a Parkerian solar wind since it reaches almost 20 nT and it has two bumps. It also presents two time ranges showing significant negative values of B_z, reaching values lower than ~ -10 nT in the second one.

In panel (b) the solar wind velocity shows a clear transition from a slow (~400 km/s) to a fast (larger than 700 km/s) solar wind, where a SIR is present (region 2, between second and third vertical dashed lines). This transition is in phase with the second bump of the large value of the IMF. A huge coronal hole, almost facing Earth in the previous days, is present in coronal images (filter of wavelength at 193 Å) obtained from observations made by Atmospheric Imaging Assembly (AIA), aboard Solar Dynamics Observatory

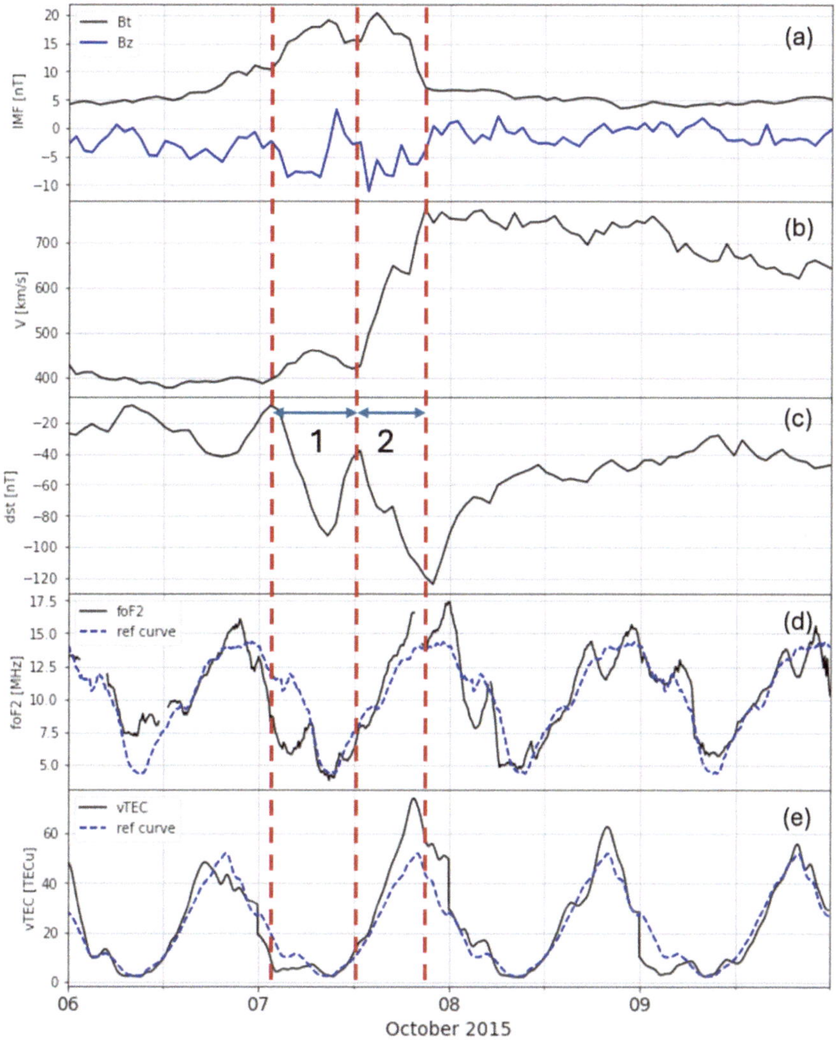

Fig. 1. Interplanetary, magnetospheric and ionospheric parameters are shown from top to bottom, respectively. From upper to bottom panel: (a) IMF intensity B_t and south component B_z (z IMF component in GSM). (b) Bulk solar wind velocity. (c) dst index. (d) Observed critical frequency foF2 shown in black solid line, and reference daily variation using the five nearest quiet days shown as blue dashed line. (e) Calibrated vTEC from RAMSAC shown in black solid line, and reference daily variation using the five nearest quiet days shown as blue dashed line. The first red vertical dashed line indicates the beginning of the IMF increase (in agreement with the first decrease of dst). The second and third red vertical dashed lines indicate the times where the solar wind velocity start/end to significantly increase.

(SDO). This coronal hole is a good candidate for the source of the fast wind observed at the rear of the analyzed SIR ([8], Fig. 2).

From interplanetary observations, [15] conclude that this event was a pure SIR. There is no ICME reported in the catalogue of [5] in a time window centered around this event and larger than 20-days length. The closest ICME before this SIR was reported in this catalogue on 21–22 September, and the closest ICME after this SIR was reported on 25–27 October. However, it is surprising that the IMF strength increases so much before the SIR (i.e., before the second dashed vertical line of Fig. 1). According with the arrival time of this first sub-structure to 1 au (first vertical line at ~02 UT of 7 October) and its bulk velocity (~ 450 km/s), assuming a constant velocity it is expected a travel time of ~3 days and 19 h, thus a launching time of the possible CME around 07 UT of 3 October. We analyzed a catalogue of solar events [16], recently described by [17] and we found that a CME is reported, launched at 06:48 UT of 3 October with a velocity of 535 km/s, a central position angle of 280° (measured from solar north counterclockwise) and an angular width of 14°.

In Fig. 2 the magnetic interplanetary conditions from ACE/MFI are shown to explore the magnetic configuration present within the region 1 of Fig. 1. We find that the time profile of the components of the magnetic field inside this region is consistent with the presence of a flux rope, as shown with the coherent change of By and Bz components. A large Bx component (directed mostly along the spacecraft crossing trajectory) is a signature of a crossing far away from the flux rope axis (i.e., with a large impact parameter). Finally, the magnetic configuration is distorted at its rear as expected from its interaction with the following SIR. The time range where the coherent rotation is observed is in agreement with region 1 since it starts around 2:30 UT of 7 October and it finishes around 19:30 UT. Then, a possible explanation is that the flank of an ICME is associated with the first structure. We conclude that the double geomagnetic and ionospheric behavior is plausibly caused by an ICME immediately followed by an SIR.

We now return to the description of Fig. 1, where the forcing and magnetospheric/ionospheric answer is shown. Panel (c) of Fig. 1 shows the dst index. The presence of two geomagnetic storms (or a two-step storm) is observed. A first storm is present from the early hours of 7 October to ~12 UT (marked as '1' in Fig. 1), reaching the dst peak value of −93 nT on 7 October at 8 UT. It is followed by a second storm that lasts until almost the end of 7 October. This second geomagnetic storm reaches its dst peak ($dst = -124$ nT) value on 7 October at 22 UT. The presence of these two geomagnetic storms is associated with the two interplanetary structures present in panels (a) and (b).

Next, in panels (d,e), we analyze the data of foF2 and vTEC obtained at Tucumán (26°51S, 65°12W). Tucumán is located near the south crest of the equatorial ionization anomaly (its dip latitude is $\approx$15°) and so, its ionization during the day is significantly higher than in other latitudes. Another characteristic of Tucumán is that it is also located close to the South Atlantic Magnetic Anomaly. Thus, due to the complex ionospheric behavior in this location, there is not a standard criterium to determine ionospheric storms from interplanetary origin.

In general, the effects of geomagnetic perturbations on the ionosphere are not easy to analyze. Furthermore, geomagnetic storms can be attenuated due to regional ionospheric conditions and can result in less (or none) evident effects on ionospheric measurable consequences. Changes in the electron concentration in different regions of the ionosphere

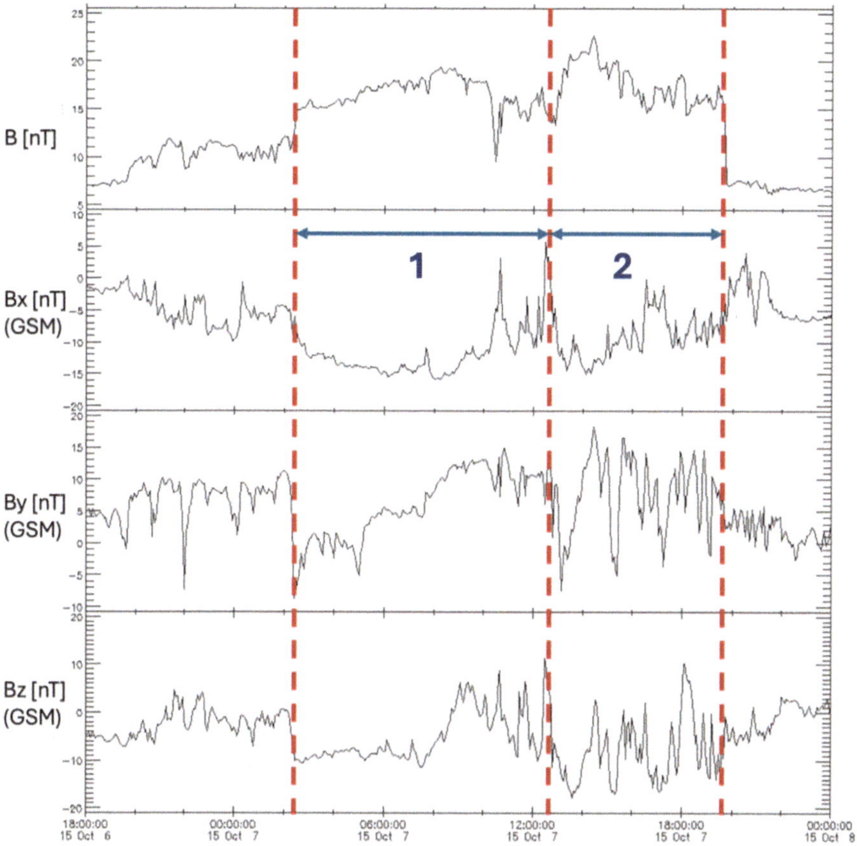

Fig. 2. Magnetic interplanetary conditions. From the upper to bottom panels: modulus of magnetic field and its three components (*Bx*, *By*, and *Bz*) in GSM. Red vertical dashed lines indicate the same times as in Fig. 1, delimiting regions 1 and 2.

are governed by different physical mechanisms, and the reaction of each region to a geomagnetic storm is variable from case to case. The perturbed state of the F2 ionospheric region can be observed as a depletion or an enhancement of the electron content [18]. A significant decrease in the electron density combined with a decrease of foF2 characterizes negative ionospheric storms, while positive storms are characterized by significant increases in the electron density and foF2 (e.g., [19]). Different thresholds can be used to determine a significant in/decrease of electron density or foF2 with respect to the reference value given by the blue dashed line in Fig. 1.d and 1.e. While some authors use a threshold of 20% (e.g., [20]), other authors use 15% (e.g., [21] and [22]). In this work we use a threshold of 20%.

During the storm duration the behavior of the main indicators of the state of the ionosphere, such as foF2 and vTEC, is a combined result of several physical processes, mainly transport/ionization/recombination. The ionospheric response is local and it depends on acting processes originated from above or/and from below. From above, a perturbed interplanetary medium and/or a distorted magnetosphere, induce for instance strong changes on the electric field, distortions of the guiding geomagnetic field, precipitation of particles, among others (see, e.g., [23]). Thermospheric gas heating, thermospheric winds, and TADs are examples of possible processes of ionospheric drivers from below [18].

Observations of foF2 (panel d) and the vTEC behavior (panel e) in region 1 indicate a negative ionospheric storm (vTEC and foF2 data are located below the reference curve). This storm starts at the beginning of 7 October during nighttime (local time is UT - 4). Its temporal evolution is almost in agreement with the main phase of the first geomagnetic storm. Later on, in region 2 both quantities (foF2 and vTEC) are in agreement with a positive ionospheric storm as they show values above the reference curve. It develops during the dayside, in the afternoon on 7 October, during the last stage of the main phase of the second geomagnetic storm (marked as '2'). This suggests a fast mechanism acting in the ionosphere which responds to the double structure observed in the interplanetary medium.

3 Conclusions

In this paper we present a case study of the impact of an interplanetary disturbance on the ionosphere. We re-visited the near Earth interplanetary observations of a previously identified SIR, observed on 7 October 2015, and the magnetospheric and ionospheric consequences. We identify a double bump of the interplanetary magnetic field intensity, with only the second one in agreement with the significant jump of solar wind bulk velocity (from ~400 km/s to ~800 km/s), showing clear signatures of an SIR. Our interpretation is that the first increase of B was caused by the flank of an ICME (reported in a catalogue, and consistent with times and velocities observed). We find that the interacting ICME/SIR caused a two-phases geomagnetic storm and two low latitude ionospheric storms, first a negative and then a positive one. Our results are in agreement with previous studies showing that the most intense geomagnetic and ionospheric perturbations originated by SIRs, are produced by a combined ICME-SIR solar wind structure.

Acknowledgements. We thank the anonymous referee for improving this manuscript. We acknowledge the use of ACE/MFI (PI N. Ness at Bartol Institute) and ACE/SWE (PI D.J. McComas at SWRI) data, and the use of CDAWeb. N.S. and C.G. are fellows of CONICET. S.D., M.G.M., and A.M.G. are a members of the Carrera del Investigador Científico, CONICET. This work was supported by the Argentinean grants PICT 2019-02754 (FONCyT-ANPCyT) and UBACyT-20020190100247BA (UBA).

References

1. Richardson, I.G.: Solar wind stream interaction regions throughout the heliosphere. Living Rev. Sol. Phys. **15**(1), 1 (2018)
2. Gosling, J.T., Hundhausen, A.J., Bame, S.J.: Solar wind stream evolution at large heliocentric distances: experimental demonstration and the test of a model. J. Geophys. Res. **81**(13), 2111–2122 (1976). https://doi.org/10.1029/JA081i013p02111

3. Hundhausen, A.J., Gosling, J.T.: Solar wind structure at large heliocentric distances: an interpretation of Pioneer 10 observations. J. Geophys. Res. **81**(7), 1436–1440 (1976)
4. Smith, E.J., Wolfe, J.H.: Observations of interaction regions and corotating shocks between one and five au: Pioneers 10 and 11. Geophys. Res. Lett. **3**(3), 137–140 (1976)
5. Richardson, I.G., Cane, H.V.: Near-earth interplanetary coronal mass ejections during solar cycle 23 (1996–2009): catalog and summary of properties. Solar Phys. **264**(1), 189–237 (2010)
6. Gulisano, A.M., Démoulin, P., Dasso, S., Ruiz, M.E., Marsch, E.: Global and local expansion of magnetic clouds in the inner heliosphere. Astron. Astrophys. **509**, A39 (2010)
7. Chi, Y., Shen, C., Luo, B., Wang, Y., Xu, M.: Geoeffectiveness of stream interaction regions from 1995 to 2016. Space Weather **16**, 1542 (2018)
8. Molina, M.G., Dasso, S., Mansilla, G., Namour, J.H., Cabrera, M.A., Zuccheretti, E.: Consequences of a solar wind stream interaction region on the low latitude ionosphere: event of 7 October 2015. Solar Phys **295**, 173 (2020)
9. Smith, C.W., L'Heureux, J., Ness, N.F., Acuña, M.H., Burlaga, L.F., Scheifele, J.: The ACE magnetic fields experiment. Space Sci. Rev. **86**, 613–632 (1998)
10. McComas, D.J., et al.: Solar wind electron proton alpha monitor (SWEPAM) for the advanced composition explorer. Space Sci. Rev. **86**, 563–612 (1998)
11. World Data Center for Geomagnetism, Kyoto. http://wdc.kugi.kyoto-u.ac.jp
12. Tsurutani, B., McPherron, R., Gonzalez, W., Lu, G., Gopalswamy, N., Guarnieri, F.: Magnetic Storms Caused by Corotating Solar Wind Streams, Geophys. Monogr. Ser. 167 (2006)
13. Alves, M.V.: Geoeffectiveness of coronating interaction regions as measured by index. J. Geophys. Res. **111**, 0148 (2006)
14. Zhang, J., Richardson, I.G., Webb, D.F.: Interplanetary origin of multiple-dip geomagnetic storms. J. Geophys. Res. **113**, 0148 (2008)
15. Grandin, M., Aikio, A., Kozlovsky, A.: Properties and geoeffectiveness of solar wind high-speed streams and stream interaction regions during Solar Cycles 23 and 24. J. Geophys. Res. **124**(6), 3871 (2019)
16. https://helio.astro.bas.bg/catalog-ars-sc24/
17. Tsvetkov, T., Nakeva, Y., Petrov, N.: Online catalog of activity events of solar cycle 24 related to active regions. Sol. Phys. **299**, 125 (2024)
18. Danilov, A.D.: Ionospheric F-region response to geomagnetic disturbances. J. Atmos. Solar-Terr. Phys. **52**(3), 343 (2013)
19. Zolesi, B., Cander, J.: Ionospheric Prediction and Forecasting. Springer, Berlin (2014)
20. Danilov, A.D.: F2-region response to geomagnetic disturbances. J. Atmos. Solar-Terr. Phys. **63**, 441 (2021)
21. Kutiev, I., Muhtarov, P.: Modelling of midlatitude F region response to geomagnetic activity. J. Geophys. Res. **106**(A8), 15501–15509 (2001)
22. Ippolitol, A., Perrone, L., Plainaki, C., Cesaroni, C.: Investigating the foF2 variations at the Ionospheric Observatory of Rome during different solar cycles minimums and levels of geomagnetic activity. J. Space Weather Space Clim. **10**, 52 (2020)
23. Schunk, R., Nagy, A.: Ionospheres – Physics, Plasma Physics, and Chemistry, 2nd edn. Cambridge University Press, Cambridge (2009)

Open Access This chapter is licensed under the terms of the Creative Commons Attribution 4.0 International License (http://creativecommons.org/licenses/by/4.0/), which permits use, sharing, adaptation, distribution and reproduction in any medium or format, as long as you give appropriate credit to the original author(s) and the source, provide a link to the Creative Commons license and indicate if changes were made.

The images or other third party material in this chapter are included in the chapter's Creative Commons license, unless indicated otherwise in a credit line to the material. If material is not included in the chapter's Creative Commons license and your intended use is not permitted by statutory regulation or exceeds the permitted use, you will need to obtain permission directly from the copyright holder.

Ensemble Simulations of Coronal Mass Ejections in Interplanetary Space with Elliptical Cone Models

Johan Muhamad[1(✉)], Tiar Dani[1,2], Muhamad Z. Nurzaman[1], Rasdewita Kesumaningrum[1], Santi Sulistiani[1], Farahhati Mumtahana[1], Gerhana P. Putri[1], Ayu D. Pangestu[1], and Ahmad Z. Utama[1]

[1] Research Center for Space, National Research and Innovation Agency (BRIN), Bandung 40135, Indonesia
johan.muhamad@brin.go.id
[2] Department of Computer Science and Electronics, Universitas Gadjah Mada, Yogyakarta 55281, Indonesia

Abstract. The estimation of CME arrival time strongly depends on the CME propagation models in interplanetary space and the geometrical aspects of the CME model. We conducted ensemble simulations of CMEs propagation with various elliptical cone shapes to study the relation between the CME speed and the optimum cone shape. We numerically searched for the best elliptical aspect ratio of the elliptical cone for each CME in our CME-ICME pair data. We found that the fast CMEs tend to have a higher elliptical aspect ratio (more circular) than the slower CMEs (flattened). Our results suggest that a fast CME gives a stronger push to all directions, which results in a more circular shape of the leading-edge. We believe that this velocity-dependent behavior is related to the different Lorentz force strengths during the early expansion of a CME.

Keywords: coronal mass ejections · cone model · drag-based model

1 Introduction

A non-constant coronal mass ejection (CME) speed model, i.e., the drag-based model (DBM), has been developed to accommodate the acceleration or deceleration of a CME in interplanetary space due to the presence of a drag force [1]. The model has undergone significant development, progressing from the fundamental one-dimensional (1D) model to more sophisticated variants including the 2D circular cone self-similar, 2D flattening cone, and the ensemble version of the 2D flattening cone DBM [2].

DBM is developed considering the aerodynamic influence of the interaction between CME and ambient plasma in the interplanetary medium [3]. CMEs moving faster than the ambient solar wind speed tend to decelerate, while those moving slower than the ambient solar wind are accelerated. By considering this interaction, for 1D DBM, the

© The Author(s) 2026
N. Gopalswamy et al. (Eds.): ISWI 2024, SPPHY 431, pp. 33–38, 2026.
https://doi.org/10.1007/978-981-95-1121-1_4

distance of a CME from the Sun as a function of time, R_t, can be calculated as [1]:

$$R_t = \frac{S}{\gamma} ln\left[S\gamma(v_0 - \omega)t + 1\right] + \omega t + R_0. \tag{1}$$

v_0 is the initial velocity, γ is the drag parameter, ω is the ambient solar wind speed, R_0 is the initial distance of a CME when the drag force is dominant compared to the Lorentz force, t is time, and S can be 1 or -1 depending on the relative speed between v_0 and ω.

The 2-D implementations typically require a CME cone model to approximate the evolving shape of the CME front during propagation. In a simple 2D DBM, which leading edge of a CME during its propagation is considered to be concentric, the distances of all elements in the leading-edge curvature and the Sun are same. This equidistance can be calculated by using Eq. (1). However, a more realistic representation incorporating an elliptical cone was later introduced to better align with observational data.

An elliptical CME evolution DBM was first developed by [4]. It was then further refined by integrating observational data acquired from STEREO/HI [5]. [6] subsequently conducted ensemble simulations utilizing this method, incorporating a broader range of ensemble parameters. Recently, a new elliptical model was developed by [7] that incorporated frontal deformation.

Our study aims to investigate the relationship between cone ellipticity and CME speed by identifying the optimal elliptical aspect ratio that best aligns with observed CME arrival times. We have developed an elliptical cone DBM with an adjustable aspect ratio for tailored fitting to observations. This study focuses on how CME arrival time depends on the ICME cone shape and initial CME velocity. We expect that this relationship between cone ellipticity and CME initial speed will lead to a simple yet effective model for predicting CME arrival times.

2 Data and Methods

2.1 Data

A set of 75 CME-ICME pairs selected from the interplanetary CME (ICME) database [8], spanning from 1997–2017, is used in our simulations. We only select CMEs that clearly have associated solar flares, neglecting CMEs from quiescent filament eruptions, multiple CME events, and stealth CMEs. The initial CME space speeds are estimated using the de-projected values provided by [9] for the corresponding dataset. The locations of CME sources are determined by the location of the corresponding active region (AR) positions when the associated flare happened. Additionally, data on initial appearance times, speeds, and central position angles of the CMEs are retrieved from the SOHO/LASCO catalog at CDAW (NASA) data center [10].

2.2 Methods

The ICME is modeled as a 2D cone model with the leading edge, which is considered as semi-elliptical shape. The center of this ellipse coincides with the intersection of two lines: one connecting the edges of the ICME legs and another connecting the CME

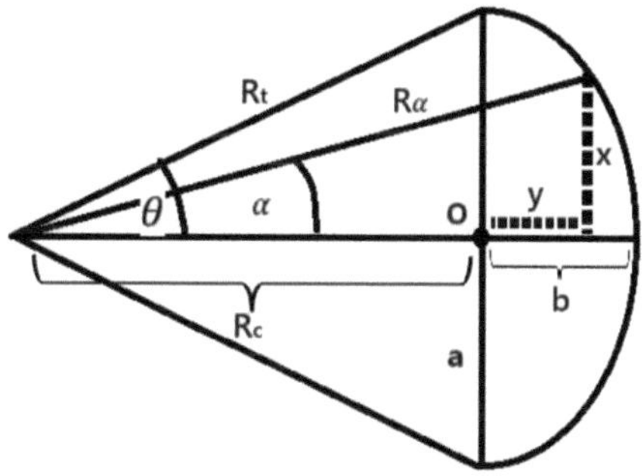

Fig. 1. ICME elliptical cone geometry in the simulations.

source with the apex (see Fig. 1). The semi-minor axis (b) acts as a free parameter, allowing for incremental adjustments to create an adaptable leading-edge that can range from a flattened shape to a semi-circular one. The lengths of the ICME legs (Rt) are equally evolving in time according to Eq. (1) [1]. The cone half-angle, θ, is determined as a function of the projected CME initial speed based on the statistical relationship proposed by [11]. Considering all the geometrical parameters in Fig. 1, the distance of any element of the elliptical leading-edge from the Sun can be calculated as

$$R_\alpha = \frac{R_c + y}{\cos\alpha}, \tag{2}$$

where y can be derived by taking advantage of elliptical formula with pre-defined semi major axis (a) and b,

$$\frac{x^2}{a^2} + \frac{y^2}{b^2} = 1, \tag{3}$$

and substituting x as,

$$x = \tan\alpha(R_c + y). \tag{4}$$

Note that our model is not identical to the model by [4], which determined the ellipse center based on the point of tangency between the ellipse and the ICME legs.

Ensemble simulations are employed to determine the optimal aspect ratio (b/a) of the ellipse for each CME event within the dataset. This optimization process aimed to achieve the best fit between the simulated arrival time and the corresponding observed arrival time. The simulation models ICME propagation by first calculating ICME legs using Eq. (1), and then deriving R_c based on the corresponding cone half angle, θ. Subsequently, it calculates R_α using Eq. (2) for all elements from the ICME apex to the ICME legs by incrementally change α. Note that the semi major axis, a, is constrained by the cone half angle, while b is a free parameter that controls the shape of the cone. This b/a variable is incrementally changed from 0.001 to 1 for flattened to semi-circular cone shape, respectively.

For each CME event, we run the 2D DBM simulation with various aspect ratios and calculate the associated arrival times until we find the one that predicts the arrival time most accurately. The corresponding aspect ratio that can predict the most accurate arrival time is considered as the optimum one. This process is repeated for all CME events.

Since the simulations are based on a pure geometrical approach, actual shapes of the observed CMEs are not fully considered. Therefore, the optimum aspect ratios in the simulations may not accurately reflect the observed CMEs' shapes. This is because the only parameter that constraints the optimization in the simulations is the observed arrival time.

The input parameters obtained from the database are v_0, central position angle, and the longitude of the corresponding AR. The drag parameter, γ, is maintained constant across all simulations at the optimal value of $\gamma = 0.2 \times 10^{-7}$ km^{-1} as suggested by [1]. Similarly, a constant ambient solar wind speed of 450 km/s is adopted in the simulations.

3 Results and Discussions

Our simulations (Fig. 2) reveal a trend towards a more circular leading-edge (higher b/a) for CMEs with higher initial velocities. This relationship strengthens when considering only front-side CMEs (source region longitude $< 30°$) $< 30°$), as depicted by the red circles in Fig. 2. The correlation coefficient (R) between the two parameters is significantly higher for front-side CMEs ($R = 0.65$) compared to the entire dataset, including near-limb CMEs, ($R = 0.28$). This may indicate that ICMEs coming from near the disc center can be estimated using a simple elliptical model during its propagation to Earth. Conversely, near-limb CMEs, due to their more complex trajectories, may require a more intricate model for accurate arrival time prediction.

We further investigate the influence of γ parameter on the relationship between b/a and v_0. Ensemble simulations are conducted using the same dataset but varying γ values. The results are presented in Fig. 3, where the left panel depicts the case with $\gamma = 0.1 \times 10^{-7}$ km^{-1} and the right panel show the case with $\gamma = 0.5 \times 10^{-7}$ km^{-1}. The corresponding correlation coefficient (R) between the two parameters for the front-side CMEs are 0.38 and 0.48, respectively.

These results demonstrate that while the overall trend of aspect ratio dependence on initial speed remains relatively consistent, the specific relationship exhibits sensitivity to the chosen γ within the simulation. Our preliminary findings suggest that a combination of $\gamma = 0.2 \times 10^{-7}$ km^{-1} and $\omega = 450$ km/s offers a sufficient basis for estimating the appropriate aspect ratio based on initial CME velocity. We realize that the calculation of arrival time in the DBM also depends on the preference of ω, which is constant in our simulations. In reality, the value of ω can vary for different CME events. However, for simplicity, we only set the constant value of ω for all cases during the simulations. This uniform ω that is applied for all CME events is useful in our simulations to isolate the influence of cone shape to the arrival time prediction, as our main focus in our current study.

4 Conclusion

Our findings consistently demonstrate that faster CMEs exhibit a more circular leading-edge on the ICME cone. This suggests a more rapid expansion in all directions for faster CMEs compared to their slower counterparts. We believe that this is related to

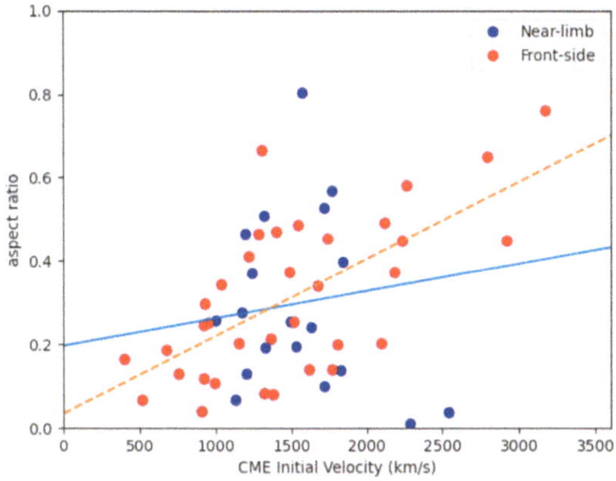

Fig. 2. Plot of the elliptical aspect ratio to the CME initial speed of our simulations for $\gamma = 0.2 \times 10^{-7}$ km^{-1} of the near limb CMEs (blue) and front-side CMEs (red). Blue (orange dashed) line represent the regression line for all (front-side) CMEs.

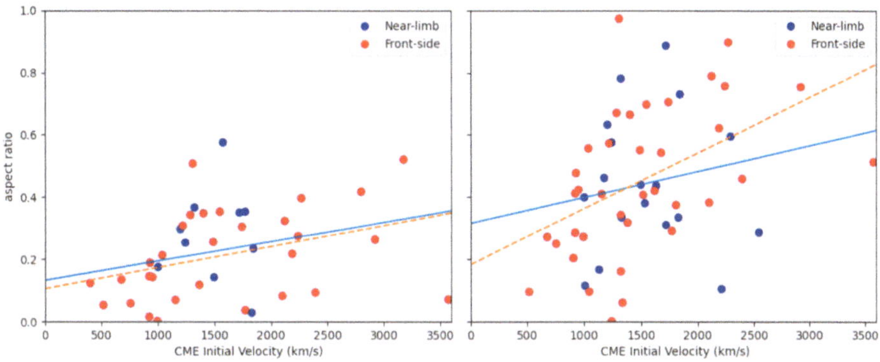

Fig. 3. Plot of the elliptical aspect ratio to the CME initial speed of our simulations for $\gamma = 0.1 \times 10^{-7}$ km^{-1} (left) and $\gamma = 0.5 \times 10^{-7}$ km^{-1} (right). Color code is the same as in Fig. 2.

the strong Lorentz force that act on a flux rope during early expansion of a CME. Our results consistent with previous study that wide CMEs are more impulsive and undergo lateral expansion lower in the corona [12]. However, it should be noted that our study only analyzes limited CME events and relies on assumptions about the constant γ and ω. These limitations may not be relevant when analyzing CMEs that have actual drag parameters and ambient solar winds speeds that largely deviate from the assumed values. Ultimately, our results unveil the potential for developing a simplified 2D DBM model that solely requires adjustments to the elliptical aspect ratio based on the CME's initial speed for CME arrival time estimation. A more extensive study using the 2D DBM is currently underway to investigate the effects of different drag parameters and ambient solar wind conditions on CME propagation, while also considering the actual shape of CMEs. The findings of this study will be published in the near future.

Acknowledgement. This CME catalog is generated and maintained at the CDAW Data Center by NASA and The Catholic University of America in cooperation with the Naval Research Laboratory. SOHO is a project of international cooperation between ESA and NASA.

References

1. Vršnak, B., Žic, T., Vrbanec, D., et al.: Propagation of interplanetary coronal mass ejections: the drag-based model. Solar Phys. **285**, 295–315 (2013)
2. Dumbović, M., Čalogović, J., Martinić, K., et al.: Drag-based model (DBM) tools for forecast of coronal mass ejection arrival time and speed. Front. Astron. Space Sci. (2021). https://doi.org/10.3389/fspas.2021.639986
3. Cargill, P.J.: On the aerodynamic drag force acting on interplanetary coronal mass ejections. Sol. Phys. **221**, 135–149 (2004)
4. Möstl, C., Rollett, T., Frahm, R.A., et al.: Strong coronal channelling and interplanetary evolution of a solar storm up to Earth and Mars 2015. Nat. Commun. **6**, 7135 (2015)
5. Rollett, T., Möstl, C., Isavnin, A., et al.: ElEvoHI: a novel CME prediction tool for heliospheric imaging combining an elliptical front with drag-based model fitting. Astrophys J **824**, 131 (2016)
6. Amerstorfer, T., Möstl, C., Hess, P., et al.: Ensemble prediction of a halo coronal mass ejection using heliospheric imagers. Space Weather **16**, 784–801 (2018)
7. Hinterreiter, J., Amerstorfer, T., Temmer, M., et al.: Drag-Based CME modeling with heliospheric images incorporating frontal deformation: ELEvoHI 2.0. Space Weather **19**, e2021SW002836 (2021)
8. Richardson, I., Cane, H.: Near-Earth Interplanetary Coronal Mass Ejections Since January 1996. https://doi.org/10.7910/DVN/C2MHTH, Harvard Dataverse, V2 (2024)
9. Napoletano, G., Foldes, R., Camporeale, E., et al.: Parameter distributions for the drag-based modeling of CME propagation. Space Weather **20**, e2021SW002925 (2022)
10. CDAW Homepage. https://cdaw.gsfc.nasa.gov/CME_list/. Last access 4 June 2024
11. Gopalaswamy, N., Yashiro, S., Michalek, G., et al.: A catalog of halo coronal mass ejections from SOHO. Sun and Geosphere. **5**(1), 7–16 (2010)
12. Balmaceda, L.A., Vourlidas, A., Stenborg, G.: On the expansion speed of coronal mass ejections: implications for self-similar evolution. Solar Phys. **295**, 107 (2020)

Open Access This chapter is licensed under the terms of the Creative Commons Attribution 4.0 International License (http://creativecommons.org/licenses/by/4.0/), which permits use, sharing, adaptation, distribution and reproduction in any medium or format, as long as you give appropriate credit to the original author(s) and the source, provide a link to the Creative Commons license and indicate if changes were made.

The images or other third party material in this chapter are included in the chapter's Creative Commons license, unless indicated otherwise in a credit line to the material. If material is not included in the chapter's Creative Commons license and your intended use is not permitted by statutory regulation or exceeds the permitted use, you will need to obtain permission directly from the copyright holder.

Models of Quasi-discontinuous Solar Wind Streams

L. Westrich[1,2(✉)], B. M. Shergelashvili[1,2,3], and H. Fichtner[1]

[1] Institute for Theoretical Physics IV, Ruhr-Universität Bochum, 44780 Bochum, Germany
lukas.westrich@ruhr-uni-bochum.de

[2] Centre for Computational Helio Studies, Faculty of Natural Sciences and Medicine, Ilia State University, 0162 Tbilisi, Georgia

[3] Evgeni Kharadze Georgian National Astrophysical Observatory, 0179 Tbilisi, Georgia

Abstract. In this work we examine the heating of the solar wind above the heliobase. Based on the discontinuous solar wind solutions of Shergelashvili et al. [1] we developed new quasi-discontinuous solar wind models in Westrich et al. [2]. First, we will present the basic concept of discontinuous solar wind solutions and the quasi-discontinuous solar wind models, which is basically equivalent to assuming a localized heating source above the heliobase in the case of a 1D quasi-adiabatic, radial expansion of a one-fluid solar wind. Furthermore, we will discuss the differences and the similarities of the discontinuous and quasi-discontinuous models. They contain strong gradients in their physical properties. Therefore, after a discussion of the characteristics of these solutions we will examine how heat conduction could weaken the discontinuities by explicitly including heat conduction in a two-fluid model. We will show that heat conduction is not strong enough to flatten the basic flow structure. As a result, the idea behind these models, i.e., that a damping of plasma waves near the transsonic point could produce such solar wind structures, remains reasonable.

Keywords: methods: numerical · Sun: corona · Sun: heliosphere · solar wind

1 Introduction

In 2020 Shergelashvilli et al. [1] introduced a new type of solar wind solutions with a discontinuity in the physical properties at the transonic point, but with a continuous Mach number. This discontinuity was discussed to be equivalent to an implicit heating at this very narrow region. This very localized heating is unprecedented in solar wind models. Therefore, Westrich et al. [2] introduced a solar wind model with a strongly localized, but spatially expanded heat source.

© The Author(s) 2026
N. Gopalswamy et al. (Eds.): ISWI 2024, SPPHY 431, pp. 39–44, 2026.
https://doi.org/10.1007/978-981-95-1121-1_5

2 Fundamental Idea of the Models

2.1 Parker Solar Wind Model

The Parker solar wind model [3] is the first and therefore the simplest solar wind model one could create. It is a one-fluid, radially symmetric, isothermal, stationary, hydrodynamical model with a central mass M and without a magnetic field. With these assumptions one can easily derive from the mass and momentum continuity the following very famous differential equation for the flow speed v:

$$\frac{dv}{dr} = \frac{v}{r} \frac{C_s^2 - \frac{GM}{r}}{v^2 - C_s^2} \tag{1}$$

Here r is the heliocentric distance, C_s is the speed of sound in the plasma and G is the gravitational constant. This equation has a critical point, where the denominator becomes equal to zero. Therefore, to obtain an acceleration of the plasma from a subsonic regime to a supersonic one the numerator must vanish at this point as well.

2.2 Discontinuous Solar Wind Solutions

The basic concept of the model of Shergelashvili et al. [1] is to make the same assumptions as the Parker model but assume that the temperature, T, behaves according to a polytropic law with the polytropic index γ. From literature (e.g. Shi et al. [4]) is it know that there is no solution for a transonic solar wind for $\gamma \geq 3/2$. However, in Shergelashvilli et al. [1] they presented an approach to solve the differential equations analytically for a polytropic index equal to $5/3$. According to this, they proposed to take two adiabatic expanding analytically solvable branches and combine them with a Mach number equal one at the critical point $r*$, so that the Mach number is continuous, but the physical quantities like number density, flow speed and temperature have a jump at this point. Shergelashvilli et al. [1] derived two quartic equations for the two branches which can be solved for different boundary conditions analytically and presented results for the slow and fast wind. They interpreted the discontinuity as the consequence of an implicit heating source at the critical point.

2.3 Quasi-discontinuous Solar Wind Models

In Westrich et al. [2] quasi-discontinuous solar wind models were introduced. The aim of this paper was to approximate the discontinuous solutions with continuous solutions with very high gradients in the physical quantities in the vicinity of the transonic point to obtain in the limit a quasi-discontinuity. Therefore, they also assumed also a polytropic index of $\gamma = 5/3$ but included an explicit heating source which was assumed to be of the following type

$$Q_H(r) = \frac{T_H}{\sqrt{2\pi\varepsilon^2}} \exp\left(-\frac{(r - r_0)^2}{2\varepsilon^2}\right) \tag{2}$$

where T_H, ε and r_0 is the heating strength, width and position of the heating function. This heating function modifies the polytropic temperature differential equation to:

$$\frac{dT}{dr} = Q_H(r) - (\gamma - 1)T\left(\frac{2}{r} + \frac{1}{v}\frac{dv}{dr}\right) \tag{3}$$

In the limit of a vanishing width the heating function becomes a delta function and is therefore a good approximation to the discontinuous solar wind solutions from Shergelashvili et al. [1]. The aim of the paper Westrich et al. [2] was to verify whether one can reproduce the discontinuous solar wind solutions with a continuous numerical model. To prove this idea, they developed a steady state model which also exhibits critical point r_c at the transonic point and another, time-asymptotic model in which a time dependent MHD-code CRONOS [5], runs long enough to yield a stable solar wind structure. A few results are presented in the next section, for more details, see Westrich et al. [2].

3 Results

We compare the following models: (D) the discontinuous model, (S) a steady-state, quasi-discontinuous model, and (T), a time-asymptotic model. Figure 1 shows two results which are derived in Westrich et al. (2024) [2].

In the upper panel the radial profile of the flow speed for slow wind conditions as determined from models (D) and (S) for different widths ε of the heating function (2). One can see that for the continuous solutions approximate the discontinuous solutions better for smaller widths of the heating function. Westrich et al. [2] present an extreme case where the differences between these two models (D and S) are negligible. In the right-hand panel of Fig. 1 one can see how well the time-asymptotic model relaxes into the stationary solution. The model (T) shows small variations in the subsonic flow region compared to model (S). These variations are caused by small fluctuations at the inner boundary which are a consequence of the open inner boundary condition for the flow speed. This open boundary condition ensures that the system can freely relax into the time-asymptotic model without big restrictions. However, these deviations are small compared to the overall flow structure. Overall, this time-dependent numerical simulation shows that the quasi-discontinuous models are stable in time. Thus, the solar wind relaxes under the influence of localized heating into these quasi-discontinuous flow structures and remains stable.

4 Two-Fluid Solar Wind with Heat Conduction

After this first success, it is necessary to check whether heat conduction could destroy these quasi-discontinuous structures. Westrich et al. [2] discuss a few indications that that should not be the case. However, a more physically consistent model is needed to verify this initial assessment. Therefore, we inserted in a first preliminary consideration a heating function, similar to the model (S) and (T), which heats both the electrons and the protons by the same energy in the two-fluid model with heat conduction taken from

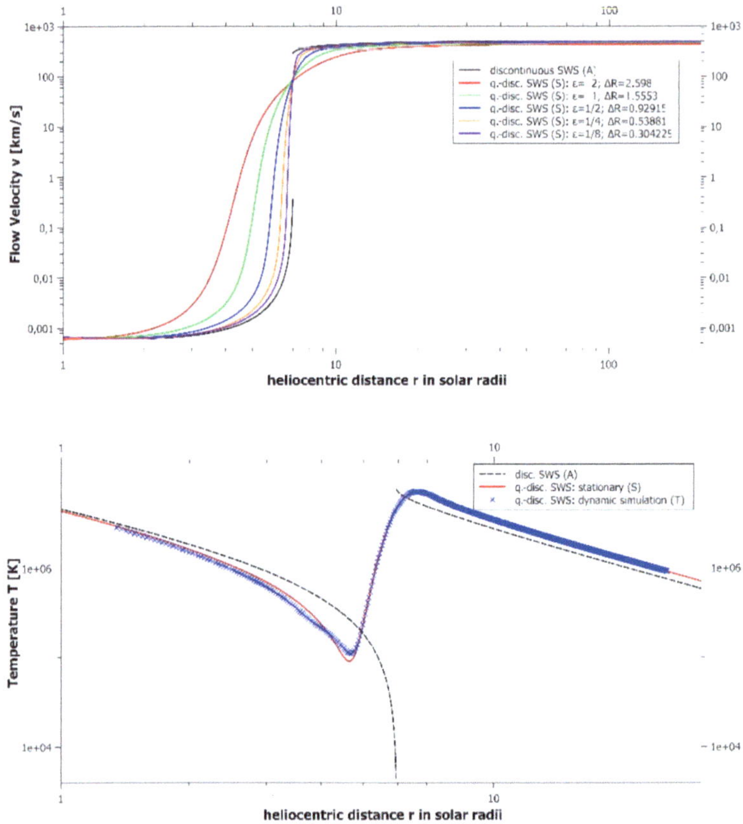

Fig. 1. The upper panel shows radial profiles of the flow speed for different slow wind models. The black curve shows the result of the discontinuous model (D), whereas the colored curves show results for the steady-state, quasi-discontinuous model (S) for different widths of the heating function. The bottom panel shows the temperature profiles for models D (black), S (red), and T (blue). Taken from Westrich et al. [2] with permission

Hartle & Sturrock 1968 [6]. Preliminary results are shown in Fig. 2 which shows that the heating causes a higher flow velocity downstream of the heating region, and a stronger acceleration within. The right-hand panel shows electron and proton temperatures for the two models. The electron temperature profile does not change much for the model with heating (orange) because heat conduction is so efficient that the additional heating is rapidly distributed over the entire simulation domain. However, protons are not as mobile as electrons, and one easily recognizes the very prominent effect of the localized heating.

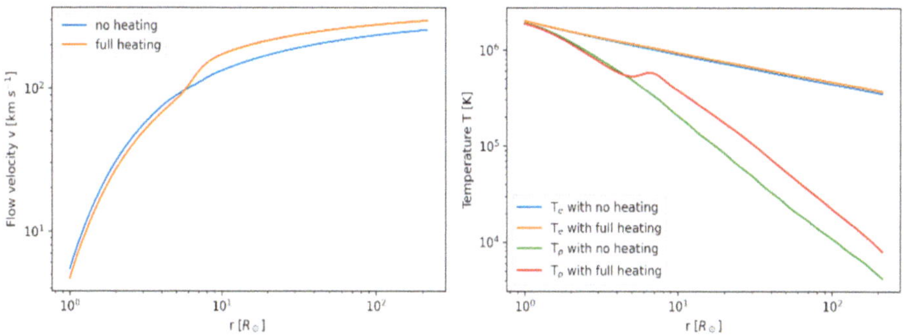

Fig. 2. Radial profiles of the flow speed (left) without (blue) & with (orange) heating and of the electron/proton temperature (right) without (blue/green) & with (orange/red) heating.

5 Conclusion and Outlook

We showed that localized heating can reproduce flow structures of discontinuous solar wind solutions and that they are stable in time. Furthermore, we found that a localized heating the proton population can cause steep gradients in proton temperature and flow speed. Therefore, heat conduction does not necessarily destroy the principal flow structure.

This result motivates the questions we want to answer in the future. We are currently working on an acoustic wave-damping pressure term (like Tu & Marsh 1995 [7] with Alfvén waves) that depends on the physical properties and their gradients in the background flow. The idea is that acoustic waves are reflected and damped near the transonic point which leads to heating of the background flow. This in turn leads to stronger gradients in, e.g., the flow speed and thus to more reflection and damping in this region. Additionally, we believe that these structures could be transient structures in the solar wind. Hence, we plan to simulate time-dependent heating functions in a more complex set-up (i.e., inclusion of magnetic fields, two dimensions, two fluids). Finally, we plan to look for possible observations or evidence of these structures. Unusual Radio Type III Bursts could be a candidate (read more about it in Shergelashvilli et al. 2020 [1]). Furthermore, comparing these solutions to actual measurements of Parker Solar Probe and Solar Orbiter below 20 solar radii could be productive. This temporal acceleration of the solar wind stream could also cause magnetic switchbacks in the solar wind [8].

References

1. Shergelashvili, B.M., Melnik, V.N., Dididze, G., et al.: A new class of discontinuous solar wind solutions. MNRAS **496**, 1023 (2020)
2. Westrich, L., Shergelashvili, B.M., Fichtner, H.: Models of quasi-discontinuous solar-wind streams. A&A **686**, A113 (2024)
3. Parker, E.N.: Dynamics of the interplanetary gas and magnetic fields. ApJ **128**, 664 (1958)
4. Shi, C., et al.: Acceleration of polytropic solar wind: Parker Solar Probe observation and one-dimensional model. Phys. Plasmas **29**, 122901 (2022)

5. Kissmann, R., Kleimann, J., Krebl, B., Wiengarten, T.: The CRONOS code for astrophysical magnetohydrodynamics. ApJS **236**, 53 (2018)
6. Hartle, R.E., Sturrock, P.A.: Two-fluid model of the Solar wind. ApJ **151**, 1155 (1968)
7. Tu, C.-Y., Marsch, E.: MHD structures, waves and turbulence in the solar wind: observations and theories. Space Sci. Rev. **73**, 1 (1995)
8. Raouafi, N.E., Matteini, L., Squire, J., et al.: Parker solar probe: four years of discoveries at solar cycle minimum. Space Sci. Rev. **219**, 8 (2023)

Open Access This chapter is licensed under the terms of the Creative Commons Attribution 4.0 International License (http://creativecommons.org/licenses/by/4.0/), which permits use, sharing, adaptation, distribution and reproduction in any medium or format, as long as you give appropriate credit to the original author(s) and the source, provide a link to the Creative Commons license and indicate if changes were made.

The images or other third party material in this chapter are included in the chapter's Creative Commons license, unless indicated otherwise in a credit line to the material. If material is not included in the chapter's Creative Commons license and your intended use is not permitted by statutory regulation or exceeds the permitted use, you will need to obtain permission directly from the copyright holder.

Flare Production Potential of Sunspot Groups Based on Their Modified Zurich and Magnetic Classifications

Ali Kilcik[1(✉)] and Atila Ozguc[2]

[1] Department of Space Science and Technologies, Faculty of Science, Akdeniz University, 07058 Antalya, Turkey
alikilcik@akdeniz.edu.tr

[2] Kandilli Observatory and Earthquake Research Institute, Bogazici University, 34684 Istanbul, Turkey

Abstract. Sunspot groups are classified into two different ways based on their morphology and evolution (Modified Zurich Classification) and magnetic properties of sunspot groups (Mount Wilson magnetic classification) systems. The modified Zurich classification system is based on three different parameters; the first parameter describes the morphology and evolution of a group, the second parameter describes the shape of the biggest spot in the group, the third one describes the distribution of spots inside a sunspot group and each parameter is divided into sub groups. Here, we used the classes in the third parameter of modified Zurich Classification system and all classes in the Mount Wilson Magnetic classification system. Then we calculated the flare production potential of all classes from the year 1996 to 2023. Our main findings are as follows: (i) More than 50% of all X flares are produced by the compact groups "c" of McIntosh classification and Beta-Gamma-Delta class of magnetic classification systems, (ii) least number of flares are observed in the "x" class of McIntosh classification and alpha class of magnetic classification system, and (iii) "x" McIntosh and alpha magnetic classes are quite compatible with each other in terms of producing flares.

Keywords: Sunspot · Solar flare · Sunspot classification

1 Introduction

Sunspots are temporary dark areas observed on the surface of the Sun due to their low temperature compared to their surroundings. These structures are indicative of a strong magnetic field. They are systematically observed for more than 400 years. Sunspots are usually observed in groups and each group shows some differences from the point of morphology, number of spots, compactness, etc. Sunspot groups are classified in different ways based on their morphology and properties since the beginning of 20th century. The first classification of sunspots made by [1], then [2] introduced Zurich classification system. To better describe the relationship between sunspots and solar flares, the Zurich classification system was modified by [3] and called the McIntosh/modified Zurich

© The Author(s) 2026
N. Gopalswamy et al. (Eds.): ISWI 2024, SPPHY 431, pp. 45–51, 2026.
https://doi.org/10.1007/978-981-95-1121-1_6

classification system. This classification system is based on three different parameters: sunspot group morphology and evolution (A, B, C, D, E, F and H), penumbral characteristic of the main sunspot (largest spot) in the group (x, r, s, a, h, and k) and the internal spot distribution / complexity of the group (x, o, i, and c). The second classification system of sunspot groups is based on their magnetic character and morphology. Mount Wilson magnetic classification system was proposed earlier by George Ellery Hale in 1919 [4] and the current scheme classifies sunspot groups into eight classes (Alpha, Beta, Gamma, Delta, Beta-Gamma, Beta-Delta, Beta-Gamma-Delta, and Gamma-Delta) based on their morphological and magnetic properties together [5].

Sunspots are strongly related to active solar events such as solar flares, coronal mass ejections. Solar flares are one of the strongest eruptive events observed on the sun. They were first discovered by Richard Carrington and Richard Hodgson independently on 1 September 1859 as an increasing brightening on the projected image of the solar disk. The relationship between sunspot activity and flare activity, as well as the physical principles underlying solar flares, has emerged as one of the significant issues [6] due to the space weather effect of these active phenomena.

Solar flares are classified in two different ways based on the observing wavelength. These are the Hα and X-ray classification systems. The Hα classification system is based on the size of bright area and the amount of brightening. According to this classification system 15 different flare classes can be described such as Sf, 2N, 4B, etc. (see, [7], for details). The X-ray classification system of solar flares is based on the peak flux in watts per square meter of soft X-rays at wavelengths 1 to 8 Å as measured by GOES satellites. There are five different classes in this system (A, B, C, M and X). Each category for X-ray flares has nine subdivisions such as B1 to B9, M1 to M9 except X classes that do not have an upper limit. In this classification C1 is 10 times stronger than B1 and X1 is 10 times stronger than M1 etc. [7].

This paper is structured as follows. The data, analysis and results are shown in Sect. 2, and our conclusions and discussion are given in Sect. 3.

2 Data, Analysis and Results

The data used in this study were downloaded from the National Oceanic and Atmospheric Administration (NOAA) Space Weather Prediction Center (SWPC) web page (https://www.swpc.noaa.gov/products/solar-region-summary). The raw data set consisted of a total of 39232 sunspot group measurements which cover the 1996–2023 time intervals. The raw data include the time information, NOAA identification numbers, coordinates, sunspot area, modified Zurich class, sunspot count, Mount Wilson magnetic class, X-ray flare class, etc. of the observed sunspot groups.

We first categorized all sunspot groups into four groups based on the third parameters of modified Zurich classification system as x, o, i, and c. We detected a total of 12302 x-class, 21027 o-class, 4041 i-class and 1862 c-class sunspot groups during the investigated time period. The daily total sunspot counts and areas of these categories and their monthly means were calculated and compared (Fig. 1).

As shown in Fig. 1, we see that: i) both monthly mean sunspot counts and sunspot areas show almost similar trend during the investigated time period; ii) the sunspot counts

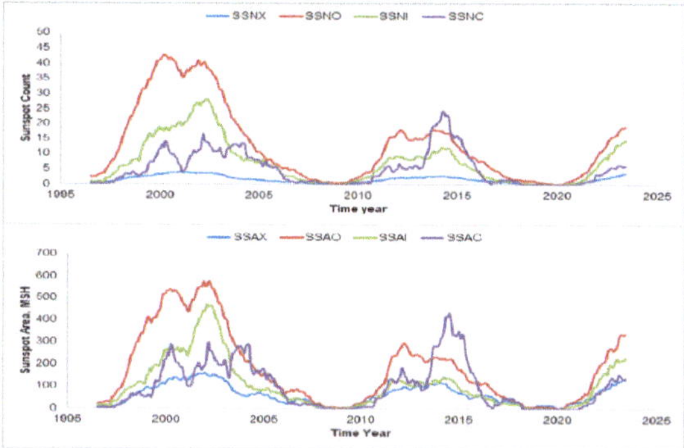

Fig. 1. Temporal variation of monthly mean sunspot counts observed in each category.

and areas generally decreased significantly during solar cycle 24 compared to cycle 23, while they show opposite behavior for the c categories in that the number of c category sunspot groups are always twice during the second peak of the cycle 24; iii) it looks like the amplitude of solar cycle 23 is approached during the solar cycle 25 (the current cycle).

Then, we compared the number of groups in each category with the total number of flare events. The flare production potential (FPP) of each category is calculated as follows:

$$FPP = \frac{\text{Total number of flaring sunspot groups}}{\text{Total number of the same} - \text{class sunspot groups}} \tag{1}$$

As a next step we calculated the average number of flares (AFN) produced by each category as follows;

$$AFN = \frac{\text{Total number of observed flares}}{\text{Total number of the same} - \text{class sunspot groups}} \tag{2}$$

Thus, the obtained results based on the third parameter of modified Zurich classification system are presented in Table 1.

Also all sunspot groups were categorized into seven groups based on their Mount Wilson magnetic classes as alpha, beta, beta-delta, beta-gamma, beta-gamma-delta, gamma and gamma-delta. Then, we performed the same analysis for these categories as shown in Table 2.

From Tables 1 and 2 we obtain following results: i) the x class in the modified Zurich classification system corresponds to the alpha class in the Mount Wilson magnetic classification system since all parameters presented in Table 1 and 2 are almost identical, ii) o class in the modified Zurich classification system shows similar behavior with beta class in the Mount Wilson magnetic classification system, iii) the other two classes (i and c) are distributed into five classes in the Mount Wilson magnetic classification system, iv) all gamma and gamma-delta classes produced at least one flare except one sunspot group out of 15. Note that two active regions (NOAA AR11430 in March 6, 2012 and 11460 in April 16, 2012) do not have magnetic classification information for these days

Table 1. FPP of sunspot groups based on the third parameter of modified Zurich classification system and some other statistics.

Third Parameter of Zurich Class	Total Number of Groups	Total Number of Flaring Groups	Total Number of Observed Flares	FPP	AFN	X-Ray Flare Class		
						C	M	X
x	12302	782	1338	0.06	0.11	1243	88	7
o	21027	3714	7342	0.18	0.35	6656	658	28
i	4041	2015	5336	0.50	1.32	4777	522	37
c	1862	1376	5447	0.74	2.93	4492	866	89

Table 2. Flare production potential of sunspot groups based on the Mount Wilson Magnetic classification system and some other statistics. Here A; alpha, B; beta, B-D; beta-delta, B-G; beta-gamma, B-G-D; beta-gamma-delta, G; gamma, and G-D; gamma-delta.

Mount Wilson Magnetic class	Total Number of Groups	Total Number of Flaring Groups	Total Number of Observed Flares	FPP	AFN	X-Ray Flare Class		
						C	M	X
A	12283	781	1336	0.06	0.11	1241	88	7
B	22261	4122	8472	0.19	0.38	7742	699	31
B-D	267	181	621	0.68	2.38	497	113	11
B-G	3308	1904	5203	0.58	1.57	4641	538	24
B-G-D	1096	885	3773	0.81	3.44	2996	690	87
G	5	5	21	1.0	4.2	19	2	0
G-D	10	9	37	0.9	3.7	32	4	1

and they have no flaring activity. To clarify the first two points we compared the two classification systems and the results are presented in Table 3.

As shown in Table 3, almost all alpha classes in the Mount Wilson magnetic classification system correspond to x classes in the modified Zurich classification system (99.7%). Similar to the alpha classes, 89 percent of all beta classes correspond to o classes. The remaining 11% in the beta classes distributed to the i (10%) and c (1%) classes in the modified Zurich classification system. Finally, classes i and c in the modified Zurich classification system are distributed to other classes in the Mount Wilson magnetic classification system.

Table 3. Comparison of the third parameter of Modified Zurich Classification and Mount Wilson Magnetic Classification Systems. Here A; alpha, B; beta, B-D; beta-delta, B-G; beta-gamma, B-G-D; beta-gamma-delta, G; gamma, and G-D; gamma-delta.

Mount Wilson Magnetic Class	Third Parameter of Modified Zurich Classes				Total
	x (12302)	o (21027)	i (4041)	c (1862)	
A	12255(99.7%)	26	0	2	12283
B	45	19903(89%)	2046(10%)	267(1%)	22261
B-D	0	78(29%)	80(30%)	109(41%)	267
B-G	1	925(28%)	1606 (49%)	776(23%)	3308
B-G-D	1	92(8%)	301(28%)	702(64%)	1096
G	0	0	3(60%)	2(40%)	5
G-D	0	1(10%)	5(50%)	4(40%)	10
Total	12302	21025	4041	1862	

3 Discussion and Conclusions

We compared the third parameter of the modified Zurich and all classes in the Mount Wilson Magnetic classification systems. Then we calculated the flare production potential of all classes from the year 1996 to 2023. Our main findings are as follows:

- Sunspot counts and areas generally decreased significantly during solar cycle 24 compared to cycle 23, while they show opposite behavior for the c categories in that the number of c category sunspot groups are always twice during the second peak of the cycle 24.
- The amplitude of solar cycle 23 is being approached during the solar cycle 25 (the current cycle).
- Almost all alpha and beta classes in the Mount Wilson magnetic classification system correspond to x and o classes in the modified Zurich classification system. Classes i and c in the modified Zurich classification system are distributed to other classes in the Mount Wilson magnetic classification system.
- FPP of compact groups i and c of the McIntosh classification and all classes except alpha/beta classes of magnetic classification systems are greater than 50%.

[8] analyzed the temporal variation of solar flare index and geomagnetic Ap, Dst, Scalar B and aa indices for the 1975–2020 time intervals. They found that all indices strongly decreased during the solar cycle 24 compared to cycle 23. Recently, [9] predicted energetic index which describe the energy supplied from the solar wind to the magnetosphere, from the data of the Lviv and Belsk geomagnetic observatories and found that the maximum of geomagnetic activity will occur around 2026 and its magnitude will be $1.5 - 2$ times higher than the maximum of the 24th cycle. Our results show good agreement with above studies. We further found that the sunspot number and areas currently higher compared to Cycle 24 except in c categories. This is consistent with the estimate of the strength of solar cycle 25 from other considerations [10].

It has been known for a long time that sunspot magnetic configurations with delta class have a tendency to produce more solar flares [11]. Later [12] found that the more complex sunspot classes produce more flare activity. This finding is statistically confirmed by different authors ([7, 13]). Some of the above authors only used magnetic classification, some others used McIntosh classification. In this study, we used both classification systems and calculated the flare production potential (FPP) based on the third parameter of McIntosh classification system and all magnetic classes. We confirm the above results for the 1996–2023 time intervals (last two and a half cycles). We further calculated the FPP and AFN produced by each category of different classification systems.

Acknowledgement. We would like to thank the referee for his constructive comments and suggestions. This study was presented with the support provided by the organizing committee.

References

1. Cortie, A.L.: On the types of sun-spot disturbances. ApJ **13**, 260 (1901)
2. Waldmeier, M.: Chromosphärische eruptionen. I. Mit 6 abbildungen. Z. Astrophys. **16**, 276 (1938)
3. McIntosh, P.S.: The classification of sunspot groups. SolPhys. **125**, 251–267 (1990)
4. Hale, G.E., Ellerman, F., Nicholson, S.B., Joy, A.H.: The magnetic polarity of sun-spots. ApJ **49**, 153 (1919)
5. Jaeggli, S.A., Norton, A.A.: The magnetic classification of solar active regions 1992–2015. ApJL **820**, L11 (2016)
6. Hudson, H.S.: Global properties of solar flares. Space Sci. Rev. **158**, 5 (2011)
7. Eren, S., et al.: Flare-production potential associated with different sunspot groups. MNRAS **465**, 68–75 (2017)
8. Ozguc, A., Kilcik, A., Yurchyshyn, V.: Temporal and periodic variations of the solar flare index during the last four solar cycles and their association with selected geomagnetic-activity parameters. SolPhy. **297**, 112 (2022)
9. Sumaruk, T.P., Sumaruk, P.V., Neska, A.: Solar and geomagnetic activity in 19–25 cycles. Odessa Astron. Public. **36**, 188–192 (2023). https://doi.org/10.18524/1810-4215.2023.36.288658
10. Gopalswamy, N., Michalek, G., Yashiro, S., Mäkelä, P., Akiyama, S., Xie, H.: What do halo CMEs tell us about solar cycle 25? ApJL **952**, L13 (2023)
11. Kunzel, H.: Die Flare-Häufigkeit in Fleckengruppen unterschiedlicher Klasse und magnetischer Struktur. AN, 285, 271 (1960)
12. Smith, S. F., Howard, R.: Magnetic Classification of Active Regions. In: Structure and Development of Solar Active Regions. Edited by Karl Otto Kiepenheuer. IAUS 35, p. 33. Dordrecht, D. Reidel (1968)
13. Guo, J., Lin, J., Deng, Y.: The dependence of flares on the magnetic classification of the source regions in solar cycles 22–23. MNRAS **441**, 2208 (2014)

Open Access This chapter is licensed under the terms of the Creative Commons Attribution 4.0 International License (http://creativecommons.org/licenses/by/4.0/), which permits use, sharing, adaptation, distribution and reproduction in any medium or format, as long as you give appropriate credit to the original author(s) and the source, provide a link to the Creative Commons license and indicate if changes were made.

The images or other third party material in this chapter are included in the chapter's Creative Commons license, unless indicated otherwise in a credit line to the material. If material is not included in the chapter's Creative Commons license and your intended use is not permitted by statutory regulation or exceeds the permitted use, you will need to obtain permission directly from the copyright holder.

Space Weather Extreme Events and Geomagnetic Storms

Extreme Space Weather Events

Natalia Buzulukova[1,2,3(✉)], Janet Kozyra[4], Timothy Fuller-Rowell[5,6], and Anton Dorodnitsyn[2,3,7]

[1] Department of Astronomy, University of Maryland College Park, College Park, MD, USA
nbuzulukova@gmail.com
[2] Partnership for Heliophysics and Space Environment Research, College Park, MD, USA
[3] NASA Goddard Space Flight Center, College Park, MD, USA
[4] NASA Headquarters, Washington, DC, USA
[5] NOAA, Boulder, CO, USA
[6] University of Colorado, Boulder, CO, USA
[7] University of Maryland, Baltimore, MD, USA

Abstract. Extreme space weather can be defined as an event with variability in solar, ionospheric, thermospheric or magnetospheric parameters that cause significant degradation of certain technological systems or pose a serious risk to human health. Space weather effects/impacts can be divided into five broad groups: 1) ionospheric effects; 2) ionizing radiation/plasma effects on s/c electronics; 3) ionizing radiation effects on human health; 4) atmospheric uplifting and satellite drag; 5) geomagnetically induced currents (GICs). Extreme space weather can be traced to their solar and interplanetary origins, including solar flares, interplanetary coronal mass ejections and shock waves, and solar energetic particles. By fitting the tail of the cumulative distributions of space weather parameters, it is possible to estimate the probability of extremes. Historical examples show that before the space age, the most common examples of extreme space weather were extreme geomagnetic storms with intense GICs in ground-based infrastructure including power grids. Advances in technologies are expanding the area of extreme space weather. With the number of space objects increasing dramatically each year, satellite drag effects are becoming a concern. With thousands of aircraft maneuvering in response to a geomagnetic storm, it could theoretically be possible to achieve extreme space weather conditions for an event that would otherwise be considered moderate.

Keywords: Carrington Event · Natural Hazards · Space Weather

1 Introduction

Space weather can be defined as a branch of science that studies the variability of solar, geomagnetic, and ionospheric-thermosphere conditions that can potentially have adverse effects on technological infrastructure, electronics, and spacecraft operations, as well as human health. An extreme space weather event can be defined as an event with such a degree of variability (e.g., variations in the magnetic field at ground level,

© The Author(s) 2026
N. Gopalswamy et al. (Eds.): ISWI 2024, SPPHY 431, pp. 55–63, 2026.
https://doi.org/10.1007/978-981-95-1121-1_7

variations in atmospheric density at 500 km altitude of the Starlink satellite, or variations in the radiation environment of the International Space Station/open space) that it can cause significant degradation, to the point of rendering certain technological systems inoperable or destroyed (e.g., power grids or spacecraft operations), or pose a serious hazard to human health during spaceflight. The last 15 years have been relatively mild in terms of extreme space weather, likely due to a mild Solar Cycle 24. On May 10–11, 2024, a very strong geomagnetic storm struck the Earth's magnetosphere, causing the hourly average Disturbance Storm Time (Dst) index to drop below -400 nT. The closest geomagnetic storm in time and intensity (as measured by Dst) occurred on November 20–21, 2003, over 20 years ago. At the request of the U.S. space weather community, the May 2024 geomagnetic storm was named the Gannon storm, in memory of a prominent American geophysicist, Dr. Jennifer Gannon, who passed away unexpectedly on May 2, 2024.

With the awareness of an approaching solar maximum, Gannon storm has been very important to monitor with the modern geospatial/heliophysics fleet and ground-based networks of magnetometers, ionosondes and radars. The evolution of mobile phone cameras over the past 20 years has made it possible to capture photos and videos of prolific auroral activity around the world, sparking interest in space weather in popular mass media and social networks and comparing the event to past extreme space events, including the famous Halloween events (October 28–November 4, 2003) and the Carrington event of September 1–2, 1859. However, the Gannon storm does not appear to have caused significant destruction like the Halloween events, although performance degradations of various services have been reported, including GPS, high-frequency (HF) radio communications, and power grid irregularities [1]. According to the Reuters report [2], the Starlink system has experienced some degradation of services, but no major issues have been reported compared to the famous case of February 3, 2022, when SpaceX lost 38 of 49 launched Starlink satellites during a very modest double-dip geomagnetic storm (in the Dst index) on February 3–4, 2022 [3].

2 Major Space Weather Impacts and Their Solar Origins

Space weather effects/impacts can be divided into five major groups: 1) ionospheric effects, including changes in ionization rate, electron density profiles, and their impact on communication and navigation systems including GNSS systems; 2) impact of ionizing radiation/plasma on s/c electronics/avionics; 3) impact of ionizing radiation on human health; 4) upper atmosphere uplifting and satellite drag; and 5) impact of geomagnetically induced currents (GICs) on ground-based infrastructure, including power grids. This division is not unique due to the multidisciplinary nature of space weather science. For more detailed information, please see [4] and the references therein.

Figure 1 shows the relationship between the major space weather effects/impacts and their solar origins. Active regions (ARs) and coronal holes (CHs) are the main sources of variability in solar space weather processes. ARs consist of groups of sunspots with varying magnetic polarities, which can lead to magnetic reconnection, a process that results in energy release and solar flares. Flares emit radiation across the entire electromagnetic spectrum, from X-rays to visible light and radio waves, causing changes in the ionospheric ionization rate and the electron density profile in the Earth's ionosphere, thereby

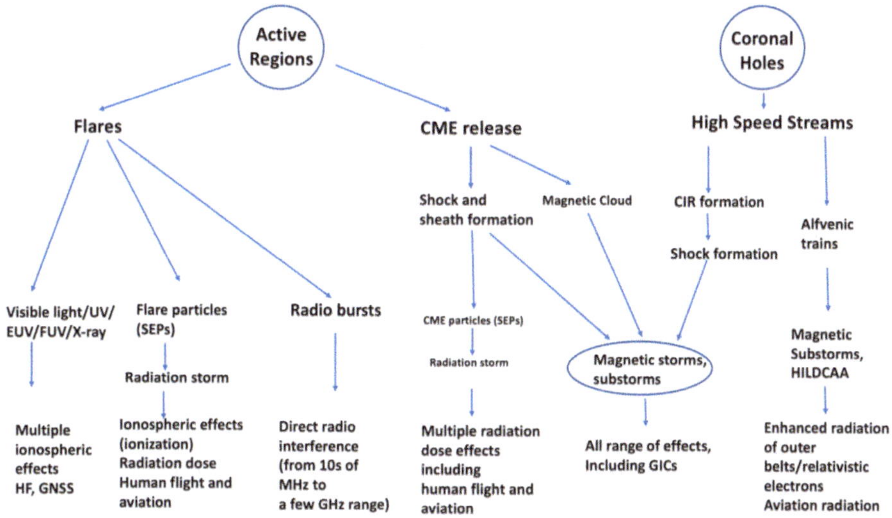

Fig. 1. Relationship between major space weather effects/impacts and their solar origins. Adapted from [4] under Creative Commons Attribution License.

affecting radio propagation (e.g., "radio blackouts") and GNSS systems, either through changes/fluctuations in the electron density profile or through direct interference with radio signal. Coronal mass ejections (CMEs), another result of magnetic reconnection, are large expulsions of plasma and magnetic fields that cause geomagnetic storms upon interaction with the Earth's magnetosphere, under a preferable orientation of the CME's magnetic field relative to the Earth's magnetic field. There is no one-to-one association between solar flares and CMEs; statistically, only ~ 20% of C-class flares are thought to be associated with CMEs, but this number is closer to 100% for X-class flares [5]. For the discussion of extreme events, it suffices to specify that all extreme solar flares (e.g., X3 and above) are eruptive, i.e., accompanied by CMEs, and that all extreme CMEs (with velocity V ~ 2000 km/s) are accompanied by solar flares. This is probably not the case for energetic stellar flares, which seem to tend to produce less energetic CMEs (with lower velocity/kinetic energy), due to a stronger background magnetic field and a greater drag effect from the ambient stellar wind [6].

A shock wave forms when an interplanetary CME (ICME) moves faster than the upstream ambient solar wind, by a speed greater than magnetosonic speed. The shock changes the properties of the ambient plasma by compressing and heating it. The region between the shock and the ICME is called the sheath. It has been shown that the increased magnetic field amplitude in the sheath can cause intense geomagnetic storms [e.g., 7] or cause double-dip storms (in terms of Dst index), where the first "dip" is caused by the sheath and the second by the ICME. When the shock strikes the Earth's magnetosphere, the increased level of plasma and magnetic field fluctuations in the sheath region produce reconfigurations in the global magnetospheric current system, causing the development of intense GICs in the Earth's infrastructure. In the case of extreme GICs, there could be power outages/blackouts in different countries/continents. Some famous and well-documented examples will be discussed below.

CMEs and flares are responsible for radiation storms on Earth, in space or on other planets (e.g. Mars), by generating solar energetic particles (SEPs) with energies up to GeV. SEPs are thought to originate either directly from flares or from CME shocks (see [8] for discussion of different SEP mechanisms). The most extreme SEPs cause ground-level enhancement (GLE) radiation at the Earth's surface and produce cosmogenic isotopes that could be used to estimate the intensity of historical extreme SEPs. It is important to note that radiation storms and geomagnetic storms are related to the same solar origins (flares and CMEs) but could occur independently on Earth.

3 What is an Extreme Geomagnetic Storm?

Geomagnetic storms are often discussed in connection with extreme space weather on Earth. Geomagnetic storms are intervals of increased energy input into the coupled magnetosphere-ionosphere-thermosphere system, caused by a southward-oriented interplanetary magnetic field (opposite to the Earth dipole), in an ICME or sheath, or corotation interaction region (CIR). When there is a significant accumulation of thermal energy in the form of energetic particles (1–500 keV) drifting around the Earth, and the diamagnetic effect of these drifts as measured by -Dst becomes greater than 50 nT, this signals the onset of a geomagnetic storm. The empirical relationship for the time derivative of the Dst index corrected for solar wind dynamic pressure can be expressed as [9]:

$$\frac{dDst(t)}{dt} = Q(t) - \frac{Dst(t)}{\tau}$$

where Q(t) is called the injection function and depends linearly on $V_{sw}B_s$; V_{sw} is the V_x component of the solar wind velocity in the GSM system; B_s is the southward component of B_z and τ is the ring current decay rate [10]. For envelope estimation, the decay term Dst(t)/τ could be omitted, and Dst_{min} becomes proportional to the integral of $V_{sw}B_s$. Dst_{min} normally defines the storm strength. How large should the Dst dropout be for an extreme storm? As noted in [11], there is no single quantification scale, some subjectivity will always be present and different authors will use different definitions for "extreme", "severe", "large", "strong". Some authors use classification based on Dst_{min} and define a 'great' storm as an event with $Dst_{min} < -350$ nT, while other authors define 'great' storm as an event occurring once per decade, or once per 100 years. How to define the strength of some historical events, when magnetometer measurements of Dst index were not available? As suggested by [11], such a measure of a storm intensity (more precisely auroral storm intensity) would correspond to auroras visible from a geomagnetic latitude of 30°. Based on an analysis of the available published literature, they concluded that a once-in-500-year event would be the 1770 event, which probably lasted up to 9 days.

4 Examples of Historic Geomagnetic Storms with Extreme Space Weather

The Carrington event of 1–2 September 1859, perhaps the most widely cited example of extreme space weather, began with a white flare visible on September 1st. The flare caused a clear "magnetic crochet" signature, or solar flare effect, SFE, of 110 nT recorded by the Greenwich magnetometer. The SFE has the same origins as ionospheric blackouts in that the flare radiation deposits energy in the Earth's ionosphere, alters ionospheric ionization and conductivity, and causes a transient current system and a characteristic trace on magnetograms. Only very intense flares can produce an SFE of >90 nT. The Carrington soft X-ray flare (SXR) class has been estimated to be X45 $\pm$ 5 [12]. About 17.5 h after the SFE was recorded, an extreme geomagnetic storm arrived at Earth. From the transit time, the average speed of the shock was estimated to be about 2380 km/s [13]. Based on the Colaba magnetometer recordings, the strength of the storm measured by Dst dropout was estimated by [13] to be -1760 nT, making the Carrington event the largest geomagnetic storm ever recorded. Currently, it is discussed that the strength of the Carrington event may be overestimated due to the ionospheric effect [12], giving estimates of Dst ~ -900 nT for the Carrington event and note that a similar Dst (in the range -825– -900 nT) was recorded for the great storm of May 1921. What made the Carrington event arguably the "worst case" space weather event on record was not the extreme Dst dropout and remarkable auroral activity at low latitudes across the globe, but the widespread disruptions and failures of telegraph systems [14], as well as reported cases of equipment overheating and fires at telegraph stations.

In this regard, the storm of 13–16 May 1921 is also considered an extreme space weather event comparable to the Carrington case, not only because of the extreme Dst dropout (~ -900 nT) and low-latitude auroral observations, but also because of multiple reports of intense GICs and their negative impacts, including outages and disruptions of telephone and telegraph services on different continents/countries, including Europe, Australia, New Zealand, Japan, the United States, Canada, and in some cases damaged equipment [15].

Another example of extreme space weather is the great geomagnetic storm of 12–13 March 1989, referred to by the mass media as "The Great Quebec Blackout", "A Scary 13th", and "The Day the Sun Brought Darkness". There were multiple solar flares from March 10 to 12, including several X-type flares, and two ICME shocks encountering Earth on March 13 [16]. The shock velocity of the second ICME was estimated at 1,350 km/s [16]. It was the arrival of the second ICME at 07:43 UT (as seen on ground magnetograms) that triggered the famous Hydro-Québec blackout at 07:44–07:45 UT [16]. Dst_{min} for the March 1989 storm was ~ -600 nT; not so extreme as for the May 1921 storm.

The famous Halloween events of October–November 2003 began on October 28 with the X17 solar flare, one of the most powerful flares ever recorded, with thc subsequent release of coronal mass ejections (CMEs) at an estimated velocity of ~2125 km/s and an Earth transit time of ~19 h (note a longer transit time than for the Carrington event). The second powerful X10-class flare occurred the next day and caused the release of a second CME at an estimated velocity of ~1948 km/s and a transit time of ~19 h [17]. The two ICMEs encountered Earth one after the other, causing a powerful composite geomagnetic storm with 3 distinct Dst dropouts, and nearly all of the types of space weather effects/impacts reflected in Table 1. As reported by NOAA [17], strong GICs and power outages occurred over Europe (see also [18]); the intensity of the radiation storm was high enough to order astronauts on the International Space Station to seek shelter; satellites and the GPS system experienced service degradation and outages;

airlines rerouted flights to avoid high radiation and communication blackout areas; and the storm likely caused a loss of the $600 million ADEOS-2 spacecraft [17]. The lowest Dst depression during the Halloween storm was −383 nT, not as extreme as the March 1989 storm and the May 1921 storm.

The solar AR 486 that produced the Halloween storm maintained its complexity, extreme size, and flare potential throughout its transit across the solar disk. The size of AR 486 was calculated in [8] and found to decrease in size by about 10% during the Halloween flares (and hence its magnetic energy), but started to grow considerably after that, culminating in the release of an extreme flare, the most powerful ever recorded, on November 4, 2003, causing saturation of the GOES X-ray detectors, with an estimated class of X28 made by NOAA, and X40 ± 5 after corrections made from observations of the D-region ionospheric absorption [19]. The extreme flare was also accompanied by an extreme CME. Fortunately, this flare occurred near the edge of the Sun and CME was not directed toward Earth, so the impact was limited.

5 Statistics of Extreme Events

Decades of observations of solar events and historical records of solar activity (e.g., active region area) allow to calculate cumulative distributions of various space weather phenomena. Fitting the tail of the distribution function with power law or other statistics could be used to estimate probabilities of extreme space weather events. Examples of distributions for CME velocity, kinetic energy, flare size, SEP peak intensity and fluence, active region area can be found in [8]. [20] provides a literature review of studies on the waiting time for geomagnetic storms. From Table 6 of [20], the waiting time for a geomagnetic storm with a Dst dropout from −589 to −663 varies from 25 to 100 years provided by different studies.

Additional data/assumptions are needed to improve the uncertainty of extreme events. As discussed in [21], knowledge of the theoretical threshold for a quantity of interest could be used to improve statistics, e.g. for geomagnetic storms. This is currently an area of active development, where first-principles numerical models could be useful. Another solution is to extend the historical data over longer periods, e.g. for extreme SEPs. It is now generally accepted that the unusually high concentration of the cosmogenic isotope ^{14}C found in Japanese cedars indicates extreme events of solar origin, for 774–775 AD [22] and 992–993 AD [23], now known as the “Miyake events”. Considering energetic solar protons $E > 200$ MeV as the main source of ^{14}C and using data for protons of fluence $E > 200$ MeV from the extreme GLE of 23 February 1956, the following estimates are given in [20] for the AD 774 flare class: X285 if the event was a single flare; X180 if there were two equal successive flares; X140 if there were 3 successive flares separated by a few months. Please see [20] for details of the methodology and discussion of uncertainties. Notably, the waiting time estimates provided by [8] for the X100 flare are about 1000 years, which is within the uncertainty range of the AD 774 event.

For the completeness we note that observations of stellar flares on solar-type and Sun-like starts could be used to expand the statistics of extreme flares. Early results suggested that an X1000-class superflare could be emitted every 800 years [24], which seems too frequent for our Sun. Subsequent analysis showed that a careful selection of Sun-like stars with $T_{rotation} > 20$ days is required, otherwise the analysis is likely dominated by fast-rotating stars with very energetic flares [20]. Currently, the ability of our Sun to emit X1000-sized flares is still in question.

6 Relationship Between Extreme Geomagnetic Storms and Extreme Space Weather

The known historical extremes of extreme geomagnetic storms are associated with the impact from intense GICs, causing telegraph/telephone/power grid systems outages. As noted above, the strength of a geomagnetic storm is defined by the integral of $V_{SW}B_S$. One can imagine an extreme storm with an extreme Dst dropout produced by smoothly increasing and then constant $V_{SW}B_S$, but without significant variations dB_S/dt or pressure pulses found in the sheath/shock region, thus likely with limited intense GIC formation. This thought experiment demonstrates that the relationship between the strength of the storm and intense GICs is not straightforward. An example to consider is the geomagnetic storm of November 20–21, 2003. The Dst dropout for the November event was −400 nT, larger than for the Halloween events, however, to our knowledge, no intense GICs were reported, and the overall impact was milder than that of the Halloween events. Notably, the ICME speed was very different for two events; according to the OMNI database for solar wind speed [25], the solar wind speed was about 2000 km/s for the Halloween event and about 750 km/s for the November event. The OMNI solar wind speed for the May 2024 Gannon storm was 700–900 km/s, making the Gannon event closer in this respect to the November event, not the Halloween event. Although more detailed study is needed, it could be hypothesized that the slower ICME speed for the Gannon storm was the reason for the overall milder GIC response, although the storm strength as measured by Dst was on the extreme side and close to the November 2003 case.

The $V_{SW}B_S$ integral is recognized as the main driving force of the magnetosphere, as it is related to the energy injected by the solar wind into the magnetosphere and ionosphere. Enhanced ionospheric drifts produce Joule heating, and this energy is transferred to the neutrals through collisions, causing uplifting of the upper atmosphere (thermosphere). This effect produces neutral density fluctuations along s/c orbits, leading to a decrease in s/c altitude due to increased satellite drag. This effect is particularly important at altitudes of 500–600 km of Starlink's low Earth orbit (LEO) trajectories. The extreme storm of March 13, 1989 caused not only power grid outages but also significant satellite drag effects; about 1,000 space objects lost track of their orbital motion for a few days [26].

Since 1989, the number of payloads launched has increased dramatically. The more space objects are launched into Earth orbit, the more difficult it would be to control them in the event of an extreme event like the 1989 storm. Important results are presented in [27] demonstrating that during the May 2024 Gannon storm, the number of s/cs performing maneuvers in LEO increased to 5,000, 5 times the pre-storm level. In contrast, for the Halloween events, this number did not change, although the level of activity in terms of Dst was similar [27]. This justifies the definition of the Gannon storm as an extreme space weather event, not in terms of Dst, but because of the total number of satellites that were affected, showing again that the definition of extreme space weather is likely to change with technological advances.

Acknowledgements. This work was partially supported by NASA LWS Grants 80NSSC19K0085.

References

1. NOAA's Current Space Weather Conditions. https://www.swpc.noaa.gov/news/g5-conditions-reached-yet-again. Last accessed 3 Oct 2024
2. Reuters. https://www.reuters.com/technology/space/starlink-experiencing-degraded-service-musk-says-satellites-under-pressure-2024-05-11/. Last accessed 3 Oct 2024
3. Hapgood, M., Liu, H., Lugaz, N.: SpaceX—Sailing close to the space weather?. Space Weather 20, e2022SW003074 (2022)
4. Buzulukova, N., Tsurutani, B.: Space Weather: from solar origins to risks and hazards evolving in time. Front. Astron. Space Sci. **9**, 1017103 (2022)
5. Yashiro, S., Gopalswamy, N., Akiyama, S., Michalek, G., Howard, R.A.: Visibility of coronal mass ejections as a function of flare location and intensity. J. Geophys. Res. **110**, A12S05 (2005)
6. Moschou, S.-P., Drake, J.J., Cohen, O., Alvarado-Gómez, J.D., Garraffo, C., F, Federico: The stellar CME–flare relation: what do historic observations reveal? Astrophys. J. **877**(2), 105 (2019). https://doi.org/10.3847/1538-4357/ab1b37
7. Meng, X., Tsurutani, B.T., Mannucci, A.J.: The solar and interplanetary causes of superstorms (minimum Dst ≤ -250 nT) during the space age. J. Geophys. Res. Space Phys. **124**, 3926–3948 (2019)
8. Gopalswamy, N.: Extreme solar eruptions and their space weather consequences. In: Buzulukova, N. (ed.) Extreme events in Geospace, Elsevier, p. 37 (2018)
9. Burton, R.K., McPherron, R.L., Russell, C.T.: An empirical relationship between interplanetary conditions and Dst. J. Geophys. Res. **80**, 4204–4214 (1975)
10. O'Brien, T.P., McPherron, R.L.: An empirical phase space analysis of ring current dynamics: solar wind control of injection and decay. J. Geophys. Res. **105**(A4), 7707–7719 (2000)
11. Knipp, D.J., Bernstein, V., Wahl, K., Hayakawa, H.: Timelines as a tool for learning about space weather storms. J. Space Weather Space Clim. **11**, 29 (2021). https://doi.org/10.1051/swsc/2021011
12. Cliver, E.W., Dietrich, W.F.: The 1859 space weather event revisited: limits of extreme activity. J. Space Weather Space Clim. **3**, A31 (2013). https://doi.org/10.1051/swsc/2013053
13. Tsurutani, B.T., Gonzalez, W.D., Lakhina, G.S., Alex, S.: The extreme magnetic storm of 1–2 September 1859. J. Geophys. Res. **108**(A7), 1268 (2003)
14. Boteler, D.H.: The super storms of August/September 1859 and their effects on the telegraph system. Adv. Space Res. **38**(2), 159–172 (2006)
15. Hapgood, M.: The great storm of May 1921: an exemplar of a dangerous space weather event. Space Weather **17**, 950–975 (2019)
16. Boteler, D.H.: A 21st century view of the March 1989 magnetic storm. Space Weather **17**, 1427–1441 (2019)
17. National Weather Service: Intense space weather storms October 19–November 7, 2003, NOAA service assessment, Silver Spring, MD (2004). https://repository.library.noaa.gov/view/noaa/6995

18. Pulkkinen, A., Lindahl, S., Viljanen, A., Pirjola, R.: Geomagnetic storm of 29–31 October 2003: Geomagnetically induced currents and their relation to problems in the Swedish high-voltage power transmission system. Space Weather **3**(8), S08C03 (2005)
19. Brodrick, D., Tingay, S., Wieringa, M.: X-ray magnitude of the 4 November 2003 solar flare inferred from the ionospheric attenuation of the galactic radio background. J. Geophys. Res. 110, A09S36 (2005)
20. Cliver, E.W., Schrijver, C.J., Shibata, K., Usoskin, I.G.: Extreme solar events. Living Rev. Sol. Phys. **19**(2) (2022)
21. Love, J.J.: Extreme-event magnetic storm probabilities derived from rank statistics of historical Dst intensities for solar cycles 14–24. Space Weather **19**, e2020SW002579 (2021)
22. Miyake, F., Nagaya, K., Masuda, K., Nakamura, T.: A signature of cosmic-ray increase in AD 774–775 from tree rings in Japan. Nature **486**, 240–242 (2012)
23. Miyake, F., Masuda, K., Nakamura, T.: Another rapid event in the carbon-14 content of tree rings. Nat. Commun. **4**, 1748 (2013)
24. Shibata, K., et al.: Can Superflares Occur on Our Sun? Publ. Astron. Soc. Jpn. **65**(3), 49 (2013)
25. https://omniweb.gsfc.nasa.gov/. Last accessed 3 Oct 2024
26. Hapgood, M.: Chapter 27 - Space Weather: What are Policymakers Seeking? In: Buzulukova, N. (ed.) Extreme Events in Geospace, Elsevier (2018)
27. Parker, W.E., Linares, R.: Satellite drag analysis during the may 2024 geomagnetic storm. ARC Tech. Notes **61**, 5 (2024)

Open Access This chapter is licensed under the terms of the Creative Commons Attribution 4.0 International License (http://creativecommons.org/licenses/by/4.0/), which permits use, sharing, adaptation, distribution and reproduction in any medium or format, as long as you give appropriate credit to the original author(s) and the source, provide a link to the Creative Commons license and indicate if changes were made.

The images or other third party material in this chapter are included in the chapter's Creative Commons license, unless indicated otherwise in a credit line to the material. If material is not included in the chapter's Creative Commons license and your intended use is not permitted by statutory regulation or exceeds the permitted use, you will need to obtain permission directly from the copyright holder.

Variations in the Ionospheric Parameters over Midlatitude Europe Region During September 6–10, 2017 Geomagnetic Storm Caused by Coronal Mass Ejections

K. D. Aksonova[1,2](✉), S. V. Panasenko[1,2], D. Burešova[2], D. V. Kotov[1], and M. O. Reznychenko[1]

[1] Institute of Ionosphere, Kharkiv, Ukraine
aksonova@ufa.cas.cz

[2] Institute of Atmospheric Physics of the Czech Academy of Science, Prague, Czech Republic

Abstract. Geomagnetic storms (GS) significantly affect ionospheric variability and state of ionization. The irregular processes occurring in the ionosphere exert influence on modern HF technologies both in outer space, for example, on satellites, and on Earth-based infrastructures such as radars, navigation systems and radio communications. Therefore, understanding and modeling the effect of GS on variations of various ionospheric parameters is extremely important for applied purposes. This study presents the response of the European midlatitude ionosphere on the intense GS of September 6–10, 2017. This GS was caused by coronal mass ejections (CMEs) launched on September 4th and 6th.

The monitoring of the ionosphere state was conducted using the data obtained from the ionosonde and Kharkiv incoherent scatter radar (ISR) (49.6ºN, 36.3ºE). The height profiles of electron temperature (T_e), variations in the F2 peak electron density (*NmF2*), height (*hmF2*) and critical frequency (*foF2*) were analyzed. Significant changes in these parameters were found. In addition, we described the wavelike events related to the propagation of traveling ionospheric disturbances (TIDs) which could be initiated by GS, as well as estimated their vertical and horizontal propagation parameters. We also discussed our results and made comparisons with the other studies of the ionospheric response to this storm.

Keywords: Geomagnetic Storm · Ionospheric Parameters · Traveling Ionospheric Disturbances

1 Introduction

It is well known that coronal mass ejections (CMEs) are solar eruptions associated with processes in the solar corona [1]. Coming out of the Sun's atmosphere, these ejections can reach the Earth's magnetosphere, causing geomagnetic storms (GS) to happen if the magnetic field in CME is oriented southward relative to the dipole magnetic field. GS is a complex process that provokes various effects not only in the magnetosphere but at the

© The Author(s) 2026
N. Gopalswamy et al. (Eds.): ISWI 2024, SPPHY 431, pp. 64–70, 2026.
https://doi.org/10.1007/978-981-95-1121-1_8

different ionospheric heights [2]. These effects depend on altitudes and latitudes and are differ in time and intensity, therefore, they have dissimilar mechanisms of action. GS has also impact on upper atmosphere (thermosphere) neutral density. In turn, the ionospheric state during GS directly affects radio communications and is the cause of electric power grids failures, satellites damage. Sharp fluctuations in the electron density can change the radio wave propagation velocity, which leads to errors in Global Positioning System and navigation systems [3]. Thus, the connections between solar phenomena and variations in the ionospheric parameters is of great applied significance for space weather. The effects of GS at ionospheric heights have been studied for a long time [4]. Whereas a specific CME-induced geomagnetic activity is the subject of fewer studies [5, 6]. Therefore, it is necessary to investigate the response for each CME event.

This study presents the response of the midlatitude ionosphere during the intense GS of September 6–10, 2017. For this purpose, we used the data obtained at the Observatory of Institute of ionosphere, Kharkiv Ukraine. We discuss local changes in the height of the peak of F2-layer, electron temperature and density, as well as wave-like disturbances and their parameters. Our results will contribute to a broader understanding of both the regional and global ionospheric response to CME storms.

2 Space Weather Conditions

According to the `LASCO CME catalog` (https://cdaw.gsfc.nasa.gov/CME_list/) three CMEs with full halo were observed: on September 4th, 6th and 10th. They were associated with the flares M5.5, X9.3 and X8.2 [7]. The second one was the most powerful in the last 12 years. The period under review occurred during the declining phase of the solar cycle with varying of $F_{10.7}$ index from 134.9 to 101.6. Figure 1 shows OMNI (https://omniweb.gsfc.nasa.gov/form/dx1.html) 5 min resolution interplanetary magnetic field (IMF) B, B_z and B_y components, interplanetary electric field (IEF) E_y component, solar wind parameters, such as velocity Vsw and temperature T and dynamic pressure P, as well as geomagnetic indices K_p, D_{st} and AE (http://wdc.kugi.kyoto-u.ac.jp/aedir/index.html) during 6–10 September 2017.

As shown in Fig. 1, from the whole period of observation only 6 September 2017 was a magnetically quiet day (K_p < 3 and D_{st} had positive values). There were no significant changes in the other parameters, except AE index which reached a peak of about 600 nT around 10 UT. A significant change in space weather conditions began to occur in the next two days. Figure 1 illustrates two abruptly enhancements in Vsw. The first one was from about 400 to 600 km/s around 0 UT on 7 September with the next slight decrease by the end of the day to about 470 km/s and the second sharp increase occurred between 7 and 8 September up to a maximum value of about 850 km/s around 8 UT on 8 September. This contributed to the K_p index increase up to 4– and 8, while D_{st} index has increased to +52 nT and reached a minimum value of –122 nT during first and second sharp increase of Vsw, respectively. At the same time for these two periods, the IMF B_z value at first increased in the north direction to 15.85 nT with the next southward rotation to –31.07 nT. The IMF B magnitude fluctuations also dramatically increased as Vsw and reached the maximum values of 17.61 nT and 32.71 nT. The period from 12 to 24 UT on 8 September is characterized by decreasing in K_p index from 8+ to 2+

and the second minimum value of D_{st} index of –109 nT around 17 UT. In this case, the values of Vsw slightly decreased from about 740 km/s to 650 km/s, the IMF B_z value rapidly decreased to –17.43 nT around 12 UT, while the IEF E_y and IMF B values sharply increased to 13.76 mV/m and 17.8 nT, respectively. In turn, AE-index indicated the amplification of auroral electrojets at each subsequent described period. It reached the maximum values of 1339 nT (around 9 UT) on 7 September, 2063 nT (around 0 UT) and 2351 nT (around 14 UT) on 8 September. While the strongest pressure P changes were observed in the time interval of 0–10 UT with the maximum value of 11.34 nPa (around 01:30 UT) on 7 September and sharp jumps of no more than 8.31 nPa (around 00:30 UT) throughout the day on 8 September. A significant gradual enhancement in temperature T up to 10^6 K occurred on 7 September. Whereas on 8 September, there was an abrupt increase to $4 \cdot 10^6$ K (around 02:20 UT).

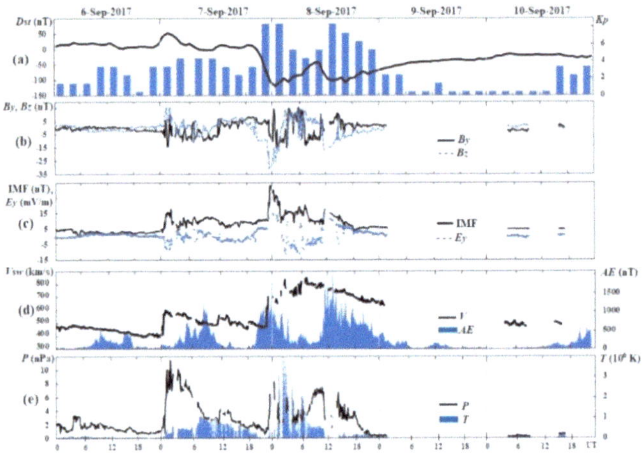

Fig. 1. Time series of (a) D_{st} and K_p geomagnetic activity indices, (b) the IMF B_z and B_y components, (c) magnitude of B component of the IMF and IEF E_y, (d) solar wind speed Vsw and AE index, (e) pressure P and temperature T during the period of 6 - 10 September 2017.

All of the above space weather parameters indicate an intense two-step GS (two minimums of D_{st} index: –122 nT and –109 nT) during the observation period.

3 Facilities

Several studies using different radiophysical methods are devoted to the solar–terrestrial interaction that occurred in September 2017 [8–10]. In this study the observation results were obtained using Kharkiv incoherent scatter radar (KhISR). It is one of the most informative sources of ionospheric parameters at midlatitude European region. KhISR is located in Ukraine (geographic coordinates: 49.6°N, 36.3°E). The main parameters and operating modes can be found in [11]. It is possible to determine the whole set of the most important parameters of the ionosphere in a wide altitude range from 100 to 1000 km by using incoherent scatter (IS) technique. Radar allows us to measure: electron density N_e, plasma temperatures (electron T_e and ion T_i), vertical component of the plasma drift velocity and ion composition. The calibration of electron density profiles is performed using ionosonde located in the vicinity of the KhISR. In addition, this method allows us to observe vertical wave-like fluctuations in IS signal power that correspond to fluctuations in the electron density at different heights, which are associated with traveling ionospheric disturbances (TIDs) [12, 13].

4 Results

Next, we will characterize the changes in the ionospheric parameters for the three periods described in Sect. 2. Figure 2 shows variations of the F2-layer peak height (*hmF2*), electron density (*NmF2*) and electron temperature (T_e) observed over Kharkiv. We used the parameters on 6 September 2017 as the quiet time references for this storm. It can be seen clearly two major changes in the T_e variations on 7 and 8 September 2017. The first one is a sharp decrease from about 3000 K to 1800 K (near 8 UT (LT = UT + 3)) and the second is the sharp jump to 4600 K (near 0 UT). It should be noted that the time of the maximum increase in T_e coincides with the first minimum of D_{st} – index.

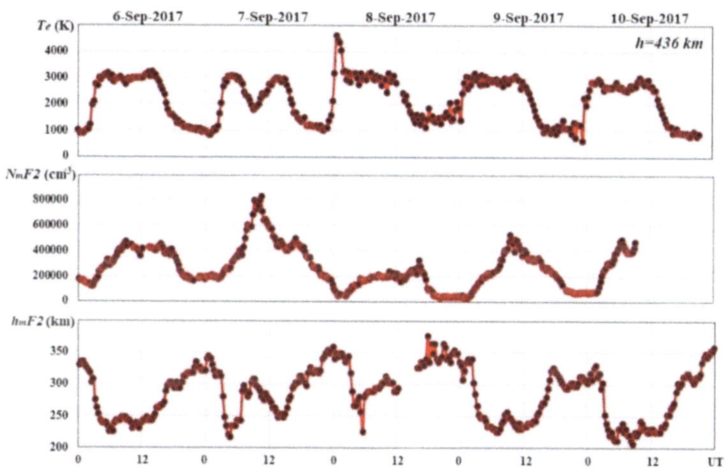

Fig. 2. Variations in T_e, *NmF2* and *hmF2* during the period of 6–10 September 2017 measured by the KhISR.

Simultaneously with the temperature decrease near 08 UT on September 7th, there is a sudden increase in *NmF2* to about $83 \cdot 10^4$ cm^{-3} accompanied by 65 km enhancement in *hmF2* compared to the previous day. While during September 8th, the lowest *NmF2* and the highest *hmF2* values, compared to other days, were recorded.

For identifying wave structures in electron density, quasi-periodic variations in the relative power (δP) of the IS signal are analyzed. Figure 3 indicates the amplitude spectrogram (δS_P) at the 200-km height and band-pass filtered variations of δP for altitude range 100–300 km. The local maximums in δS_P values were interpreted as a possible manifestation of TIDs. As can be seen from Fig. 3, the oscillations in the period range of 100–200 min had the greatest amplitudes. Based on this, the appropriate subrange was chosen for further bandpass filtration in the range of 120–240 min, which corresponds to large-scale structures. The dominant periods of TIDs varied from day to day in the range of 150–180 min and were revealed near the solar terminator passages. Such long-period perturbations correspond to large-scale (LS) TIDs. Our results show that maximum values of δP vary from 14 to 30%. The heights at which they were found range from 200–225 km. In a whole, it is clearly seen, that they are higher on 7 September than on the other days by a factor of 1.5 and 2 (compared with 6 September). It should be noted that the nighttime anomaly (from 15 to 23 UT) observed on September 8 at altitudes above 230 km is associated with the polar wall of the main ionospheric trough, which had significantly shifted into middle latitudes. For determining the TID propagation parameters we used cross-correlation analysis. The vertical components of the phase velocity V_z and wavelength of LSTIDs were estimated. We found that $V_z =$ 35–50 m/s and $\Lambda_z = 450$–900 km. Also, in most cases, the downward phase progression is clearly visible, which indicates that the sources of TIDs are located at lower heights. Only for two events, on September 7 and 9, the propagation of V_z is upwards. This indirectly indicates that they were most likely caused by variations in the electric field at high latitudes.

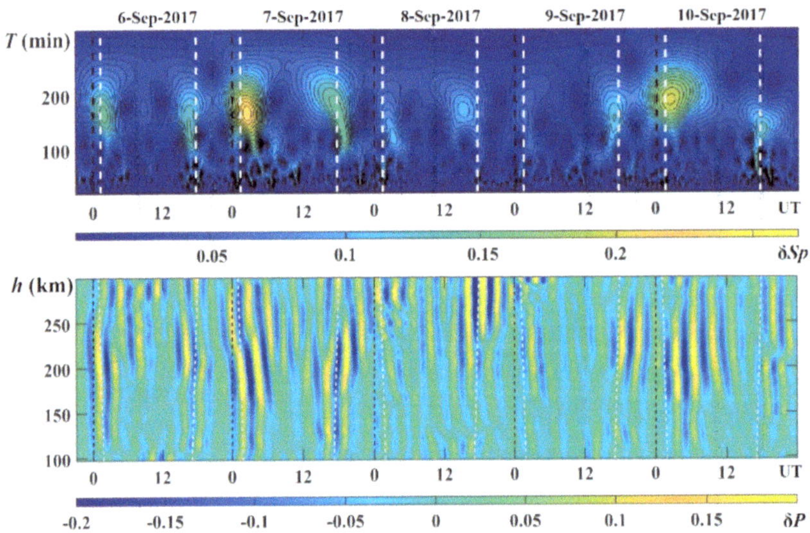

Fig. 3. Amplitude spectra of relative variations in IS signal power δP at the altitude of 200 km (top panel) and altitude-time dependence of δP band-pass filtered in the range of 120–240 min (bottom panel). Dotted white lines show the local solar terminator passage times at the studied heights.

5 Conclusions

In this study we analyzed the ionospheric behavior over midlatitude European region during the period of the intense geomagnetic storm 6–10 September 2017, caused by CMEs. A strong increase of *NmF2* accompanied by a sharp T_e decrease and significant *hmF2* uplift were found at ~8 UT on September 7th. On September 8th, it was found that a sharp increase in T_e is associated with the first minimum of D_{st} – index. This was an unexpectedly strong change of the topside electron temperature (from ~1000 K to ~5000 K). The probable cause of this was an increased ring current. Along with a sharp decrease in T_e, on September 7th, there was an increase in *NmF2* to about 5000 cm^{-3} and *hmF2* by 65 km. In addition, LSTIDs with the periods of 120–240 min as well as their vertical parameters were studied. Intense fluctuations were observed in the morning and evening hours, which indicated the solar terminator as the most likely additional source of their generation.

References

1. Zhang, J., Dere, K.P., Howard, R.A., Kundu, M.R., White, S.M.: On the temporal relationship between coronal mass ejections and flares. Astrophys J. **559**(1), 452 (2001)
2. Danilov, A.D., Lastovicka, J.: Effects of geomagnetic storms on the ionosphere and atmosphere. Int. J. Geomagn. Aeron. **2**(3), 209–224 (2001)
3. Basu, S., et al.: Large magnetic storm-induced nighttime ionospheric flows at midlatitudes and their impacts on GPS-based navigation systems. J. Geophy. Res. Space Phys. **113**(A3) (2008)
4. Laštovička, J.: Effects of geomagnetic storms in the lower ionosphere, middle atmosphere and troposphere. J. Atmos. Terr. Phys. **58**(7), 831–843 (1996)
5. Fujiwara, H., et al.: Extreme ion heating in the dayside ionosphere in response to the arrival of a Coronal Mass Ejection on 12 March 2012. Ann. Geophys. **32**(7), 831–839 (2014)
6. Matamba, T.M., Habarulema, J.B.: Ionospheric responses to CME-and CIR-driven geomagnetic storms along 30 E–40 E over the African sector from 2001 to 2015. Space Weather **16**(5), 538–556 (2018)
7. Redmon, R.J., Seaton, D.B., Steenburgh, R., He, J., Rodriguez, J.V.: September 2017's geoeffective space weather and impacts to Caribbean radio communications during hurricane response. Space Weathe **16**(9), 1190–1201 (2018)
8. Mosna, Z., et al.: Ionospheric storm of September 2017 observed at ionospheric station Pruhonice, the Czech Republic. Advances in Space Research **65**(1), 115–128 (2020)
9. Zhang, S.R., et al.: Subauroral and polar traveling ionospheric disturbances during the 7–9 September 2017 storms. Space Weather **17**(12), 1748–1764 (2019)
10. Oikonomou, C., et al.: Investigation of the negative ionospheric response of the 8 September 2017 geomagnetic storm over the European sector. Adv. Space Res. **70**(4), 1104–1120 (2022)
11. Domnin, I., et al.: Kharkiv incoherent scatter facility, Bulletin of the National Technical University "Kharkiv Polytechnic Institute". Series: Radiophysics and ionosphere **47**(1089), 28–32 (2014)
12. Panasenko, S.V., Goncharenko, L.P., Erickson, P.J., Aksonova, K.D., Domnin, I.F.: Traveling ionospheric disturbances observed by Kharkiv and Millstone Hill incoherent scatter radars near vernal equinox and summer solstice. J. Atmos. Solar Terr. Phys. **172**, 10–23 (2018)

13. Aksonova, K.D., Sopin, A.O., Burešová, D., Zalizovski, A.V., Domnin, I.F.: Synchronous observations of traveling ionospheric disturbances by the multipoint Doppler sounding, ionosonde and the incoherent scatter radar: Case study. Adv. Space Res. **73**(9), 4414–4425 (2024)

Open Access This chapter is licensed under the terms of the Creative Commons Attribution 4.0 International License (http://creativecommons.org/licenses/by/4.0/), which permits use, sharing, adaptation, distribution and reproduction in any medium or format, as long as you give appropriate credit to the original author(s) and the source, provide a link to the Creative Commons license and indicate if changes were made.

The images or other third party material in this chapter are included in the chapter's Creative Commons license, unless indicated otherwise in a credit line to the material. If material is not included in the chapter's Creative Commons license and your intended use is not permitted by statutory regulation or exceeds the permitted use, you will need to obtain permission directly from the copyright holder.

Effect of Geomagnetic Storm in Ionosphere Over Nepal

Drabindra Pandit[1(✉)], Christine Amory-Mazaudier[2], Rolland Fleury[3], Narayan Prasad Chapagain[4], and Binod Adhikari[1]

[1] Department of Physics, St. Xavier's College, Kathmandu, Nepal
drabin2076@gmail.com

[2] LPP, Polytechnique, Sorbonne Universite´s UPMC Paris 06, Paris, France

[3] Lab-STICC, UMR 6285, Institut Mines-Telecom Atlantique, Site de Brest, France

[4] Amrit Campus, Tribhuvan University, Thamel, Kathmandu, Nepal

Abstract. This study examines the deviation in ionospheric total electron content (TEC) over Nepal during 49 geomagnetic storms of solar cycle 24. For this investigation, the daily RINEX files downloaded from the UNAVCO website are used to calculate the TEC on GPS measurements only. The study shows the increase and decrease in TEC during the event days. The increase or decrease in percentage TEC is compared to a reference quiet day. Among 49 selected storms in 26 stormy days the TEC has increased and in 23 stormy days the TEC has decreased. The maximum deviation in TEC is found during the recovery phase geomagnetic storm of 2010-08-04 and 2012-07-16. During storm on 2010-08-04 it is 101.97% increase and during storm of 2012-07-16 it is 50.68% decrease. The outcomes of this study enhance the understanding of ionospheric dynamics over Nepal under geomagnetically disturbed conditions.

Keywords: Geomagnetic Storms · Ionosphere · Total Electron Content (TEC)

1 Introduction

Interactions between the solar wind and Earth's magnetic field produce geomagnetic storms. The ionosphere may get disturbed as a result of these storms (Gosling, 1993). TEC is an essential parameter for assessing ionospheric changes, refers to the total number of electron content in unit column cross-section, 1 TECu = 10^{16} el/m^2 (Lean et al., 2016). Positive and negative ionospheric storms are produced when a substantial amount of energy from the solar wind enters the polar ionosphere over a few hours to several days (Horvath & Lovell, 2015). During geomagnetic storms, there can be positive ionospheric storms, characterized by an increase in TEC, and negative ionospheric storms, characterized by a decrease in TEC. Positive and negative ionospheric storms can be observed during different phases of geomagnetic storms (Rishbeth, 1963). Bagiya et al. (2009) studied the variation in the TEC during the geomagnetic storm of 24 August 2005. On 24 August, the TEC increased due to the rapid penetration of the electric field. On 25 August, there was a depletion of the TEC which could be created by Joule heating

© The Author(s) 2026
N. Gopalswamy et al. (Eds.): ISWI 2024, SPPHY 431, pp. 71–78, 2026.
https://doi.org/10.1007/978-981-95-1121-1_9

in the auroral zone. In fact, this type of depletion is very often associated with Joule heating. Joule heating causes a thermal expansion of the atmosphere which modifies temperatures, pressures and thermospheric winds, and provokes a mass transport from the pole to the equator. This produces a significant modification of the composition of the atmosphere. Trivedi et al. (2011) observed deviation in the vertical total electron vTEC (up to − 90%) during geomagnetic storm of July 17, 2005.

During four intense geomagnetic storms of solar cycle 24, Pandit et al. (2022) compared TEC derived from GPS data with TEC obtained from Center for Orbit Determination in Europe CODG and International GNSS service group IGSG and found a higher degree of discrepancy between the TEC derived from GPS measurements and the models. This paper presents comparison of TEC obtained from local modeling for 49 geomagnetic storms over Nepal. Due to its constant operation and vast network of global receivers, GPS system is an extremely useful tool for studying the ionosphere, particularly during periods of magnetic disturbance when the magnetosphere-thermosphere is undergoing dynamics and energy dissipation. This research contributes to improve the comprehension of the consequences of space weather on Earth's atmosphere and its occurrence: during geomagnetic storms, the impact of space weather events on the Earth's environment causes changes in the TEC which disturb the GNSS signal.

2 Datasets and Methodology

For this study four ground-based GPS stations JMSM, DLPA, GRHI and KKN4 of Nepal are used (see Fig. 1). Table 1 displays geomagnetic, geographic and dip latitudes of the stations selected. The RINEX (Receiver Independent Exchange) data has format v2.1 and has a temporal resolution of fifteen minutes is taken from www.unavco.org website. This data is processed using Fleury's software (http://www.girgea.org), which runs in the Windows operating system to obtain the required vertical TEC which has the same time resolution. The list of geomagnetic storms has taken from https://www.spaceweatherlive.com/en/auroral-activity/top-50-geomagnetic-storms/solar-cycle/24.html. The Kp index, which characterizes the strength of these geomagnetic storms has obtained from the website: https://spaceweather.com/archive.php?day=13&month=09&year=2014&view=view. The data solar wind and geomagnetic parameters has been downloaded from https://omniweb.gsfc.nasa.gov/form/omni_min.html. The quiet days for each storm are taken as the quietest day of the same month from the website https://datapub.gfz-potsdam.de/download/10.5880.Kp.0001/Quiet_Disturbed_Days/. Table 2 summarizes the characteristics of the storms such as Kp-index, SSC (Sudden Storm Commencement), the minimum value of SYMH with time, the date and the maximum deviation in TEC from the reference value along with the percentage change in deviation of TEC. The Kp and Dst geomagnetic indices are used to categorize varying levels of geomagnetic activity. On the basis of Dst index Geomagnetic storm is moderate ($50 \leq |Dst| < 100$ nT), intense ($100 \leq |Dst| < 250$ nT), and super-storms ($|Dst| \geq 250$ nT) (Gonzalez et al., 1994) and using Kp index classification of geomagnetic storm is as follows: mild ($0 \leq Kp \leq 5$), moderate ($5 < Kp \leq 7$), and intense ($7 \leq Kp \leq 9$) (Bartels et al., 1940).

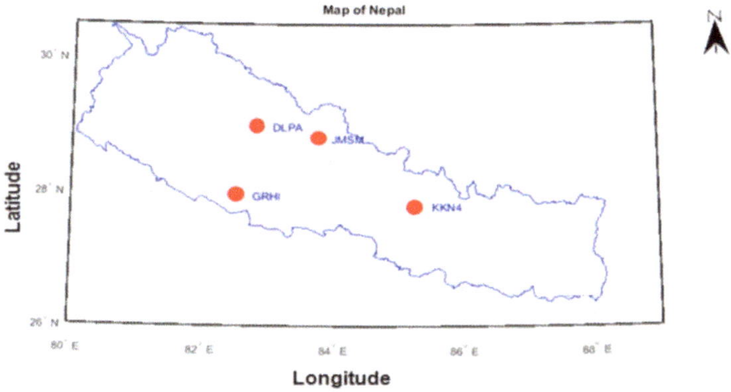

Fig. 1. Map of Nepal showing GPS stations (bold circle) data used in this study (Pandit et al., 2021).

Table 1. The selected GPS stations and their coordinates, the data of which is used in this study.

Station ID	location	Geomagnetic latitude	Geomagnetic longitude	Dip latitude	Local time LT
KKN4	Kakani Nepal	18.62°	159.41°	43.86°	UT + 05:45 h
GRHI	Gorahi Nepal	18.94°	156.82°	44.25°	UT + 05:45 h
JMSM	Jomsom Nepal	19.71°	156.06°	45.31°	UT + 05:45 h
DLPA	Dolpa Nepal	19.94°	157.21°	46.03°	UT + 05:45 h

3 Results and Discussion

Figures 2 and 3 represents variation in TEC (first panel) during quiet and disturbed day and variation SYMH (second panel) along with the day of August 2010 and July 2012 respectively. The vertical black lines in both panels indicate sudden storm commencement (SSC) in universal time (UT). For each selected day we determine the amplitude of the ionospheric positive or negative deviation on TEC (DTEC) by comparison with a magnetic quiet reference taken as the average of the five quietest days of the same month, see Fig. 2 and 3. In this figure the storms selected exhibit the major impact on DTEC observed. The plots show increase in TEC during 4 August 2010 and decrease in TEC during 16 July 2012. The maximum deviation in TEC during 4 August 2010 obtained is 101.97% and its value during 16 July 2012 is 50.65%. Various physical processes influence the mid and low latitudes during magnetic storms. 1) There is the penetration of the electric field of magnetospheric convection, which affects high and low latitudes almost simultaneously (Nishida, 1968; Vasyliunas, 1970). 2) There is the thermal expansion of the atmosphere, due to Joule heating in the auroral zone, which causes changes in pressure, temperature and composition due to molecular mass transport (Fuller-Rowell et al., 1994, 1996). And 3) there are disturbed thermospheric winds which generate the disturbed ionospheric dynamo (Blanc and Richmond, 1980). At the start of the geomagnetic storm, the penetration of the magnetospheric electric convection field is the only process influencing the middle and low latitudes. After a few hours, the thermal expansion of the atmosphere, which strongly affect the composition, and the disturbed ionospheric dynamo are present at mid and low latitudes. Fuller Rowell et al., 1996 showed that the thermal expansion of the thermosphere produces a negative ionospheric storm in the Northern hemisphere.

In Fig. 2 we observe two positive ionospheric storms (increase of VTEC) during the main phases of the two geomagnetic storms produced by two CME (space weather.com). Joule Heating occurred during 2 short periods and this is not sufficient to produce a negative ionospheric storm related to change in composition and disturbance dynamo. In Fig. 3 during the main phase of the storm we observe an ionospheric positive storm (increase of VTEC) and during the recovery phase the negative ionospheric storm.

The Table 2 shows the characteristics features of strong geomagnetic storms and their impacts on TEC in ionosphere over Nepal. The first, second, third, fourth, fifth, sixth, seventh, eighth, nine and tenth column of the table shows event day, kp-index, SSC date, SSC time, minimum SYMH, date and time of minimum SYMH, peak value or minimum value of TEC on the day of event, peak value or minimum value of TEC on quiet reference day, deviation of TEC from quiet reference day and the percentage deviation in TEC in compare to the quiet reference day respectively.

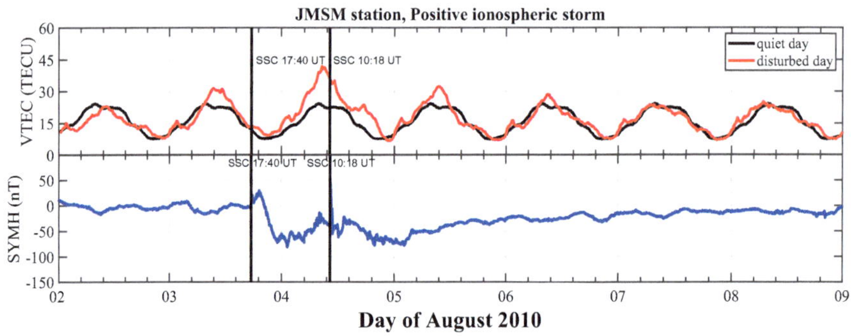

Fig. 2. Positive ionospheric storm showing increase in VTEC during geomagnetic storm on 2010-08-04 (red curve) and quiet reference day taken as average of five quietest days of the same month (black curve). The lower panel shows the corresponding SYHM value (blue curve). The time on x-axis is universal time.

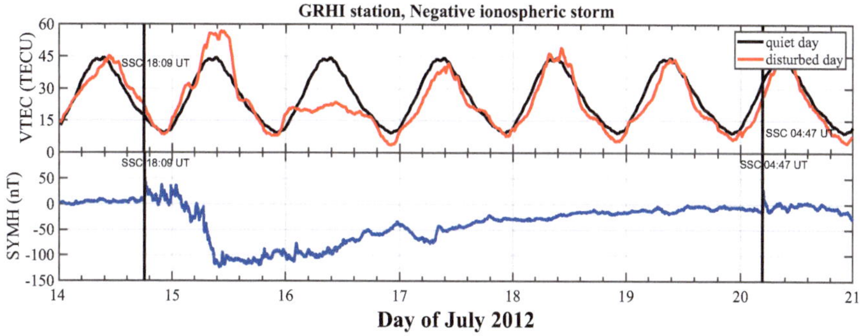

Fig. 3. Negative ionospheric storm showing decrease in VTEC during geomagnetic storm on 2012-07-16 (red curve) and quiet reference day taken as average of five quietest days of the same month (black curve). The lower panel shows the corresponding SYMH value (blue curve). The time on x-axis is universal time.

The significant deviations from the quiet reference day TEC level are the hallmark of these storms. The largest TEC rise observed during the positive ionospheric storm on August 4, 2010, was greater than 100%, whereas the maximum TEC decrease observed during the negative ionospheric storm on July 16, 2012, was greater than 50%. The two major ionospheric storms presented in Figs. 2 and 3, 04.08.2010 (positive storm) and 16.07.2012 (negative storm), are underlined in red in the Table 2. Among the 49 cases, there are 26 cases classified positive and 23 cases classified negative. The distribution (±) is therefore balanced. The direction of variation (±) does not depend on the maximum value of the SYM-H index (column 5).

Table 2. Characteristics features of strong geomagnetic storm of solar cycle 24 and its impact on TEC in ionosphere over Nepal.

Date of storm	Kp	SSC Date	SSC time hh:mm	Min SYMH (nT) Disturbed	Min. SYMH day and time in hr	TEC disturbed day (TECU)	TEC quiet reference day (TECU)	δTEC = disturbed TEC – quiet reference TEC (TECU)	%δTEC = disturbed TEC – quiet reference TEC)/ quiet reference TEC x100%
08.09.17	8+	07.09.17	23:00	−140	8; 1:10	35.96	43.93	−7.97	−18.14
22.06.15	8+	22.06.15	18:35	−210	23; 4:20	45.28	43.08	2.2	5.10
09.03.12	8	09.03.12	1:30	−149	9; 8:15	101.54	74.54	27	36.22
17.03.15	8-	17.03.15	4:45	−230	17; 22:40	71.82	78.55	−6.73	−8.56
23.06.15	8-	22.06.15	18:40	−210	23; 4:20	42.07	43.08	−1.01	−2.34
02.10.13	8-	02.10.13	4:30	−90	2; 6:10	66.48	50.89	15.59	30.63
05.04.10	8-	05.04.10	8:25	−90	6; 7:40	25.51	40.36	−14.85	−36.79

(continued)

Table 2. (*continued*)

Date of storm	Kp	SSC Date	SSC time hh:mm	Min SYMH (nT) Disturbed	Min. SYMH day and time in hr	TEC disturbed day (TECU)	TEC quiet reference day (TECU)	δTEC = disturbed TEC – quiet reference TEC (TECU)	%δTEC = disturbed TEC – quiet reference TEC)/ quiet reference TEC x100%
05.08.11	8-	05.08.11	20:00	−130	6; 3:25	36.03	30.54	5.49	17.97
07.09.17	8-	07.09.17	23:00	−145	8; 1:10	47.02	43.93	3.09	7.03
07.10.15	7+	No SSC		−125	7; 18:50	51.95	48.27	3.68	7.62
26.08.18	7+	No SSC		−205	26; 7:10	25.20	24.45	0.75	3.06
25.10.11	7+	24.10.11	18:35	−160	25; 1:20	73.49	71.54	1.95	2.72
15.07.12	7	14.07.12	18:10	−125	15; 10:00	56.80	47.75	9.05	18.95
11.09.15	7	No SSC		−90	11; 14:20	34.49	58.15	−23.66	−40.68
28.05.17	7	27.05.17	15:15	−140	28; 7:10	34.37	57.66	−23.29	−40.39
20.09.15	7	20.09.25	6:05	−85	20; 11:05	48.57	58.15	−9.58	−16.47
24.10.11	7	24.10.11	18:30	−160	25; 1:20	59.44	71.45	−12.01	−16.80
17.03.13	7-	17.03.13	6:00	−130	17; 20:30	84.76	72.54	12.22	16.84
20.12.15	7-	20.12.15	3:30	−170	20; 21:55	42.62	25.17	17.45	69.32
28.09.17	7-	26.09.17	23:55	−75	28; 5:55	48.16	43.93	4.23	9.62
09.10.12	7-	No SSC		−115	9; 2:10	56.17	79.83	−23.66	−29.63
09.07.12	7-	No SSC		−80	9; 21:50	66.34	47.75	18.59	38.93
24.04.12	7-	23.04.12	3:15	−130	24; 3:30	50.87	85.74	−34.87	−40.66
21.12.15	7-	20.12.15	3:30	−170	21; 2:00	31.72	25.17	6.55	26.02
01.10.12	7-	30.09.12	11:30	−135	1; 4:00	72.59	79.83	−7.24	−9.06
03.08.10	7-	03.08.10	17:40	−80	4; 23:00	31.39	20.71	10.68	51.56
08.05.16	6+	No SSC		−110	8; 8:20	51.96	31.37	20.59	65.63
25.10.16	6+	No SSC		−80	25; 23:00	23.56	24.47	−0.91	−3.71
27.08.15	6+	No SSC		100	27; 20:30	43.29	41.51	1.78	4.28
29.06.13	6+	No SSC		−110	29; 6:30	38.68	44.63	−5.95	−13.33
04.08.10	**6+**	**04.08.10**	**10:20**	**−78**	**5; 2:10**	**41.83**	**20.71**	**21.12**	**101.97**
27.03.17	6+	No SSC		−85	27; 16:10	30.55	31.02	−0.47	−1.51
13.10.16	6+	No SSC		−115	13: 23:55	29.50	22.47	7.03	31.28
28.05.11	6+	No SSC		−90	28; 11:10	44.18	43.12	1.06	2.45
08.11.17	6+	No SSC		−90	8; 1:10	22.97	24.18	−1.21	−5.00
07.09.15	6+	No SSC		−90	8; 1:50	42.62	58.15	−15.53	−26.70

(*continued*)

Table 2. (*continued*)

Date of storm	Kp	SSC Date	SSC time hh:mm	Min SYMH (nT) Disturbed	Min. SYMH day and time in hr	TEC disturbed day (TECU)	TEC quiet reference day (TECU)	δTEC = disturbed TEC – quiet reference TEC (TECU)	%δTEC = disturbed TEC – quiet reference TEC)/ quiet reference TEC x100%
19.02.14	6+	No SSC		−130	19; 8:20	88.44	71.89	16.55	23.02
16.07.12	**6+**	**14.07.12**	**18:10**	**−125**	**15; 10:00**	**23.55**	**47.75**	**−24.2**	**−50.68**
08.10.12	6+	No SSC		−115	9; 10:20	50.22	79.83	−29.61	−37.09
15.08.15	6+	15.08.15	8:30	−95	16; 7:40	39.82	41.51	−1.69	−4.07
08.06.14	6+	No SSC		−72	8; 6:50	68.62	36.08	32.54	90.18
26.09.11	6+	26.09.11	12:30	−120	26; 21:20	60.90	61.89	−0.99	−1.59
27.09.17	6+	No SSC		−55	28; 5:55	34.00	43.93	−9.93	−22.60
12.09.14	6+	12.09.14	16:00	−95	12; 23:00	87.22	69.48	17.74	25.53
14.11.12	6+	No SSC		−120	14; 7:30	66.25	48.07	18.18	37.81
15.03.12	6+	12.03.12	9:20	−80	15; 19:50	74.57	74.54	0.03	0.04
07.11.17	6+	No SSC		−90	8; 0:50	19.93	24.18	−4.25	−17.57
12.03.12	6+	12.03.12	9:10	−67	12; 17:00	48.16	74.54	−26.38	−35.39
03.08.10	5	03.08.10	17:40	−70	3; 23:00	14	9	5	55.55

4 Conclusion

In this research the total electron content (TEC) deviation in the ionosphere over Nepal during the 49 geomagnetic storms of solar cycle 24 is investigated using GPS TEC RINEX data from the UNAVCO website. The study indicates that TEC increases and decreases on event days. Twenty-six stormy days out of forty-nine selected storms show rise in TEC, whereas twenty-third stormy days showed drop in TEC. During the recovery phase of geomagnetic storms on 2010-08-04 and 2012-07-16 had the greatest deviation in TEC. 101.97% more TEC was detected during the storm on 2010-08-04 compared to the reference quiet day, while 50.68% less TEC was found during the storm on 2012-07-16. The findings of this investigation contribute to our knowledge of the dynamics of the ionosphere above Nepal in the presence of geomagnetically induced disruption. The values of the magnetic indices (Kp, SYM-H) are not sufficient to classify the type of deviation.

References

Bagiya, M.S., Joshi, H.P., Iyer, K.N., Aggarwal, M., Ravindran, S., Pathan, B.M.: TEC variations during low solar activity period (2005–2007) near the equatorial ionospheric anomaly crest region in India. Ann. Geophys. **27**(3), 1047–1057 (2009)

Bartels, J., Heck, N.H., Johnston, H.F.: Geomagnetic three-hour-range indices for the years 1938 and 1939. Terr. Magn. Atmos. Electr. **45**(3), 309–337 (1940)

Blanc, M., Richmond, A.D.: The ionospheric disturbance dynamo. J. Geophys. Res. **85**(A4), 1669–1686 (1980)

Fuller-Rowell, T.J., Codrescu, M.V., Rishbeth, H., Moffett, R.J., Quegan, S.: Response of the thermosphere and ionosphere to geomagnetic storms. J. Geophys. Res. **99**(A3), 3893–3914 (1994)

Fuller-Rowell, T.J., Codrescu, M.V., Rishbeth, H., Moffett, R.J., Quegan, S.: On the seasonal response of the thermosphere and ionosphere to geomagnetic storms. J. Geophys. Res. **101**(A2), 2343–2353 (1996)

Gonzalez, W.D., et al.: What is a geomagnetic storm? J. Geophys. Res. **99**(A4), 5771–5792 (1994)

Gosling, J.T.: The solar flare myth. J. Geophys. Res. Space Physics **98**(A11), 18937–18949 (1993)

Horvath, I., Lovell, B.C.: Positive and negative ionospheric storms occurring during the 15 May 2005 geomagnetic superstorm. J. Geophys. Res. Space Physics **120**(9), 7822–7837 (2015)

Lean, J.L., Meier, R.R., Picone, J.M., Sassi, F., Emmert, J.T., Richards, P.G.: Ionospheric total electron content: Spatial patterns of variability. J. Geophys. Res. Space Physics **121**(10), 10–367 (2016)

Nishida, A.: Geomagnetic DP2 fluctuations and associated magnetospheric phenomena. J. Geophys. Res. **73**(5), 1795–1803 (1968)

Pandit, D., Amory-Mazaudier, C., Fleury, R., Chapagain, N.P., Adhikari, B.: VTEC observations of intense geomagnetic storms above Nepal: comparison with satellite data, CODE and IGSG models. Indian J. Phys. **97**(3), 701–718 (2022)

Pandit, D., Ghimire, B., Amory-Mazaudier, C., Fleury, R., Prasad Chapagain, N., Adhikari, B.: Climatology of ionosphere over Nepal based on GPS total electron content data from 2008 to 2018. Ann. Geophys. **39**(4), 743–758 (2021)

Rishbeth, H.: Ionospheric storms and the morphology of magnetic disturbances. Planet Space Science **11**(1), 31–43 (1963)

Trivedi, R., Jain, A., Jain, S., Gwal, A.K.: Study of TEC changes during geomagnetic storms occurred near the crest of the equatorial ionospheric ionization anomaly in the Indian sector. Adv. Space Res. **48**(10), 1617–1630 (2011)

Open Access This chapter is licensed under the terms of the Creative Commons Attribution 4.0 International License (http://creativecommons.org/licenses/by/4.0/), which permits use, sharing, adaptation, distribution and reproduction in any medium or format, as long as you give appropriate credit to the original author(s) and the source, provide a link to the Creative Commons license and indicate if changes were made.

The images or other third party material in this chapter are included in the chapter's Creative Commons license, unless indicated otherwise in a credit line to the material. If material is not included in the chapter's Creative Commons license and your intended use is not permitted by statutory regulation or exceeds the permitted use, you will need to obtain permission directly from the copyright holder.

Daily Variations of Inter-Hemispheric Field-Aligned Currents During Severe Geomagnetic Storms Using Equatorial Magnetic Field Measurements from Davao Station, Philippines

Manjula Ranasinghe[1](✉), Akiko Fujimoto[2], Akimasa Yoshikawa[3], and Chandana Jayaratne[1]

[1] Astronomy and Space Science Unit, Department of Physics, University of Colombo, Colombo-03, Sri Lanka
manjula@phys.cmb.ac.lk

[2] Faculty of Computer Science and Systems Engineering, Kyushu Institute of Technology, 680 4, Kawazu, Iizuka 820 8502, Fukuoka, Japan

[3] International Research Center for Space and Planetary Environmental Science(I-SPES), Kyushu University, Motooka, Nishi-Ku, Fukuoka 744819-0395, Japan

Abstract. An examination of geomagnetic data collected in Davo, Philippines, over a 20-year period encompassing solar cycles 23 and 24 resulted in the identification of attributes linked to the Inter-Hemispheric Field-Aligned Currents (IHFACs). The study focused on the analysis of the normalized D component of geomagnetic field (ΔD). Specifically, characteristics were investigated in the dawn, noon, and dusk sectors, revealing the identification of both long-term features across solar cycle 23 and 24. During the declining phases of solar cycles, peak-to-peak amplitudes of ΔD in the dusk sector show enhancements, indicating increased sensitivity to geomagnetic variations as solar activity wanes. The variation of ΔD during storm periods was studied and compared to quiet time ΔD variations to identify the changes in IHFACs during high-intensity geomagnetic storms. During the main phase of severe geomagnetic storms, significant enhancements in the negative component of ΔD occur around local noon, while at night, the positive component of ΔD shows marked increases. During the recovery phase, the positive component's daily amplitude gradually decreases over several days, indicating a slow return to quieter geomagnetic conditions, whereas the negative component remains relatively stable compared to quiet periods. It is evident that the response of IHFACs to solar disturbances significantly depends on the phase of the geomagnetic storm and the time of day.

Keywords: Inter-hemispheric field-aligned currents (IHFACs) · Ground-based magnetometer data · Equatorial magnetic field · Solar cycle 23 · Solar cycle 24 · MAGDAS/CPMN · geomagnetic storm

© The Author(s) 2026
N. Gopalswamy et al. (Eds.): ISWI 2024, SPPHY 431, pp. 79–87, 2026.
https://doi.org/10.1007/978-981-95-1121-1_10

1 Introduction

Asymmetry in the hemispheres within the solar quiet (Sq) current system at middle to low latitudes is the driving force behind the IHFACs. This initial idea was proposed by Van Sabben [1]. IHFACs are significantly influenced by seasonal changes in the Earth's atmosphere. The direction of the current reverses with the transition between winter and summer seasons in the hemispheres, reflecting the asymmetrical heating and atmospheric dynamics [2]. Furthermore, the intensity of the current is also affected by these seasonal variations. Fukushima suggested a theoretical model and Fukushima's IHFAC model is distinguished by several key features: IHFACs flow from the summer hemisphere to the winter hemisphere in the dawn sector and the other way in the noon sector, the polarity of IHFACs between the noon and dusk sectors is in-phase, the absolute intensity of IHFACs is stronger in both the dawn and noon sectors compared to the dusk sector [3]. Some studies have investigated IHFACs behavior in relation to solar activities. However, the analysis periods in these studies were often too short to demonstrate the solar cycle dependence of IHFACs, typically covering less than the full 11-year solar cycle [4, 5]. Shinbori and the team utilized long-term ground-based magnetometer data spanning 59 years to investigate the long-term characteristics of the Sq current [6]. Additionally, they compared the intensity of IHFACs between periods of high and low solar activity. IHFACs are highly dynamic and significantly influenced by solar activity. This study focuses on analyzing the intensity change patterns of IHFACs during periods of high solar activity, along with a detailed investigation of IHFAC behaviors during high-intensity geomagnetic storms.

The westward ΔD (daily variation of the D-component relative to the midnight-midnight baseline) magnetic field corresponds to northward Inter-Hemispheric Field-Aligned Currents (IHFACs), while the eastward ΔD magnetic field corresponds to southward IHFACs. In our previous study, we identified significant differences in IHFAC behavior between Solar Cycles 23 and 24, revealing the general characteristics of IHFACs [7]. In the present study, we investigate IHFAC behavior during intense geomagnetic events across two solar cycles, with a focus on understanding how IHFACs respond to different solar dynamics, including high-intensity geomagnetic storms. The analysis spans two solar cycles to account for factors such as solar cycle variability, including differences in solar activity levels, solar wind conditions, and interplanetary magnetic field (IMF) configurations. Additionally, this approach enables the identification of long-term trends and assesses the potential impact on space weather forecasting. This study examines the long-term behavior of IHFACs during solar cycles 23 and 24 revealing both long-term and short-term features. By selecting several intense geomagnetic storms, a comparative analysis of IHFACs during solar disturbed and quiet periods was conducted to identify the characteristics of IHFACs during intense geomagnetic storms.

Recent advancements in identifying IHFAC features have been made through the analysis of various satellite data. Swarm satellite data has been employed to examine IHFAC characteristics, with peak patterns that align with our findings [8, 9]. Studies using CryoSat-2 and GRACE-FO data have further investigated IHFAC features, revealing the source regions of the current systems and clearly demonstrating reversal patterns [10]. Additionally, IHFAC climatology has been explored using satellite data, showing that during equinoxes, IHFAC behavior is similar to that of the June solstice, while the

December solstice displays distinct, stand-alone characteristics. Although reversal patterns are consistent with our observations, deviations are noted during the December solstice [11]. Total Electron Content (TEC) is a critical indicator of ionospheric dynamics. During solar events, an increase in TEC reflects enhanced particle precipitation into the ionosphere [12]. The source of IHFAC is ionospheric imbalance, which arises due to the Sq current. This current is significantly influenced during storm conditions. The magnetic field component linked to the disturbed ionospheric current exhibits anti-Sq variations, which are driven by the prompt penetration electric field and the disturbance dynamo electric field [13]. TEC increases during the main phase of a storm, with a reduction observed in the initial recovery phase relative to the TEC in the main phase [14].

2 Data and Methods

The study is based on 20 + years of geomagnetic data with one-hour resolution, spanning from 1998 to 2020, were obtained from the Davao station in the Philippines, provided by the International Research Center for Space and Planetary Environmental Science(i-SPES), at Kyushu University [15, 16]. The study focused on analyzing the geomagnetic east-west component, known as the D-component, to investigate the characteristics of Inter-Hemispheric Field-Aligned Currents (IHFACs). Since the geomagnetic field at Davao is approximately horizontal due to its proximity to the magnetic equator, we assume that the D-component variations are primarily caused by IHFACs. To isolate the effect of IHFACs on the D-component, we first subtracted a baseline calculated from the midnight-averaged values from each hourly data point (ΔD). Given that the source of the daytime ionospheric wind dynamo is primarily solar radiation, its magnetic effects are generally negligible during nighttime hours [17]. Since northward and southward IHFACs induce westward and eastward magnetic field variations on the ground, respectively, the D-component was particularly relevant for this analysis. Given that the Davao station is located near the magnetic equator (Geomagnetic latitude −2.22°N, Geomagnetic longitude 197.9°E, dip latitude −0.24°), it was assumed that the D-component variations were due to magnetic field changes caused by IHFACs [15]. The term "northward IHFACs" refers to currents flowing from the southern hemisphere to the northern hemisphere. Accordingly, a westward ΔD magnetic field corresponds to northward IHFACs, while an eastward ΔD magnetic field corresponds to southward IHFACs. The daytime ionospheric wind dynamo is driven by solar radiation, with its magnetic effects generally negligible at night. The daily baseline was derived by linearly interpolating between two consecutive midnights. The daily variation of the D-component, ΔD, was calculated relative to this midnight-to-midnight baseline [7].

The resulting ΔD variation patterns were analyzed to identify the characteristics of IHFACs. The Dst and Kp indices were used to detect and measure the intensity of solar disturbances, while the F10.7 index was used to measure solar activity intensity across solar cycles [18, 19]. The monthly quietest day was determined by averaging the daily geomagnetic variations of the five quietest days provided by World Data Center for Geomagnetism, Kyoto [19]. Variations in disturbance-time IHFACs were then compared to this monthly quietest daily variation. The variation of ΔD in intense geomagnetic storms was examined to study respective IHFACs behavior.

3 Results and Discussion

3.1 Long-Term Variation

Averaged values (smoothed by 61-day moving average) of each sector of the daily ΔD, revealing significant variations across different sectors are shown in Fig. 1. Previous study has revealed some notable characteristics of IHFACs based on the time of the day [7]. Specifically, the following characteristics were notable: 1) the amplitudes in the dawn and noon sectors were approximately 2–3 times greater than those observed in the dusk sector; 2) there was a marked decrease in the yearly amplitude of ΔD during solar cycle (SC24) in both the dawn and noon sectors; 3) and peak-to-peak amplitudes of ΔD in the dusk sector showed enhancements during the declining phases of both solar cycles. Responses of ΔD during the declining phases and solar max underscore the complex and dynamic nature of geomagnetic and IHFACs responses to solar influences and Milan and colleagues have demonstrated that the solar wind speed increased, and the Interplanetary Magnetic Field (IMF) intensified during the declining phases of Solar Cycle 23 and Solar Cycle 24, with the effects being more pronounced in Solar Cycle 23 [20]. The aforementioned characteristics (2 and 3) could be attributed to these changes in the IMF and solar wind speed.

It is well known that solar cycle 23 (SC23) was more active than solar cycle 24 (SC24). To study IHFACs behavior during high-intensity geomagnetic storms, we selected four intense geomagnetic storm events (E1, E2, E3, and E4) from SC23 (see Fig. 1). These events will be described in detail in the next section. Wave patterns, exemplified by F1 and F2 in Fig. 1 around the September equinox, are mainly observed in the noon sector, with similar patterns visible in the dawn sector as well. Further studies are being conducted to investigate the causes of these notable patterns and identify any possible characteristics associated with IHFACs.

3.2 IHFACs During High-Intensity Geomagnetic Storm Events

Figure 2 show the variations in geomagnetic field components ΔH, ΔD, and ΔZ during high-intensity geomagnetic storm events E1, E2, E3, and E4 (see Fig. 1 and Table 1). These four events were selected from the solar maximum and declining phase of SC23, ensuring that all events occurred in the same month of different years. Label for each component X (i.e. H, D, Z) "ΔX_d" represents disturbance time variation while "ΔX_q" represents quiet time variation. Histogram inside the graphs represent difference between disturbed time and quiet time variation (ΔX_d – ΔX_q).

The polarity pattern of the IHFAC in November generally exhibits a positive trend during the morning, leading to a negative ΔD. At noon, the polarity becomes negative, resulting in a positive ΔD, and by evening, it transitions to a weakly positive state [7]. During the solar cycle maximum, ΔD_d exhibits positive enhancements at noon, while a significant negative ΔD_d is observed in the evening (Fig. 2a, b) highlighting the temporal complexity of geomagnetic disturbances and IHFACs characteristics. These patterns are the opposite of the typical IHFAC ΔD behavior in November. During the declining phase of the solar cycle, positive ΔD_d enhancements are observed at both noon and dusk (Fig. 2c and d), except during the main phase of the second storm (Fig. 2d).

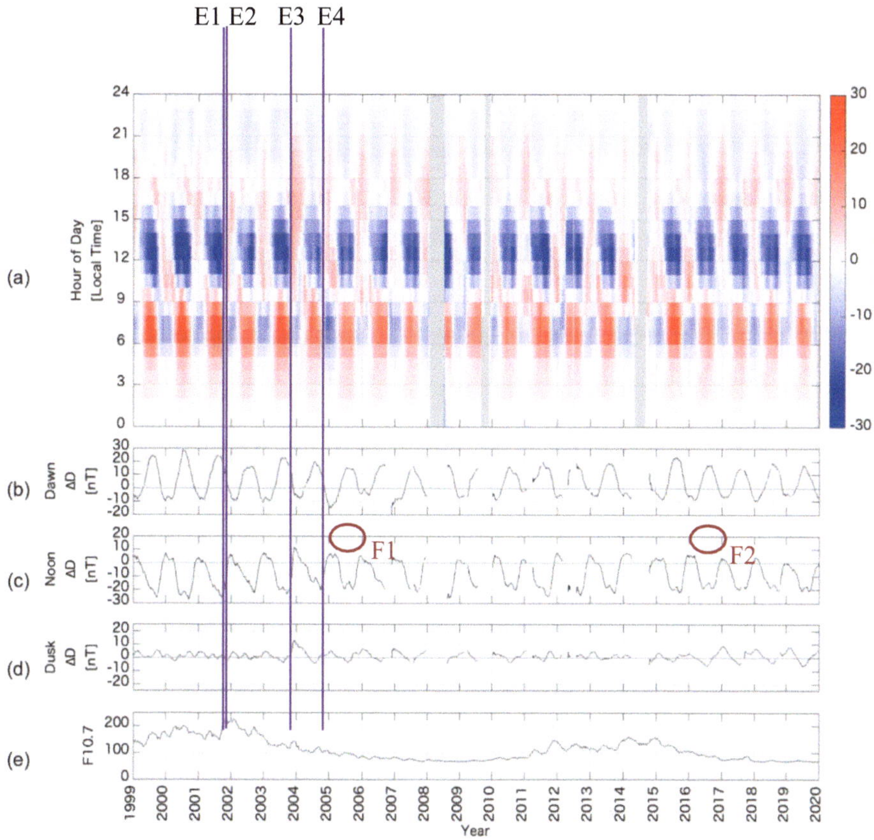

Fig. 1. Yearly ΔD variation in dusk, noon, and dawn sectors in solar cycles 23–24. Adapted from Ranasinghe et al., 2021[7].

During the geomagnetic storms where the recovery phase started around midnight, the positive daily amplitude of ΔD gradually decreased over several days (e.g., Fig. 2, Panels (b) and (c)), while no such pattern was observed on the negative side. Similarly, for two events where the recovery phase started around noon, a similar gradual decline in the negative amplitude of ΔD was observed (eg. Figure 2. Panel (a), (d)). This gradual decline suggests a slow return to quieter geomagnetic conditions as the storm's effects wane. It underscores the significant temporal variations and differential impacts on the positive and negative components of ΔD, providing insights into the IHFACs response to such extreme space weather events. In equatorial regions, ΔH primarily reflects the contribution of Sq and EEJ currents. It is evident that during geomagnetic storms, not only are the Sq and EEJ currents influenced, but IHFACs are also affected. Daytime deviations in IHFACs are likely due to changes in the Sq current, which can be attributed to the strengthening of the ring current during these events. Further investigation is needed to identify features associated with F1 and F2, which have not yet been thoroughly studied.

Table 1. Geomagnetic storm events under the study.

Event	Graph	Min Dst	Min Dst at (UT)	Min Dst at (LT)	Solar cycle phase
E1	Figure 2. (a)	−292	2001-11-06 06:00	2001-11-06 14:00	Solar max
E2	Figure 2. (b)	−221	2001-11-24 16:00	2001-11-25 00:00	Solar max
E3	Figure 2. (c)	−422	2003-11-20 20:00	2003-11-21 04:00	Declining
E4	Figure 2. (d)	−374	2004-11-08 06:00	2004-11-08 14:00	Declining

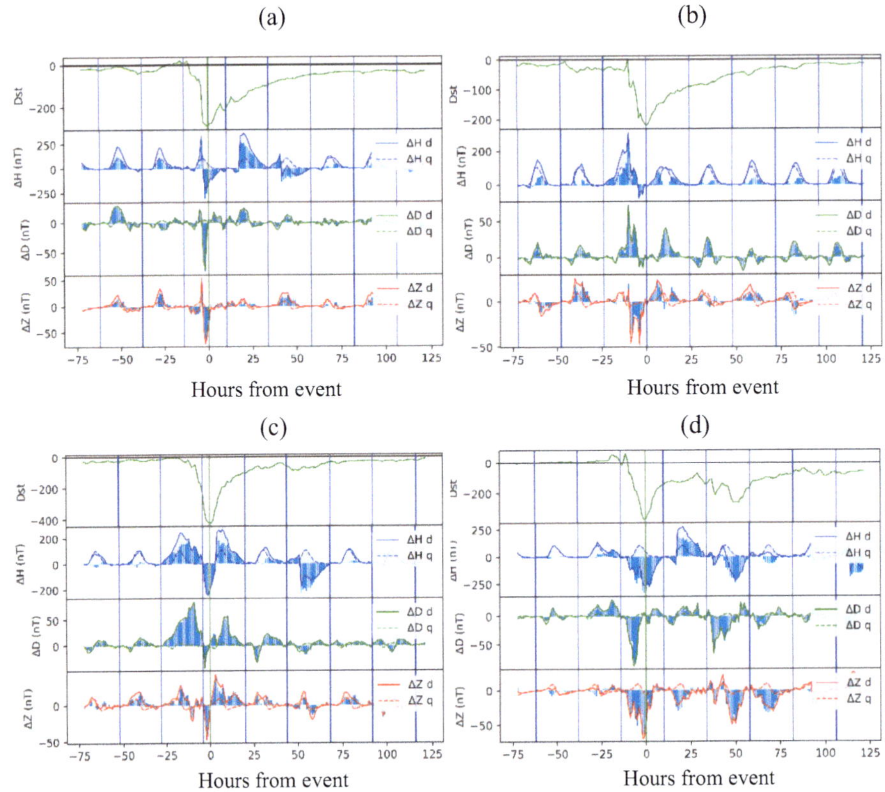

Fig. 2. Geomagnetic ΔH, ΔD, ΔZ component variations in extreme geomagnetic storm events.

Additionally, the dusk sector enhancements observed during the declining phase in Fig. 1 present an area for future research. Ionospheric conductivity is influenced by solar irradiation, which drives ionospheric dynamics [21]. Moreover, magnetospheric and ionospheric current systems are affected by the Interplanetary Magnetic Field (IMF) and solar wind conditions [22, 23]. Together, these factors shape the dynamic behavior of IHFACs during high-intensity geomagnetic storms.

4 Conclusion

The study examined geomagnetic data collected over a 20-year period from the MAGDAS station in Davao, Philippines, spanning solar cycles 23 and 24, to investigate characteristics related to inter-hemispheric field-aligned currents (IHFACs). It was found that the peak-to-peak amplitude values of the yearly ΔD in the dusk sector were approximately twice as high during the declining phases of the solar cycles compared to the rising and maximum phases. During the main phase of these geomagnetic storms, there was a notable increase in the negative component of ΔD around noon. The findings clearly indicate that the response of IHFACs to solar disturbances is significantly influenced by the geomagnetic storm phase and the time of day.

In conclusion, the study highlights several key findings: amplitudes in the dawn and noon sectors were significantly greater than in the dusk sector, with marked amplitude decreases in Solar Cycle 24. The dusk sector showed enhanced peak-to-peak amplitudes during the declining phases of both solar cycles, likely linked to increased solar wind speed and intensified Interplanetary Magnetic Field (IMF). The geomagnetic disturbances and IHFACs responses revealed distinct temporal patterns, with gradual recovery observed post-storm. These findings underscore the complex interactions between geomagnetic activity, solar influences, and IHFACs, warranting further investigation, particularly in equatorial regions and the dusk sector.

Acknowledgements. Manila University, for collecting the magnetometer data at Davao observatory of MAGDAS network. NRC of Sri Lanka (16–098) for supporting MR. JSPS (15253005, 18253005) for supporting MAGDAS, JSPS KAKENHI Grant (17J40136, 19K03956, 21K03646, 21H04518, 23K18467) for supporting AF, JSPS KAKENHI Grant (19K03956, 15H05815, JP20H01961, 21H04518, 22K21345, 22K03707) for supporting AY.

References

1. Sabben, V.: Magnetospheric currents, associated with the N-S asymmetry of Sq. J. Atmos. Terr. Phys. **28**(10), 965–982 (1966)
2. Fukushima, N.: Electric potential difference between conjugate points in middle latitudes caused by asymmetric dynamo in the ionosphere. J. Geomagn. Geoelectr. **31**, 401–409 (1979)
3. Fukushima, N.: Some topics and historical episodes in geomagnetism and aeronomy. J. Geophys. Res. **99**(A10), 19113–19142 (1994)
4. Park, J., Lühr, H., Min, K.: Climatology of the inter-hemispheric field-aligned current system in the equatorial ionosphere as observed by CHAMP. Ann. Geophys. **29**, 573–582 (2011)
5. Owolabi, O., Bolaji, O., Adeniyi, J., Oyeyemi, E., Rabiu, A., Habarulema, J.: Excursions of interhemispheric field-aligned currents in Africa. JGR Space Phy **123**, 6042–6053 (2018)
6. Shinbori, A., Koyama, Y., et al.: Characteristics of seasonal variation and solar activity dependence of the geomagnetic Sq daily variation. JGR Spa Phy. **122**, 10796–10810 (2017)
7. Ranasinghe, M., Fujimoto, A., Yoshikawa, A., Jayaratne, C.: Seasonal variation of inter-hemispheric field-aligned currents deduced from time-series analysis of the equatorial geomagnetic field data during solar cycle 23–24. Earth Planets Space **73**, 146 (2021)
8. Wang, F., Lühr, H., Xiong, C., Zhou, Y.: Improved field-aligned current and radial current estimates at low and middle latitudes deduced by the Swarm dual-spacecraft. JGR Space Physics **127**(6), e2022JA030396 (2022)

9. Lühr, H., Kervalishvili, G. N., Stolle, C., Rauberg, J., Michaelis, I.: Average characteristics of low-latitude interhemispheric and F region dynamo currents deduced from the swarm satellite constellation. JGR Space Phys. 124 (2019)
10. Park, J., Stolle, C., Yamazaki, Y., et al.: Diagnosing low-/mid-latitude ionospheric currents using platform magnetometers: CryoSat-2 and GRACE-FO. Earth Planets Space **72**, 162 (2020)
11. Park, J., Yamazaki, Y., Lühr, H.: Latitude dependence of interhemispheric field-aligned currents (IHFACs) as observed by the Swarm constellation. JGR Space Phys. **125**, e2019JA027694 (2020)
12. Akala, A.O., et al.: Solar origins of August 26, 2018 geomagnetic storm: Responses of the interplanetary medium and equatorial/low-latitude ionosphere to the storm. Space Weather **19**, e2021SW002734 (2021)
13. Yumoto, K.: 210° Magnetic observation group: the STEP 210° magnetic meridian network project. J. Geomag. Geoelectr. **48**(11), 1297–1310 (1996)
14. Kashcheyev, A., et al.: Multivariable comprehensive analysis oftwo great geomagnetic storms of 2015. JGR Space Phys. **123**(6), 5000–5018 (2018)
15. Yumoto, K.: 210° Magnetic observation group: the STEP° 210 magnetic meridian network project. J. Geomag. Geoelectr. **48**(11), 1297–1310 (1996)
16. Yumoto, K.: CPMN Group: Characteristics of Pi 2 magnetic pulsations observed at the CPMN stations: a review of the STEP results. Earth Planets Space **53**, 981–992 (2001)
17. Yamazaki, Y., Maute, A.: Sq and EEJ—a review on the daily variation of the geomagnetic field caused by ionospheric dynamo currents. Space Sci. Rev. **206**(1–4), 299–405 (2017)
18. OMNIWeb Data Service. http://omniweb.gsfc.nasa.gov/. Last accessed 01 May 2024
19. Kyoto, M., Nose, T., Iyemori, M., Sugiura, T., Kamei: World Data Center for Geomagnetism. Dst. Index and International Q-Days (2015). https://doi.org/10.17593/14515-74000
20. Milan, S.E., Imber, S.M., Fleetham, A.L., Gjerloev, J.: Solar cycle and solar wind dependence of the occurrence of large dB/dt events at high latitudes. JGR Space Phys. 128(4), e2022JA030953 (2023)
21. Green, D.L., Waters, C.L., Anderson, B.J., Korth, H.: Seasonal and interplanetary magnetic field dependence of the field-aligned currents for both Northern and Southern Hemispheres. Ann. Geophys. **27**(4), 1701–1715 (2009)
22. Ganushkina, N.Y., et al.: Defining and resolving current systems in geospace. Ann. Geophys. **33**(11), 1369–1402 (2015)
23. Korth, H., Zhang, Y., Anderson, B.J., Sotirelis, T., Waters, C.L.: Statistical relationship between large-scale upward field-aligned currents and electron precipitation. JGR Space Phys. **119**(8), 6715–6731 (2014)
24. Imtiaz, N., Younas, W., Khan, M.: Response of the low- to mid-latitude ionosphere to the geomagnetic storm of September 2017. Ann. Geophys. **38**(2), 359–372 (2020)

Open Access This chapter is licensed under the terms of the Creative Commons Attribution 4.0 International License (http://creativecommons.org/licenses/by/4.0/), which permits use, sharing, adaptation, distribution and reproduction in any medium or format, as long as you give appropriate credit to the original author(s) and the source, provide a link to the Creative Commons license and indicate if changes were made.

The images or other third party material in this chapter are included in the chapter's Creative Commons license, unless indicated otherwise in a credit line to the material. If material is not included in the chapter's Creative Commons license and your intended use is not permitted by statutory regulation or exceeds the permitted use, you will need to obtain permission directly from the copyright holder.

Ground Level Muon Detectors Response to Extreme Space Weather Events During Solar Cycle 25

Nikola Veselinović(✉), Mihailo Savić, Aleksandar Dragić, Dimitrije Maletić, Dejan Joković, Radomir Banjanac, Miloš Travar, and Vladimir Udovičić

Institute of Physics, Pregrevica 118, Belgrade, Serbia
veselinovic@ipb.ac.rs

Abstract. With the increasing activity of the Sun due to the rising phase of solar cycle 25, there is a greater probability of extreme solar weather events accompanied by coronal mass ejections, geomagnetic storms, and modulation of cosmic rays. Cosmic ray flux measured with ground (and underground) muon detectors is sensitive to the energy of primary cosmic rays higher than neutron monitors and space-borne detectors, and they can complement the study of solar-terrestrial coupling. In this work, the variation of cosmic ray flux detected with ground-based detectors during recent extreme events was discussed. These various detectors have different median primary rigidities, which allow monitoring the impact of space weather on different energy ranges of primary cosmic rays. In addition, these data are compared with conditions measured in situ in interplanetary space around Earth in order to assess implications for solar-terrestrial coupling processes.

Keywords: Cosmic rays · muon detectors · Forbush decrease · space weather extreme events

1 Introduction

Cosmic rays (CR), originated mostly from outside Solar System. As high-energy charged particles, CR interact [1] with heliosphere, a region of space where interplanetary magnetic field (IMF) is carried by a solar wind (SW), a stream of supersonic plasma blowing outward from the Sun. With increasing activity eruptive phenomena from solar active regions [2, 3] such as Solar Flares (SFs), Coronal Mass Ejections (CMEs), their interplanetary counterpart (ICMEs) appear more often. These transient, accompanied with interplanetary shocks created due to interaction of faster ICME with slower SW, can additionally modulate CR [4, 5]. This modulation of CR, can be detected on Earth as non-recurrent Forbush decreases (FDs), a sudden offset in CR flux followed by a gradual recovery phase that last several days [6, 7 and references therein]. In the last couple of decades, cosmic ray flux has been detected using detectors on space probes, both outside and inside the heliosphere, as well as in the near-Earth environment,. At ground level, the most widely used method for detecting CR is by using Neutron Monitors (NM)

© The Author(s) 2026
N. Gopalswamy et al. (Eds.): ISWI 2024, SPPHY 431, pp. 88–93, 2026.
https://doi.org/10.1007/978-981-95-1121-1_11

[8], but secondary CR muons can also be used for monitoring the solar modulation of primary CR. Ground-level secondary CR muons originate from primary CR with higher energies than primary CR detected by Neutron Monitors (NM), and even higher energies than primary CR detected with detectors onboard space probes outside Earth's magnetic field, in the heliosphere. Therefore, the energy range of primary CR observed by muon detectors extends and complements the energy range detected by the NM network and solar probes, while still being sensitive to CR modulation within the heliosphere. Ground detectors primarily measure the integral flux over different energy ranges and are characterized by cut-off rigidity (R_c), the minimum magnetic rigidity that primary CR must have to penetrate the geomagnetic (and IMF) field to reach the detector. Another key parameter is the median rigidity (R_m) of the detected primary CR. These characteristics of each detector depend on its geomagnetic location, as well as its elevation, asymptotic direction, and response function.

The CR station at the Low-Background Laboratory for Nuclear Physics at the Institute of Physics in Belgrade, Serbia, consists of two identical detector setups that measure CR muon flux at ground level (GLL) and shallow underground level (UL). Details of the experimental setup, as well as the calculated response function of the detectors, are presented in [9].Time series of 5-min time resolution are presented in Fig. 1. The rising activity of the Sun during ascending phase of solar cycle 25 is visible in GLL time series as a gradual decrease of yearly count rate but not so in UL which detect muons related to primary CR with higher energy.

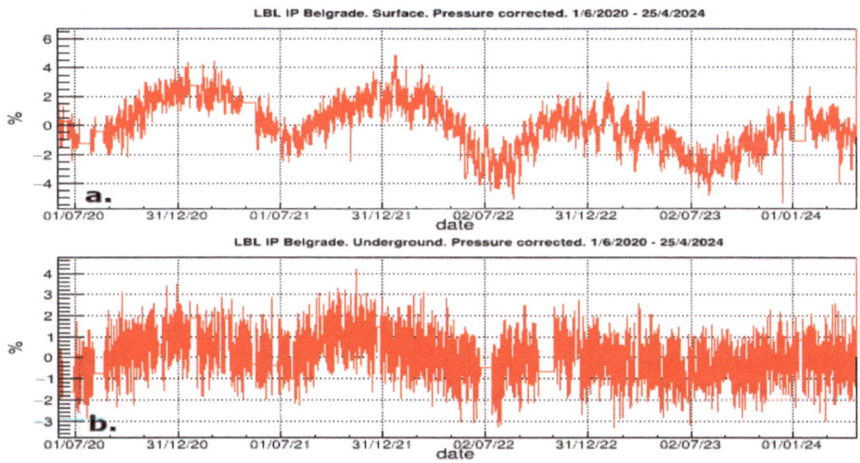

Fig. 1. Time series of muon variation with 5-min resolution detected in GLL (a) and UL (b) at Belgrade CR station. The flux is corrected for barometric effect but not for temperature effects.

The variations presented in Fig. 1 have already been corrected for atmospheric depth using barometric (atmospheric pressure) correction, so the remaining variation is due to heliospheric and temperature effects. The temperature effects can addressed using both empirical and theoretical methods. The temperature of the entire atmospheric column affects the muonic component of secondary cosmic rays (CR) at the ground in two ways: positive and negative [10 and references therein]. Secondary pions, produced by the interactions of primary CR particles with atmospheric nuclei, either decay into muons or interact with air nuclei. At higher temperatures, the air density decreases, leading to a lower probability of pion interaction and, consequently, higher muon production. However, higher temperatures also raise the altitude at which muons are produced, resulting in a longer path for the muons. As lower energy muons originate mainly from lower energy primary CR hence, so these muons are more sensitive to negative temperature effect then higher energy muons, capable to penetrate through ground. A yearly variation, mostly due to temperature effect is also visible more in pressure corrected GLL time series because negative effect of temperature on production of muons in atmosphere is more pronounce for lower energy primary CR.

2 Four Prominent FD Events During Rising Phase of Solar Cycle 25

The ongoing solar cycle 25, which began in December 2019, is still in its rising phase and has produced numerous strong X-ray solar flares accompanied by halo CMEs directed towards Earth. However, in this study, we will focus our analysis on four events detected by the Belgrade Cosmic Ray Station, as well as by the NM detector network. These four events were initiated by a series of solar flares, followed by CMEs, and were accompanied by sudden storm commencements, geomagnetic field disturbances, and auroras.

On November 5, 2021, the first significant FD of solar cycle 25 was observed. It was caused by a fast CME directed towards Earth and was accompanied by the first Ground Level Enhancement (GLE) of solar cycle 25 [11]. A GLE is a rare high-energy Solar Energetic Particle (SEP) event observed by the NM network, and in this case, it occurred on October 28.

Second event occurred on April and May 2023. Solar Active region AR 13283 produced a halo CME on April 21, 2023 which also generated a sever geomagnetic storm [12] Such intense geomagnetic storms are unexpected, since the CME was launched from a weak magnetic field region on the Sun [13]. Activity lasted during end of April and beginning of May which lead to two FDs in succession.

A National Solar Observatory telescope in Australia on March 23, 2024 recorded SF originate from two active regions AR3614 and AR3615 which produced CME directed toward Earth. On Sunday, March 24, CME triggered a severe G4-class geomagnetic storm, the most potent solar storm since 2017.

From May 3 through May 9, 2024, NASA's Solar Dynamics Observatory observed 82 solar flares. These flares originated mainly from two solar active regions: AR 13663 and AR 13664. Solar storm started on May 7th, followed by seven CMEs directed toward Earth. As a result, G5-class geomagnetic storm, a strongest since 2003, followed.

3 Cosmic Rays Activity Registered by Belgrade Station During Prominent FD Events

All of the mentioned extreme events produced FDs, which were simultaneously recorded by cosmic ray (CR) detectors worldwide, including the worldwide NM network and the Belgrade CR station. For example, the time series of detected flux during the first two events, for the Belgrade station (GLL, 44.85°N, 20.38°E, 75 m, $R_c = 5.3$ GV, $R_m = 63$ GV, and UL $R_c = 12$ GV, $R_m = 122$ GV) and, for comparison, one of the NM detectors at Thule (THULE, 76.5°N, 68.7°W, 26 m, $R_c = 0.3$ GV, $R_m = 12.6$ GV), for events in April and May 2023, are presented in Fig. 2. The properties of the near-Earth solar wind magnetic field and some plasma parameter data from several spacecraft were acquired from the OMNI database [14] and are also provided in the same figure.

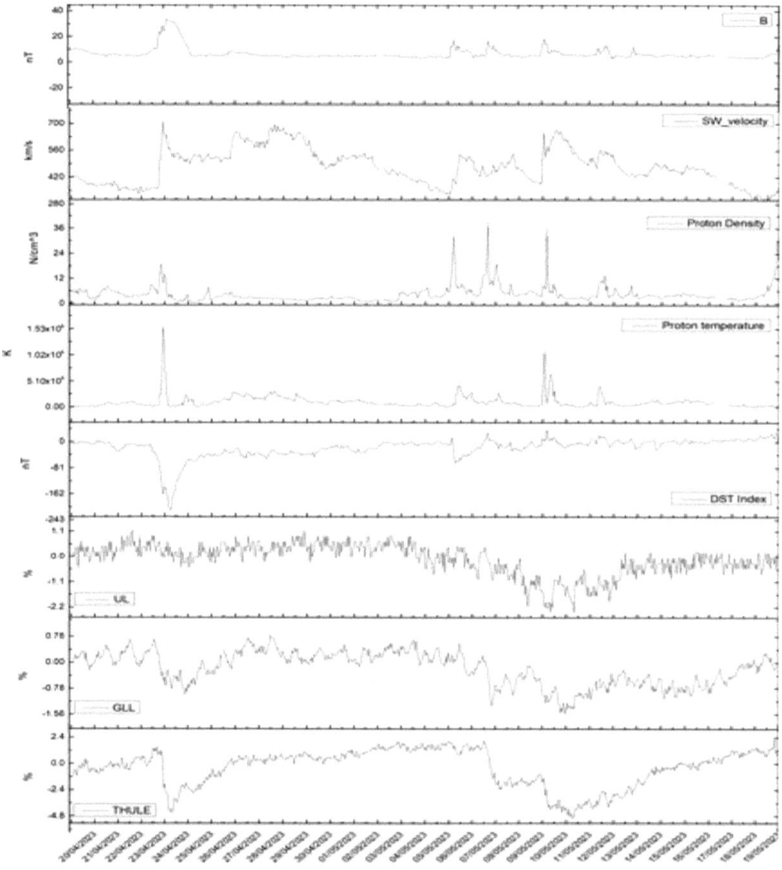

Fig. 2. Hourly-averaged variation in CR intensity measured at ground station (Thule, Belgrade CR station ground (GLL) and underground level (UL) detectors, Dst index,, Solar wind proton temperature, proton density and velocity (SW_velocity), magnitude of interplanetary magnetic field B during end of April and beginning of May 2023.

Cross-correlation Pearson coefficients between CR time series measured by Belgrade muon station, Thule NM and selected space weather parameters during these extreme events show (significance < 0.0001) that there is a anti-correlation between SW speed and CR count for lower energy CR (-0.6398 for Thule and -0.5032 for GLL) for April-May 2023 events and no correlation between SW speed and UL count. Another strong anti-correlation is for average SW magnetic field B (-0.6039 for Thule and -0.3916 for

GLL) and correlation with Dst index (0.7095 for Thule and 0.4937 for GLL) and again no correlation with UL. Data from OMNI database, for the two latest strong FDs during 2024, has not been publicly available so these FDs are not analyzed.

To determine the amplitude of the FD for each station and each event, a baseline was established. One of the methods, applied in this study is to use average hourly count several days before the event when solar activity was lower and more stable. We used a 7-day interval, immediately prior the activity. This longer interval was chosen due to unusually high solar activity during the periods of interest. The results for the amplitude of the FD are given in Table 1.

Table 1. The maximal amplitude of the FD during the extreme events in May 2024, March 2024, May 2023, April 2023 and November 2021.

FD amplitude %	10.5. 2024	24.3. 2024	11.5. 2023	24.4.2023	4.11. 2021
Thule NM	9.4	9.1	5.1	4.2	7.8
GLL	4.0	4.9	2.2	1.6	4.6
UL	2.6	2.8	2.0	-	1.5

4 Conclusion and Discussion

In this work, we studied several strong FDs that occurred during solar cycle 25, using data from the Belgrade muon station and other multiple sources. The series of extreme solar activities, followed by a series of sometimes overlapping CMEs, affected CR and caused strong geomagnetic storms. The energy range of the affected primary CR was wide enough to have a measurable effect, observed as a decrease in measured flux by NM network and muon stations. It is shown, based on the measured amplitude of FDs at different stations and correlations between time series of solar wind characteristics and CR flux, that higher energy CRs were less affected by these events. To better understand the extent of solar modulation, which is also connected with solar-terrestrial coupling, more data from future FDs as well as data from other muon stations are needed.

References

1. Heber, B., Fichtner, H., Scherer, K.: Solar and heliospheric modulation of galactic cosmic rays. Space Sci. Rev. **125**(1), 81–91 (2006)
2. Rotti, S.; and Petrus C. Martens P. C. Analysis of SEP Events and Their Possible Precursors Based on the GSEP Catalog 2023 ApJS , 267, 40 (2023)
3. Reames, D.V.: The Two Sources of Solar Energetic Particles. Space Sci. Rev. **175**, 53–92 (2013)
4. Belov, A., Abunin, A., Abunina, M., et al.: Coronal mass ejections and non-recurrent Forbush decreases. Sol. Phys. **289**, 3949–4396 (2014)
5. Belov, A., Shlyk, N., Abunina, M., Belova, E., Abunin, A., Papaioannou, A.: Solar energetic particle events and forbush decreases driven by the same solar sources. Universe **8**, 403 (2022)

6. Cane, H.V.: Coronal mass ejections and forbush decreases. Space Sci. Rev. **93**, 55 (2000)
7. Gopalswamy, N.: History and development of coronal mass ejections as a key player in solar terrestrial relationship. Geosci. Lett. **3**(8) (2016)
8. NMBD homepage, https://www.nmdb.eu/nest/. Last accessed 06 June 2024
9. Veselinović, N., Dragić, A., Savić, M., et al.: An underground laboratory as a facility for studies of cosmic-ray solar modulation, Nuclear Instruments and Methods in Physics Research Section A: Accelerators, Spectrometers, Detectors and Associated Equipment, 875 (2017)
10. Savić M., Dragić A., Maletić D., et al.: A novel method for atmospheric correction of cosmic-ray data based on principal component analysis, Astroparticle Physics 109 (2019)
11. Papaioannou, A., Kouloumvakos, A., Mishev, A., et al.: The first ground-level enhancement of solar cycle 25 on 28 October 2021. A&A **660**, L5 (2022)
12. Habarulema, J.B., et al.: Absence of high frequency echoes from ionosondes during the 23–25 April 2023 geomagnetic storm; what happened? Journal of Geophysical Research: Space Physics 129 (2024)
13. Vemareddym, P.: ApJ **961**, 199 (2024)
14. OMNI Homepage, https://omniweb.gsfc.nasa.gov/ow.html. Last accessed 06 June 2024

Open Access This chapter is licensed under the terms of the Creative Commons Attribution 4.0 International License (http://creativecommons.org/licenses/by/4.0/), which permits use, sharing, adaptation, distribution and reproduction in any medium or format, as long as you give appropriate credit to the original author(s) and the source, provide a link to the Creative Commons license and indicate if changes were made.

The images or other third party material in this chapter are included in the chapter's Creative Commons license, unless indicated otherwise in a credit line to the material. If material is not included in the chapter's Creative Commons license and your intended use is not permitted by statutory regulation or exceeds the permitted use, you will need to obtain permission directly from the copyright holder.

Solar Energetic Particles and the Associated Phenomena Such as Coronal/Interplanetary Radio Bursts

Solar Energetic Particle Events and Radio Bursts

Nat Gopalswamy(✉)

NASA Goddard Space Flight Center, Maryland, USA
nat.gopalswamy@nasa.gov

Abstract. Solar Energetic Particles (SEPs) and radio bursts are indicators of particle acceleration on the Sun and in the heliosphere. The accelerated particles have energies significantly higher than that of thermal particles up to several orders of magnitude. SEPs are detected directly by particle detectors on Earth and in space. Understanding SEPs is important from both science and application points of view because they are poorly understood and present space weather hazard to humans and their technology in space. SEPs accompany energetic flares, coronal mass ejections (CMEs), and intense radio bursts, which help us understand particle properties such as intensity, spectra, and time evolution. This paper summarizes how SEP properties are closely related to solar eruptions and the associated solar radio bursts.

Keywords: Solar Energetic Particles · Solar Radio Bursts · CMEs · Flares

1 Introduction

Solar energetic particle (SEP) events and radio bursts are indicative of particle energization in the corona and interplanetary (IP) medium. During solar eruptions, the Sun releases large amounts of stored energy in solar magnetic regions such as active regions and quiescent filament regions. Solar flares and coronal mass ejections (CMEs) are two of the key manifestations of solar eruptions. Eruptive prominences and fast mode magnetohydrodynamic (MHD) shocks form the inner core and outermost structure of CMEs, respectively. A significant amount of the released energy is converted into the kinetic energy of the energetic particles that manifest as SEPs and various types of radio bursts. SEPs were first observed in 1942 by ground-based instruments used for detecting cosmic rays [1]; they are now routinely observed by spaceborne detectors. The ground level enhancement (GLE) events are SEP events that indicate tens of GeV particles. In space, SEP ions are observed at energies down to the solar wind particle energies [2, 3]. SEPs also consist of electrons, which are commonly observed at energies up to a few MeV [4], although they have been observed up to a few 100 MeV [5]. Indirect observations of energetic particles are via their electromagnetic signatures. Interestingly, nonthermal radio emission from the Sun were also first detected in 1942 [6]. While all radio emission from the Sun is due to electrons, energetic ions produce gamma-ray emission when the particles interact with the dense solar atmosphere. Particles are energized in the solar

© The Author(s) 2026
N. Gopalswamy et al. (Eds.): ISWI 2024, SPPHY 431, pp. 97–114, 2026.
https://doi.org/10.1007/978-981-95-1121-1_12

corona and propagate toward and away from the Sun. Electrons propagating away from the Sun generate various types of radio bursts [7] at wavelengths ranging from decimeters to kilometers. Electrons flowing toward the Sun produce microwave bursts from cm to mm wavelengths. Some of these electrons also produce hard X-ray bursts and continuum emission extending to 100s of MeV [8]. The energetic electrons responsible for the radio bursts and SEP events have a common source of energization. Therefore, radio bursts are often used to infer SEP events in advance. While SEP events significantly alter Earth's space environment resulting in adverse impact on humans and their technology in space, radio bursts generally do not have direct space weather effect. There is one exception: occasional intense microwave emission can interfere with spacecraft signals (GPS) and radar signals (airport operations) [9, 10]. Extensive discussion on solar energetic particles and radio bursts can be found in the monographs [11, 12]. Recent reviews on SEPs can be found in [13–15]. This article focuses on large SEP events relevant for space weather (those with a $>$ 10 MeV flux of at least 10 pfu (particle flux units)), and the associated radio bursts.

2 Why Do We Care About SEPs?

SEPs represent one of the major radiation hazards in space, both inside Earth's magnetosphere and outside in the IP space. According to NOAA's Space Weather Prediction Center, large SEP events are of concern for (i) high frequency (HF) radio communication systems, (ii) satellite operations, and (iii) biological systems [16]. SEP events are called "Solar Radiation Storms" whose severity is classified on a scale of 1–5 denoted by S1-S5 to indicate $>$ 10 MeV proton intensity increasing by a factor 10 from10 pfu to 10^5 pfu. S1 events have a minor impact on HF radio in the polar regions with little impact on satellite operations and biological systems. The severity of the impact progressively increases for S2, S3, S4, and S5 levels on all aspects (i)–(iii). In the extreme case of ~ 10^5 pfu (S5), a complete HF radio blackout is possible; wide-ranging problems may arise in satellite operations–loss of mission, loss of control, permanent damage to solar panels, and image degradation due to noise created in the detectors; high radiation risk to passengers and crew in polar aircraft and astronauts performing extravehicular activities.

There is plenty of anecdotal evidence showing extensive damage to space missions or subsystems due to particle radiation. The widespread impact of the 2003 Halloween storms has been well documented [17]. The Martian Radiation Environment Experiment (MARIE) on the Mars Odyssey mission was designed to assess the radiation environment of Mars. MARIE succumbed to the SEP event that occurred on 28 October 2003 [18]. Damage to solar panels of satellites has also been well documented, the main concern being the reduction of efficiency. Two powerful SEP events that occurred on 29 September 1989 and October 19 each produced a step-like decrease in the expected current from the GOES-7 solar panels by ~ 5–10% [19]. Apart from these super-intense SEP events, regular SEP events result in frequent spacecraft anomalies. Statistical studies have shown that the frequency of spacecraft anomalies depends on the intensity and energy range of SEP events, the spacecraft altitude, and spacecraft inclination [20]. A notable result is that the number of spacecraft anomalies increases after SEP events and peaks 4–5 days following the start of an SEP event. Even though there is not much advance warning for GLE events, one can take precautions when an SEP event occurs owing to the possibility of spacecraft anomalies over the following few days. These results illustrate the importance of predicting the intensity, temporal evolution, and spectrum of SEP events [21].

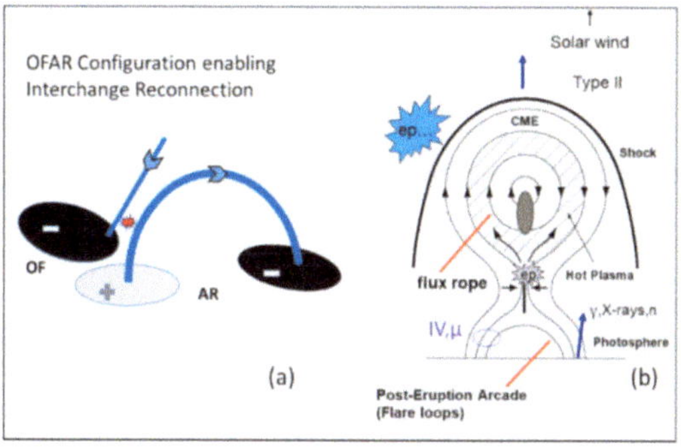

Fig. 1. Particle acceleration sites in the corona: (a) interchange reconnection region formed between an open field (OF) region and a closed magnetic field region such as an active region (AR). Such a magnetic configuration is known as an OFAR region. The interchange reconnection region is denoted by the red patch. (b) Flare reconnection in a closed magnetic region that produces post eruption arcade and a coronal mass ejection (CME). Particles are energized in the reconnection region and at the shock (ep denotes acceleration of electrons and protons). Electrons accelerated in the flare reconnection region produce microwave (μ) type IV bursts, and hard X-ray bursts ("X-rays"). Accelerated protons precipitating from the flare site result in impulsive gamma-ray bursts ("γ") and neutron ("n") emission due to interaction with the solar atmospheric particles. The shock-accelerated electrons and ions are observed as SEPs in space; the lower energy electrons produce type II bursts via the plasma emission mechanism.

3 Sources of Energetic Particles

There are many sources of energetic particles in the heliosphere [22]. Here we discuss those sources involving energy release at the Sun. There are other sources of energetic particles that we do not discuss here: reconnection exhausts in the solar wind [23], stream interaction regions [24, 25], and planetary atmospheres [26, 27].

The commonly accepted sources of energetic particles from the Sun are shown schematically in Fig. 1, viz., interchange reconnection between open and closed field lines (Fig. 1a) and reconnection between closed field lines (Fig. 1b). In both cases a closed magnetic region such as an active region (AR) or a quiescent filament region is involved. The magnetic region resulting in interchange reconnection has an open filed (OF) + AR (OFAR) magnetic configuration. OFAR configuration is thought to be responsible for type III storms, which consist of short duration type III-like bursts that occur in rapid succession for days, often lasting for more than one solar rotation [28–31]. The bursts in type III storms are caused by energetic electrons accelerated in the interchange reconnection region. A similar configuration is involved in causing impulsive SEP events, but the configuration is more dynamic, involving jets [32–34]. These events may also appear as narrow CMEs in the coronagraph field of view [35]. The helium isotope ^{3}He is significantly enhanced in these events, and hence are known as ^{3}He-rich events. One of the primary characteristics of impulsive SEP events is the association with type III radio bursts. These are regular type III bursts that are more intense and last longer than the type III storm bursts. The type III burst association indicates that electrons are also accelerated in the source region of impulsive SEP events.

Figure 1b schematically shows particle energization sites during solar eruptions. The energy release is thought to be due to the so-called flare reconnection in the current sheet below the rising flux rope. The flare reconnection results in two magnetic structures, viz., the post eruption arcade (PEA) at the Sun and the ejected flux rope. The flux rope accelerates while the reconnection is in progress and can attain high speeds, occasionally exceeding 3000 km/s. When the flux rope speed exceeds the local magnetosonic speed, a fast mode MHD shock is formed ahead of the flux rope. The shock can accelerate protons and electrons to very high energies [36, 37]. Large SEP events are thought to be primarily due to shock acceleration [2], although there is an ongoing debate on the contribution from the associated flare [38, 39]. The two sites of particle acceleration in Fig. 1b are evidenced by confined flares (or compact flares [40]) that do not have an associated CME and CME-driven shocks from quiescent filament region that have only weak flare signatures but are associated with large SEP events.

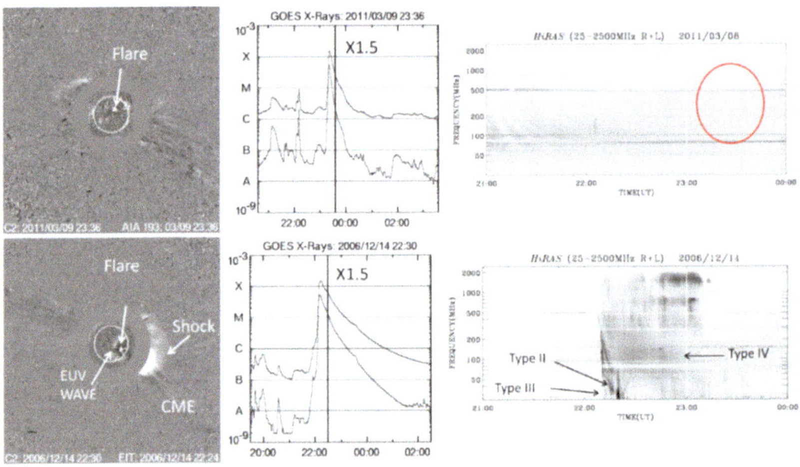

Fig. 2. (top left) Coronagraph image showing a flare brightening in an EUV difference image (SDO/AIA 193 Å) superposed on a LASCO/C2 difference image. (top middle) GOES soft X-ray light curve in the 1–8 Å band showing the X1.5 flare. The vertical line marks the time of the coronagraph image. (top right) Radio dynamic spectrum from the Hiraiso radio spectrograph (HiRAS) showing no emission in the expected time window (marked by the red circle). (Bottom) The panels are similar to the top ones, except for an eruptive flare of the same size, but accompanied by a shock-driving CME and meter wave radio bursts of type II, type III, and type IV.

Figure 2 compares a confined flare with an eruptive flare of the same size (X1.5). The confined flare has a slightly shorter duration. The soft X-ray flares are due to energy deposition by electrons accelerated in the flare reconnection streaming towards the Sun. In both cases, 17 GHz microwave emission was observed (not shown), indicating sunward propagating nonthermal electrons. No radio emission is observed in the metric radio band indicating that there is no upward flow of nonthermal electrons in the confined flare case. On the contrary, the eruptive flare is associated with a shock-driving CME. The shock also manifests as an EUV wave as can be seen in the SDO/AIA 193 Å difference image. In addition, type II, type III, and type IV bursts can be found in the metric radio dynamic spectrum. While type III and type IV bursts are thought to be due to electrons accelerated in the flare reconnection region, the type II burst is due to electrons accelerated in the shock front. Furthermore, the eruptive event was associated with a large (215 pfu) SEP event (see Fig. 3). The SEP event, metric type II burst, and the IP type II burst are all due to the CME-driven shock. The triple plots in Fig. 3 available online are useful in identifying the underlying CME and flare for SEP events.

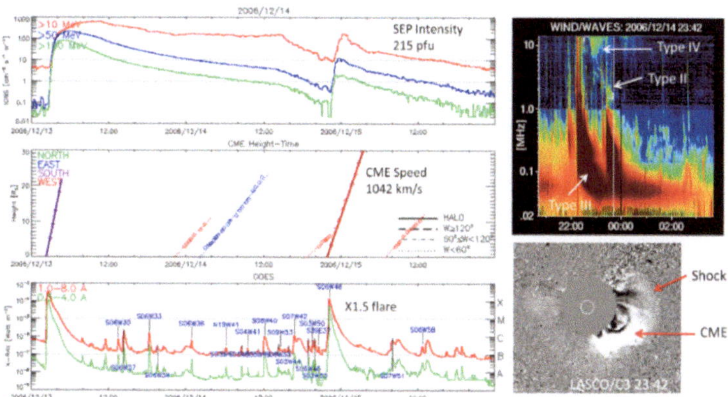

Fig. 3. (left) A composite plot showing the GOES SEP flux in pfu in three integral channels, the CME height-time plots for all CMEs occurring in a 3-day window, and the GOES soft X-ray light curve in two channels 1–8 Å and 0.5–4 Å, where the heliographic locations of the individual flares are indicated in blue. The X1.5 flare from S06W48 (cf. Figure 2) is towards the end of 2006 December 14, associated with a halo CME (sky plane speed = 1042 km/s) and a large SEP event (intensity = 215 pfu). Plot from https://cdaw.gsfc.nasa.gov. (top right) A Wind/WAVES radio dynamic spectrum showing IP type II, Type II, and type IV bursts associated with the SEP event. (bottom right) SOHO LASCO CME and the shock towards the end of the type II burst.

CMEs originating from quiescent filament regions accelerate slowly and attain super-Alfvenic speeds at relatively large distances from the Sun, and the SEPs, if present, are purely from shock acceleration [41, 42]. Also, when a CME-driven shock remains strong until 1 au and intercepted by a spacecraft with particle detectors, an energetic storm particle event is observed. These particles are locally accelerated and hence represent purely shock-accelerated particles with little flare contribution. One of the outstanding problems in SEP studies is the relative contribution of particles originating from flare and shock sources in a given event [43, 44]. Nevertheless, CME-driven shocks and type II bursts seem to be the best indicators of large SEP events [45].

4 Intensity of SEP Events

Among the observed properties of SEP events, the intensity, energy spectra, and temporal evolution are the ones that decide the space weather impact of these events. These properties are controlled by the CME/shock characteristics as well as the conditions in the ambient medium in which the shock propagates.

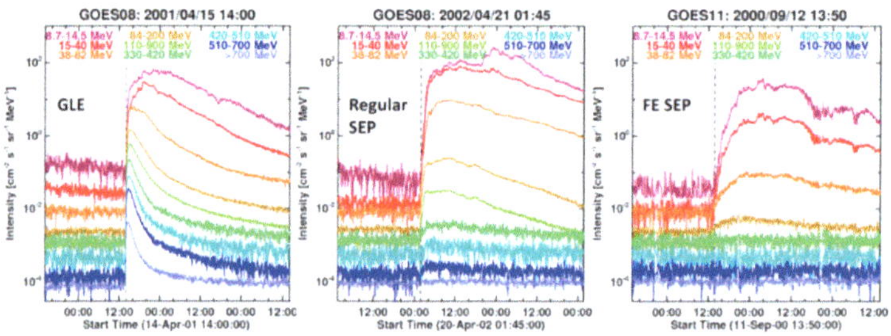

Fig. 4. Intensity vs. time profiles of three SEP events: (left) an SEP event with ground level enhancement (GLE) showing significant intensity increase in all energy channels, (middle) a regular SEP event with intensity signals in all channels except the highest one, (right) an SEP event due to a CME from a quiescent filament region (FESEP) that has intensity signal only in the lowest few channels. Clearly, spectral hardness decreases as one goes from left to right.

Figure 4 shows the time evolution of three large SEP events observed by GOES satellites. The 2001 April 15 event is a GLE event, indicating that particles are accelerated to GeV energies. The GOES light curves show a large peak in the > 700 MeV channel. The 2002 April 21 event is a regular SEP event with significant particle intensities up to ~ 200 MeV. At higher energies, the intensity is relatively low, with no increase in the > 700 MeV channel. The 2000 September 12 event has no intensity increase above 200 MeV. While the first two events are from an active region, the third event is from a quiescent filament region (FESEP event). One can infer that the spectra become progressively softer as one goes from GLE to regular SEP to FESEP events. The intensity and spectra of large SEP events are related to the properties of underlying CMEs as we illustrate in the next section.

The requirement of high CME speed for producing SEP events was reported a while ago [46, 47]. A moderate correlation was found between CME speed and SEP intensity, which increased when shock speed was used in place of the CME speed [48]. It was pointed out that the SEP intensity varied by 3–4 orders of magnitude for a given CME speed [49], probably due to (i) spectral variation of individual events, and (ii) presence of seed particles in the heliosphere. These correlations are obtained using sky-plane speeds, which need to be revised using 3-D speeds. An issue related to the seed particles is the result of higher SEP intensity when CMEs happen in quick succession from the same active region [49]. We discuss some of these points in more detail below.

5 Which CMEs Produce Large SEP Events?

CMEs occur at the rate of one every other day during solar minima and several per day during solar Maxima. There have been about 30,000 CMEs recorded in the SOHO/LASCO CME catalog since 1996 when SOHO started routinely observing CMEs until mid-2024. This corresponds to an average rate of over 1000 CMEs per year. However, the number of CMEs that produce SEP events is much smaller–only a fraction of a percent. Figure 5 shows the speed, width, and solar source locations of CMEs that produced large SEP events since 1996. The speed and width are measured in the sky plane and the source locations are heliographic coordinates of the eruption. We see that the CME speed has an approximate normal distribution, with an average speed of ~ 1545 km/s. This is almost a factor of 4 higher than the average speed of the general population (~ 400 km/s). The lowest bin of the speed distribution (600 km/s) has ~ 10 CMEs, indicating that SEP-producing CMEs need to have above-average speed to drive a shock. The number of CMEs with speeds > 1000 km/s drops rapidly and only a couple of CMEs have speeds exceeding 3000 km/s, limited by the amount of energy that can be stored/released in an active region. The width distribution shows that most (~ 82%) of the CMEs are halos. Halo CMEs are more energetic than regular CMEs and are very wide. The small number of non-halo CMEs have an average width of ~ 185° in the sky plane, which is quite large compared to the average width of regular CMEs, ~ 40°. Thus, only fast and wide CMEs possess enough energy to accelerate particles to energies > 10 MeV. Figure 5c shows that the SEP-producing CMEs are concentrated in the latitude range $\pm$ 30° corresponding to the active region belt. Furthermore, the source locations are heavily concentrated in the western hemisphere. This is because an Earth observer is magnetically connected to the western hemisphere (Parker spiral field lines) and energetic particles stream along those field lines to reach Earth. Energetic eastern hemispheric CMEs do produce SEP events but at locations behind the east limb. For example, the STEREO mission has detected large SEP events due to eastern hemispheric CMEs in Earth view, even though they did not produce a significant event at Earth [50].

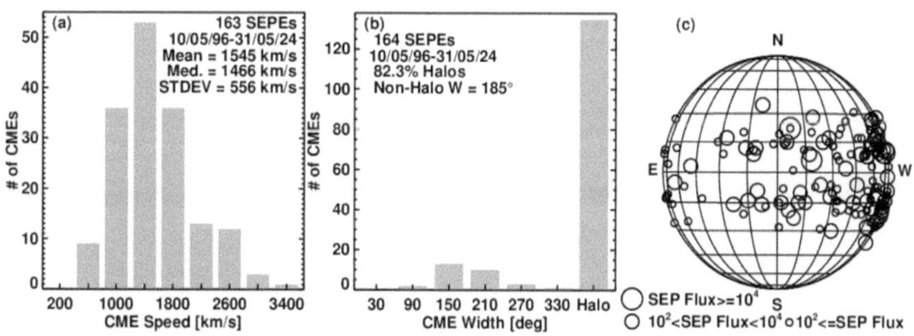

Fig. 5. Speed (a), width (b), and source location (c) distributions of CMEs associated with large SEP events observed by GOES since 1996, when SOHO started routinely observing CMEs. In (c), the size of the circle represents the SEP event size (peak intensity).

The speed used in Fig. 5 is an average within the coronagraph field of view (FOV, 2.5–32 Rs) measured in the sky plane. CMEs initially accelerate, reach a peak speed, and then decelerate [51]. The initial acceleration can range from a few m s^{-2} to ~ 10 km s^{-2}. Some CMEs finish accelerating very close to the Sun, while others continue to accelerate over large distances. Therefore, one expects that the high mean speed in Fig. 5 can be attained at various heliocentric distances. The ambient magnetic field significantly declines with distance from the Sun and the ambient Alfven speed also declines following an initial peak around 3 Rs. While the shock strength depends on its speed and the ambient Alfven speed, the particle acceleration efficiency of shocks depends on the ambient magnetic field, and hence declines with heliocentric distance.

Table 1. Kinematics and spectral properties of events in Fig. 4

SEP date	Ip pfu	V_{CME} km/s	a_{CME} m/s^2	V_{CMEi} km/s	a_{flr} km/s^2	Flare Size	Δt min	γ	Source Location
2001/04/15	951	1199	−35.9	1697	2.50	X14.0	8	2.33	S20W85
2002/04/21	2520	2393	−1.39	2088	0.59	X1.5	68	2.77	S14W84
2000/09/12	321	1550	+58.2	938	0.38	M1.0	50	3.81	S19W06

Table 1 lists CME properties of the three SEP events shown in Fig. 4. The average CME speed in Fig. 5a is 1545 km/s, which is above, similar, and below the CME speed (V_{CME}) in the GLE event (2001 April 15: 1199 km/s), FESEP event (2000 September 12: 1550 km/s), and regular SEP event (2002 April 21: 2393 km/s), respectively. Clearly the > 10 MeV SEP intensity (Ip, column 2) is not ordered by the sky-plane speed averaged over the coronagraph FOV. The average CME acceleration within the coronagraph FOV (a_{CME}, column 4) indicates different kinematics of the underlying CMEs (rapid deceleration in the GLE event, small deceleration in the regular SEP event, and continued acceleration in the FESEP event). The initial speed of CMEs (V_{CMEi}, column 5) computed from the first two height-time data points clearly orders the SEP intensities. A high initial speed implies shock formation close to the Sun. $V_{CMEi} > V_{CME}$ in the GLE event, whereas $V_{CMEi} < V_{CME}$ in the regular SEP and FESEP events. This means that the GLE CME finished accelerating close to the Sun, whereas the other two events continued to accelerate in the coronagraph FOV. This is indicated by the initial acceleration (a_{flr}, column 6) obtained as $V_{CME}/\Delta t$, where Δt (column 8) is the rise time of the associated soft X-ray (SXR) flare (column 7). The power law index of the fluence spectrum (γ, column 9) indicates the spectral hardness of the SEP events: the GLE event has the hardest spectrum and the FESEP event has the softest spectrum with the regular SEP event in-between. We see an inverse relationship between a_{flr} and γ. Finally, the SXR flare size orders the SEP spectral hardness, but not the SEP intensity.

6 Shock Formation Heights, SEP Spectra, and CME Kinematics

It is well known that the earliest indicator of a CME-driven shock is the onset time of a type II radio burst. Type II bursts typically start at a frequency of ~ 150 MHz. For a fundamental plasma emission, this frequency corresponds to an electron density of ~ 2.8×10^8 cm^{-3}, which typically occurs at a heliocentric distance of ~ 1.5 Rs [52]. There are bursts which start at higher frequencies (i.e., closer to the Sun) [53–56] or lower frequencies (i.e., farther from the Sun) [57] depending on the local plasma conditions and the CME kinematics. CMEs accelerating impulsively and attaining high speeds close to the Sun (within a couple of Rs) result in high-energy SEPs. This is because the particle acceleration efficiency of shocks is higher when the ambient magnetic field is higher [58]. Once a shock forms, it typically takes several minutes before the high-energy particles are released from the shock. For instance, in GLE events, the shock formation height is typically ~ 1.5 Rs and the particle release occurs when the CME is at a height of ~ 2–3 Rs [55, 59]. Thus, the CME initial acceleration, initial speed, shock formation height, and height of particle release are the key parameters that are of interest to space weather research connected to energetic particles (see Fig. 6).

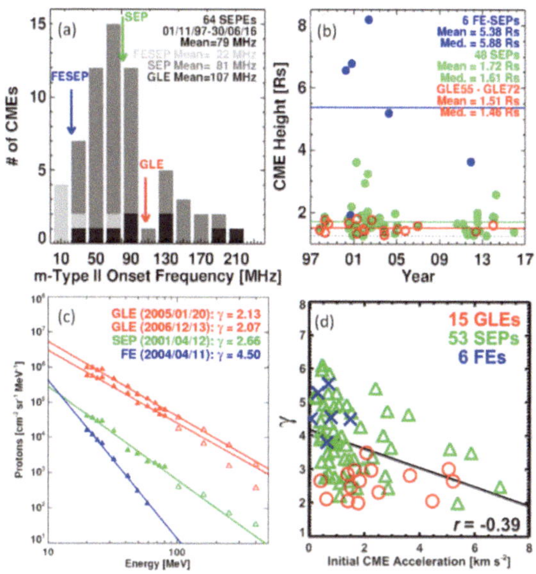

Fig. 6. (a) Histograms of starting frequencies of type II radio bursts associated with GLE events (dark bars), regular SEP events (dark gray bars), and FESEP events (light gray bars). The means of the distributions are shown on the plot and pointed by arrows. (b) Shock formation heights obtained as the CME leading-edge height at the time of the type II burst onset for the three groups of SEP events. (c) Representative 10–100 MeV fluence spectra of two GLE events (red), a regular SEP event (green), and an FESEP event (blue). The spectral indices (γ) are noted on the plot. (d) Variation of the fluence spectral index as a function of CME initial acceleration distinguished by the event group. The initial acceleration was computed as $a_{flr} = V_{CME}/\Delta t$, from the average CME speed (V_{CME}) in the coronagraph field of view and the flare rise time Δt. The correlation is moderate but statistically significant (the Pearson critical correlation coefficient is ~ 0.376 for p = 0.0005, the probability that the reported correlation is by chance).

Figure 6a shows that Type II bursts occur at highest frequencies in GLE events (~107 MHz), followed by regular SEP events (~81 MHz) and FESEP events (22 MHz). A similar pattern can be recognized in the shock formation height obtained as the CME leading-edge height at the time of type II burst onset (see Fig. 6b). Although there is some overlap, the mean shock formation height is ~ 1.51 Rs for GLE events, 1.72 Rs for regular SEP events, and 5.38 Rs for FESEP events. Accordingly, the hardness of the 10–100 MeV fluence spectrum progressively decreases from GLE to regular SEP to FESEP events as illustrated in Fig. 6c. Power-law fits to the particle fluence (Fp) data of the form Fp ~ $E^{-\gamma}$ show that the GLE events have the hardest spectra (γ ~ 2.0), regular SEP events have moderate hardness (γ ~ 2.66), and the FESEP events have the softest spectra (γ ~ 4.50). The progressive spectral hardness is further illustrated in Fig. 6d: the spectral index is negatively correlated with the CME initial acceleration with a correlation coefficient of ~ –0.39.

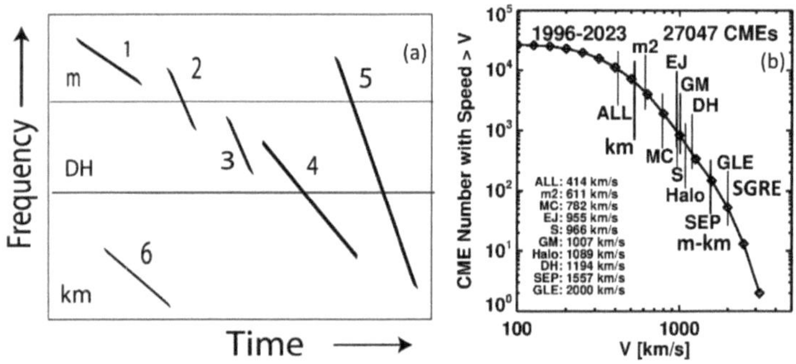

Fig. 7. (a) Type II radio bursts in a schematic dynamic spectrum: pure metric (m, 1), metric to decameter-hectometric (m-DH, 2), pure DH (3), DH-kilometric (DH-km, 4), m-km (5), and pure km (6). Cases 2–5 are sometimes combined as DH for simplicity. (b) average speeds of various CME populations associated with metric (m2), km, DH, and m-km type II bursts; other populations included for comparison are those associated with magnetic clouds (MC), non-cloud ejecta (EJ), interplanetary shocks (S), intense geomagnetic storms (GM), SEP events (SEP), ground level enhancements (GLE) in SEP events, and sustained gamma-ray emission (SGRE) events; the average speed of halo CMEs is also shown for comparison (Halo).

7 Shock Formation Heights, SEP Spectra, and CME Kinematics

While the starting frequency of type II bursts indicate the height of shock formation, the ending frequency indicates the shock strength. Type II bursts have been observed to start and end at various wavelength domains from meter waves (m) to kilometer waves (km) with the intervening decameter-hectometric (DH) waves (see Fig. 7). Type II bursts can start and end in the m domain and are associated with CMEs with an average speed of ~ 600 km/s (case 1 in Fig. 7a). If we combine cases 2, 3, and 4, the underlying CMEs have an average speed of ~ 1200 km/s. CMEs producing type II emission components at all wavelengths from m to km (case 5) have the highest average speed (~1500 km/s). Finally, CMEs producing type II bursts only in the km domain (case 6) have the lowest average speed (~500 km/s). These average values are marked and compared with the speeds of other energetic CME populations in Fig. 7b. CMEs in all cases except case 6 have an average deceleration within the coronagraphic field of view, while those in case 6 continue to accelerate and become super-Alfvenic only at tens of Rs from the Sun and hence they produce type II bursts only in the km domain. The hierarchical relationship between CME speed (or kinetic energy) and the wavelength range of type

II bursts is an indication of the close connection between shock strength and duration of particle acceleration [57, 60]. The m-km type II bursts indicate the strongest shocks that propagate to 1 au and even beyond.

The fact that type II bursts extending beyond the metric domain indicate stronger shocks has been illustrated in terms of SEP association [61]: only 25% of the western hemispheric m type II bursts (similar to case 1 in Fig. 7a) were associated with large SEP events, increasing to 90% when the metric type II bursts were accompanied by DH type II bursts. Their DH type II bursts include all type II events with ending frequency below 14 MHz. As noted, m-km type II bursts are associated with very fast CMEs, so they must have the best association with SEP events. Figure 8 compares the speed, width, and residual acceleration of CMEs associated with SEP events and m-km type II bursts. The CME properties are the same because the same shock accelerates electrons to produce type II bursts and ions that constitute SEP events.

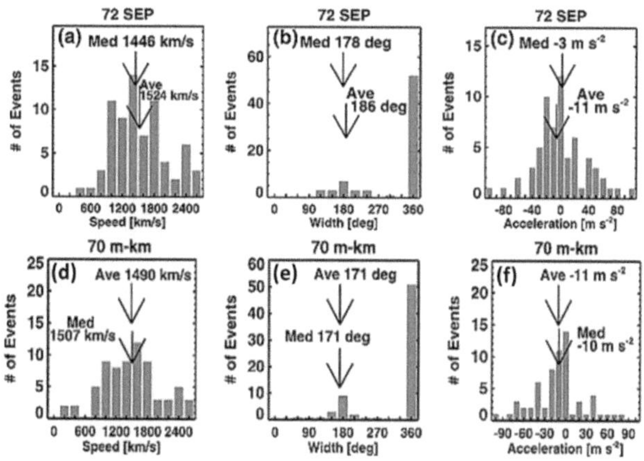

Fig. 8. Distributions of CME speed, width, and residual acceleration of CMEs associated with large SEP events (top) and type II radio bursts with emission components from metric to kilometric (m-km) wavelengths (bottom). Note that the CME properties are nearly the same.

The SEP association rate of DH type II bursts (cases 2–5 in Fig. 7a combined) from the western hemisphere steadily increases with CME speed, attaining 100% for $V_{CME} \geq 1800$ km/s [62], consistent with the results in Figs. 7 and 8. The SEP-DH type II burst association is further illustrated in Fig. 9, which plots the solar sources of CMEs associated with DH type II bursts. We see that the source locations of type II bursts are clustered around two latitude ranges corresponding to the active region belt, and uniformly distributed in longitude (type IIs are electromagnetic emission and hence are not affected by the IP magnetic field). Type II bursts originating mainly from the western longitudes are associated with SEP events because of the required magnetic connectivity to detect an SEP event at Earth. A small fraction (~21%) of DH type II bursts from the eastern hemisphere do have SEP association. A closer look at these events finds that they are associated with extremely fast and CMEs [63]. Such energetic CMEs drive extended shocks whose flanks that have magnetic connectivity to Earth.

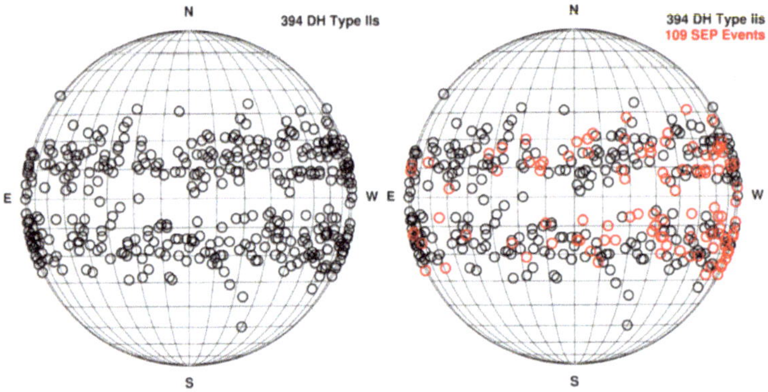

Fig. 9. Solar source locations of CMEs that produced a type II radio burst observed by the Wind/WAVES instrument (left). All type II bursts with an emission component in the DH domain are included (cases 2–5 in Fig. 7a). Heliographic coordinates of 394 CME sources are plotted. The grid spacing is 10 degrees in longitude and latitude. The sources of 109 CMEs are plotted in red indicating that they are associated with large SEP events (right). From [63].

What are the space weather implications of the type II burst–SEP association? Type II bursts can isolate the small number of energetic CMEs that drive a shock. A western hemispheric type II burst at the Sun indicates a high probability of occurrence of a large SEP event. Ten MeV protons have a speed of ~ 43470 km/s, so they can travel to Earth in ~ 75 min (assuming a Parker spiral path length of ~ 1.3 AU) if they are not scattered significantly during propagation. Normalizing to the arrival of electromagnetic emission at Earth, this represents an advance warning of about an hour. However, the higher energy particles travel much faster and hence the type II bursts cannot provide meaningful warning. For example, GeV particles (in GLE events) arrive at Earth only about ~ 3 min behind the electromagnetic emissions. The onset times of a type II burst or an SEP event may still be useful because spacecraft anomalies peak a couple of days after the onset of SEP events [20].

The scatter-free propagation noted above is not realistic because the real solar wind has turbulence superposed on the Parker spiral field lines. Therefore, a proper understanding of the transport of energetic particles in the ambient solar wind is needed to predict their properties at Earth. Solar wind properties are typically measured at 1 au, so a model is needed to characterize the solar wind/ambient medium from 1 au to the shock vicinity where the particles are released. This model needs to be coupled to another model that describes the transport of the particles. There are several such models developed recently with this general procedure of combining a solar wind model with a transport model [64–68]. A full list of SEP models, their current state of development, and validations can be found in [69].

8 Solar Cycle Variation of SEP Events

Figure 10 illustrates the solar cycle (SC) variation of peak SEP intensity (from GOES) and the source latitudes during five solar cycles since 1976. A higher concentration of SEP events near solar maxima is evident, although a significant number of events can be found during the declining phases (more pronounced in cycles 21 and 23). All cycles show events with NOAA scales at least S3 (i.e., 10^3 pfu, see Sect. 2), except for the partial cycle 25. SCs 22 and 23 have many S4 (10^4 pfu) events, but SC21 and 23 have S3 events, but no S4 events. Even though SC 21 has the highest sunspot number (SSN), the SEP intensities are relatively low, indicating that SSN alone is not an indicator of SEP intensities. This is because some large active regions are CME poor and hence are not associated with SEP events. Furthermore, some active regions emerge close to the end of a cycle, resulting in large SEP events. For example, there were four large SEP events during February-May of 1986 from ARs 4711, 4713, 4717, and 4717. AR 4711 was reported to be very active during its disk passage (e.g., [70] and references therein).

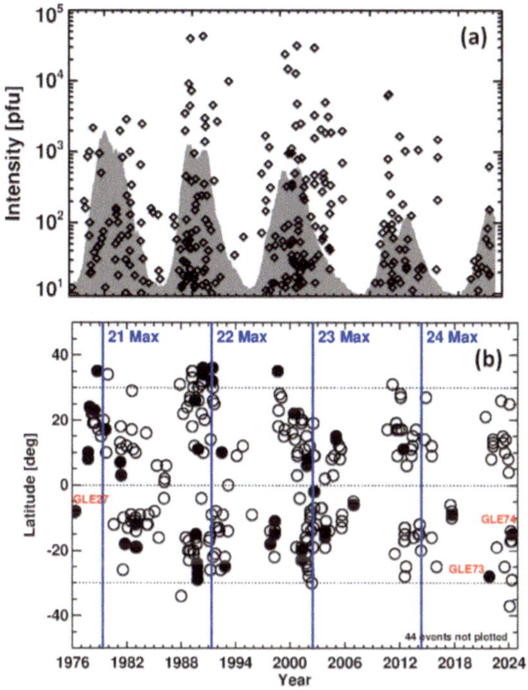

Fig. 10. (a) Maximum intensity (in pfu) of large SEP events as a function of time as observed by GOES in the > 10 MeV integral channel. The Y-axis corresponds to the NOAA radiation storm scale S1-S5 (see Sect. 2). The sunspot number is shown in gray for reference (from WDC-SILSO, Royal Observatory of Belgium, Brussels). The maximum intensity often occurs when the associated shock arrives at 1 au indicating that the peak is due to energetic storm particle (ESP) events. (b) Latitude vs. time plot of the solar source locations of large SEP events that occurred since 1976. Filled circles denote GLE events. Three GLE events at the beginning and end are identified. The blue lines mark the time of maximum of cycles 21–24.

Figure 10b shows that SEPs occur in clusters in both northern and southern hemispheres and on either side of the SSN maximum. The GLE events have a distribution similar to that of SEP events. In cycles 21–23, there were about a dozen GLE events in each cycle. The GLE occurrence rate drastically dropped in SC 24 that had only two

GLE events. The lack of high-energy SEP events in SC 24 has been attributed to the weakened heliospheric state in that cycle [71]. SC 25 is in its maximum phase but has only two GLEs in the first 4.5 years indicating that this cycle is similar to SC 24.

9 Summary

SEPs result from fundamental physical processes that convert free energy stored in solar magnetic fields to the kinetic energy of particles in the corona and IP medium. SEPs are also of great practical importance since they pose radiation hazard to humans and their technology in space. They degrade solar panels of satellites, cause satellite anomalies (deep dielectric discharges, single event upsets) in high and low latitude satellites, result in polar cap absorption of radio waves, modify Van Allen belt particle population, deplete ozone in Earth's atmosphere, and provide radiation dose to crew and passengers in aircraft flying in polar routes. Radio bursts are electromagnetic phenomena accompanying solar eruptions that result in SEPs. Long-duration type III bursts are indicative of electron acceleration in solar flares associated with CMEs. Low-frequency type IV bursts are indicative of flare-accelerated electrons in post eruption arcades and CME flux ropes. Type II bursts are due to electrons accelerated in CME-driven shocks that also energize ions and hence are a key electromagnetic signature of SEP events. Fast and wide CMEs that attain high speeds close to the Sun are responsible for the large SEP events. The intensity, spectra, and time evolution SEP events determine the ensuing space weather impact, but it is difficult predict these because of the high energies involved. These SEP properties are closely related to the properties of underlying CMEs such as early kinematics, shock formation distance from the Sun, and magnetic connectivity of the source region to the observer.

Acknowledgments. Work supported by NASA's Living With a Star Program and the STEREO project.

References

1. Forbush, S.E.: Phys. Rev. **70**, 771 (1946)
2. Reames, D.V.: Particle acceleration at the sun and in the heliosphere. Space Sci. Rev. **90**, 413 (1999)
3. Reames, D.V.: Solar Energetic Particles. Lecture Notes in Physics, vol. 932. Springer International Publishing AG (2017)
4. Dresing, N., Kouloumvakos, A., Vainio, R., Rouillard, A.: On the role of coronal shocks for accelerating solar energetic electrons. Astrophys. J. Lett. **925**(2), L21 (2022)
5. Lin, R.P.: Energetic solar electrons in the interplanetary medium. Solar Phys. **100**, 537–561 (1985)
6. Hey, J.: Solar radiations in the 4–6 metre radio wave-length band. Nature **157**, 47–48 (1946)
7. Wild, J.P., Smerd, S.F., Weiss, A.A.: Solar Bursts. Ann. Rev. Astron. Astrophys. **1**, 291 (1963)
8. Ramaty, R., Mandzhavidze, N.: Solar flares: gamma rays. In: Murdin, P., (eds.) Encyclopedia of Astronomy and Astrophysics. Article id. 2292; Institute of Physics Publishing: Bristol, UK, pp. 1–4 (2001).

9. Cerruti, A.P., Kintner, P.M., Gary, D.E., et al.: Effect of intense December 2006 solar radio bursts on GPS receivers. Space Weather **61**, S10D07 (2008)
10. Marque, C., et al.: J. Space Weather Space Clim. **8**(A42), 13 (2018)
11. Gary, D.E., Keller, C.U.: Solar and Space Weather Radiophysics - Current Status and Future Developments, astrophysics and Space Science Library, vol. 314. Kluwer Academic Publishers, Dordrecht (2004)
12. Gopalswamy, N., Mewaldt, R.A., Torsti, J.: Solar Energetic Particles and Plasmas. Geophys. Monogr. Ser. **165**, 385 (2006)
13. Desai, M., Giacalone, J.: Large gradual solar energetic particle events. Living Rev. Sol. Phys. **13**, 3 (2016)
14. Cliver, E.W., Schrijver, C.J., Shibata, K., Usoskin, I.G.: Extreme solar events. Living Rev. Sol. Phys. **19**(1), 2 (2022). https://doi.org/10.1007/s41116-022-00033-8
15. Gopalswamy, N.: The sun and space weather. Atmosphere **13**(11), 1781 (2022). https://doi.org/10.3390/atmos13111781
16. https://www.swpc.noaa.gov/noaa-scales-explanation
17. Gopalswamy, N., Barbieri, L., Cliver, E.W., Lu, G., Plunkett, S.P., Skoug, R.M.: Introduction to violent sun-earth connection events of October-November 2003. J. Geophys. Res. **110**(A9), A09S00 (2005)
18. Zeitlin, C., Boynton, W., Mitrofanov, I., et al.: Mars Odyssey measurements of galactic cosmic rays and solar particles in Mars orbit, 2002–2008. Space Weather **8**, S00E06 (2010)
19. Marvin, D.C., Gorney, D.J.: Solar proton events of 1989 - effects on spacecraft solar arrays. J. Spacecraft & Rockets **28**(6), 713–719 (1991)
20. Iucci, N., Levitin, A.E., Belov, A.V., et al.: Space weather conditions and spacecraft anomalies in different orbits. Space Weather **3**, 01001 (2005)
21. Reames, D.V.: SEPs: space weather hazard in interplanetary space. In: Song, P., Singer, H.J., Siscoe, G.L. (eds.) Space Weather, Geophysical Monograph Series, vol. 125, pp. 101–107. AGU, Washington, D.C (2001)
22. Mewaldt, R.A., Mason, G.M., Gloeckler, G., et al.: Long-term fluences of energetic particles in the heliosphere. In: Wimmer-Schweingruber, R.F. (Ed.) Joint SOHO/ACE Workshop "Solar and Galactic Composition", ASP Conference Series, vol. 598, pp. 165-170 (2001).
23. Khabarova, O.V., Zank, G.P.: Energetic particles of keV-MeV energies observed near reconnecting current sheets at 1 au. Astrophys. J. **843**(1), 4–16 (2017)
24. Barnes, C.W., Simpson, J.A.: Evidence for IP acceleration of nucleons in corotating interaction regions. Astrophys. J. **210**, L91–L96 (1976)
25. Bucık, R., Mall, U., Korth, A., Mason, G.M.: On acceleration of <1 MeV/n He ions in the corotating compression regions near 1 AU: STEREO observations. Ann. Geophys. **27**, 3677 (2009)
26. Elliott, S.S., Gurnett, D.A., Kurth, W.S., et al.: The acceleration of electrons to high energies over the jovian polar cap via whistler mode wave-particle interactions. J. Geophys. Res. **123**(9), 7523–7533 (2018)
27. Mauk, B., Clark, G.B., Gladstone, R., et al.: Energetic particles and acceleration regions over jupiter's polar cap and main aurora; a broad overview. J. Geophys. Res. **125**(3), e27699 (2020)
28. Bougeret, J.-L., Fainberg, J., Stone, R.G.: Interplanetary radio storms. I - Extension of solar active regions through the interplanetary medium. Astron. Astrophys. **136**(2), 255 (1984)
29. Morioka, A., et al.: Micro–type III radio bursts. Astrophys. J. **657**, 567 (2007)
30. Del Zanna, G., Aulanier, G., Klein, K.-L., Török, T.: A single picture for solar coronal outflows and radio noise storms. Astron. Astrophys. **526**, A137 (2011)
31. Gopalswamy, N., Mäkelä, P., Yashiro, S., Akiyama, S., Xie, H.: The relation between type III radio storms and CIR energetic particles. In: 3rd URSI Atlantic and Asia Pacific Radio Science Meeting (AT-AP-RASC), Gran Canaria, Spain, pp. 1-4 (2022)

32. Reames, D.V., Stone, R.G.: Astrophys. J. **308**, 902 (1986). https://doi.org/10.1086/164560
33. Sterling, A.C., Moore, R.A., Falconer, D.A., Adams, M.: Small-scale filament eruptions as the driver of X-ray jets in solar coronal holes. Nature **523**(7561), 437–440 (2015). https://doi.org/10.1038/nature14556
34. Bucik, R.: Space Sci. Rev. **216**, 24 (2020)
35. Kahler, S.W., Reames, D.V., Sheeley, N.R., Jr.: Coronal mass ejections associated with impulsive solar energetic particle events. Astrophys. J. **562**(1), 558–565 (2001)
36. Decker, R.B.: Space Sci. Rev. **48**(3–4), 195–262 (1988)
37. Vainio, R.: Particle acceleration and turbulence transport in heliospheric plasmas. In: Universal Heliophysical Processes, Proceedings of the International Astronomical Union, IAU Symposium, vol. 257, pp. 413–23 (2008).
38. Kahler, S.W., McAllister, A.H., Cane, H.V.: A search for interplanetary energetic particle events from solar posteruptive arcades. Astrophys. J. **533**(2), 1063 (2000)
39. Cane, H.V., Mewaldt, R.A., Cohen, C.M.S., von Rosenvinge, T.T.: Role of flares and shocks in determining solar energetic particle abundances. J. Geophys. Res. **111**(A10), A06S90 (2006)
40. Pallavicini, R., Serio, S., Vaiana, G.S.: A survey of soft X-ray limb flare images: the relation between their structure in the corona and other physical parameters. Astrophys. J. **216**, 108–122 (1977)
41. Gopalswamy, N., Mäkelä, P., Akiyama, S., et al.: Large solar energetic particle events associated with filament eruptions outside of active regions. Astrophys. J. **806**, 8 (2015)
42. Kahler, S.W., Cliver, E.W., Cane, H.V., et al.: Solar filament eruptions and energetic particle events. Astrophys. J. **302**, 504 (1986)
43. Cane, H.V., Richardson, I.G., von Rosenvinge, T.T.: A study of solar energetic particle events of 1997–2006: their composition and associations. J. Geophys. Res. **115**(A8), A08101 (2010)
44. Bucik, R., Mason, G.M., Gomez-Herrero, R., et al.: The first gradual solar energetic particle event with an enhanced 3He abundance on solar orbiter. A&A **669**, A13 (2023)
45. Reames, D.V.: How do shock waves define the space-time structure of gradual solar energetic particle events? Space Sci. Rev. **219**(1), 14 (2023)
46. Kahler, S.W.: Coronal mass ejections and solar energetic particle events, high energy solar physics. AIP Conf. Proc. **374**, 61–77 (1996)
47. Kahler, S.W.: The correlation between solar energetic particle peak intensities and speeds of coronal mass ejections: effects of ambient particle intensities and energy spectra. J. Geophys. Res. **106**(20), 947 (2001)
48. Reames, D.V., Kahler, S.W., Ng, C.K.: Spatial and temporal invariance in the spectra of energetic particles in gradual solar events. Astrophys. J. **91**, 414 (1997)
49. Gopalswamy, N., Yashiro, S., Krucker, S., Stenborg, G., Howard, R.A.: Intensity variation of large solar energetic particle events associated with coronal mass ejections. J. Geophys. Res. **109**(A12), A12105 (2004)
50. Gopalswamy, N., Xie, H., Akiyama, S., Mäkelä, P.A., Yashiro, S.: Major solar eruptions and high-energy particle events during solar cycle 24. Earth Planets Space **66**, 104 (2014)
51. Zhang, J., Dere, K.P.: A statistical study of main and residual accelerations of coronal mass ejections. Astrophys. J. Lett. **649**, 1100–1109 (2006). https://doi.org/10.1086/506903
52. Gopalswamy, N., Lara, A., Kaiser, M.L., Bougeret, J.-L.: Near-sun and near-earth manifestations of solar eruptions. J. Geophys. Res. **106**(A11), 25261–25278 (2001)
53. Pohjolainen, S., van Driel-Gesztelyi, L., Culhane, J.L., Manoharan, P.K., Elliott, H.A.: CME propagation characteristics from radio observations. Solar Phys. **244**(1–2), 167–188 (2008)
54. Cho, K.-S., Gopalswamy, N., Kwon, R.-Y., Kim, R.-S., Yashiro, S.: A high-frequency type II solar radio burst associated with the 2011 February 13 coronal mass ejection. Astrophys. J. **765**(2), 148 (2013)

55. Gopalswamy, N., Xie, H., Yashiro, S., Akiyama, S., Mäkelä, P., Usoskin, I.G.: Properties of ground level enhancement events and the associated solar Eruptions during solar cycle 23. Space Sci. Rev. **171**, 23 (2012)
56. Gopalswamy, N., Xie, H., Mäkelä, P., et al.: Height of shock formation in the solar corona inferred from observations of type II radio bursts and coronal mass ejections. Adv. Space Res. **51**(11), 1981–1989 (2013)
57. Gopalswamy, N.: Coronal mass ejections and type ii radio bursts. In: Gopalswamy, N., Mewaldt, R., Torsti, J. (eds.) Geophysical Monograph Series, vol. 165, pp. 81–94. AGU Publications, Washington (2006)
58. Kirk, J.G.: Particle acceleration. In: Kirk, J.G., Melrose, D.B., Priest, E.R. (eds.) Plasma Astrophysics, p. 215. Springer Verlag, Berlin (1994)
59. Reames, D.V.: Solar release times of energetic particles in ground-level events. Astrophys. J. **693**, 812–821 (2009)
60. Gopalswamy, N., Aguilar-Rodriguez, E., Yashiro, S., Nunes, S., Kaiser, M.L., Howard, R.A.: Type II radio bursts and energetic solar eruptions. J. Geophys. Res. **110**(A12), A12S07 (2005)
61. Cliver, E.W., Kahler, S.W., Reames, D.V.: Coronal shocks and solar energetic proton events. Astrophys. J. **605**, 902 (2004)
62. Gopalswamy, N., Yashiro, S., Akiyama, S., et al.: Coronal mass ejections, type II radio bursts, and solar energetic particle events in the SOHO era. Ann. Geophys. **26**(10), 3033–3047 (2008)
63. Gopalswamy, N., Mäkelä, P., Yashiro, S.: A Catalog of Type II radio bursts observed by Wind/WAVES and their statistical properties. Sun Geosph. **14**, 111–121 (2019)
64. Kozarev, K.A., Evans, R.M., Schwadron, N.A., et al.: Global numerical modeling of energetic proton acceleration in a coronal mass ejection traveling through the solar corona. Astrophys. J. **778**, 43 (2013)
65. Hu, J., Li, G., Ao, X., Zank, G.P., Verkhoglyadova, O.: Modeling particle acceleration and transport at a 2-D CME-Driven shock. J. Geophys. Res. **122**, 10938 (2017)
66. Luhmann, J.G., Mays, M.L., Odstrcil, D., et al.: Modeling solar energetic particle events using ENLIL heliosphere simulations. Space Weather **15**, 934 (2017)
67. Manchester, W.B., IV., van der Holst, B., Tóth, G., Gombosi, T.I.: The coupled evolution of electrons and ions in coronal mass ejection-driven shocks. Astrophys. J. **756**, 81 (2012)
68. Wijsen, N.: PARADISE: A Model for Energetic Particle Transport in the Solar Wind. Ph.D. dissertation, KU Leuven and Univ. Barcelona, Belgium, Spain (2020).
69. Whitman, K., Egeland, R., Richardson, I.G., et al.: Review of solar energetic particle prediction models. Adv. Space Res. **72**(12), 5161–5242 (2020)
70. Gopalswamy, N., Kundu, M.R.: A slowly moving plasmoid associated with a filament eruption. Sol. Phys. **122**(1), 91–110 (1989)
71. Gopalswamy, N., Akiyama, S., Yashiro, S., Xie, H., Mäkelä, P., Michalek, G.: Anomalous expansion of coronal mass ejections during solar cycle 24 and its space weather implications. Geophys. Res. Lett. **41**, 2673–2680 (2014)

Open Access This chapter is licensed under the terms of the Creative Commons Attribution 4.0 International License (http://creativecommons.org/licenses/by/4.0/), which permits use, sharing, adaptation, distribution and reproduction in any medium or format, as long as you give appropriate credit to the original author(s) and the source, provide a link to the Creative Commons license and indicate if changes were made.

The images or other third party material in this chapter are included in the chapter's Creative Commons license, unless indicated otherwise in a credit line to the material. If material is not included in the chapter's Creative Commons license and your intended use is not permitted by statutory regulation or exceeds the permitted use, you will need to obtain permission directly from the copyright holder.

Forecasting and Analysis of Solar Energetic Particle Radiation Storms

Olga Malandraki[1(✉)], Michalis Karavolos[1], Arik Posner[2,3], Kostas Tziotziou[1], Henrik Droege[4], Bernd Heber[4], and Patrick Kuehl[4]

[1] National Observatory of Athens, IAASARS, Athens, Greece
omaland@noa.gr
[2] NASA Headquarters, Washington, USA
[3] NASA Johnson Space Center, SRAG, Houston, USA
[4] Christian-Albrechts-Universität zu Kiel, Kiel, Germany

Abstract. Solar Energetic Particle (SEPs) with energies ranging from tens of keV to a few GeV, are a significant component in the description of the space environment. In this work, the HESPERIA REleASE product is emphasized, which, based on the Relativistic Electron Alert System for Exploration (REleASE) forecasting scheme, generates real-time predictions of the proton flux (30–50 meV) at L1, making use of relativistic and near-relativistic electron measurements by the SOHO/EPHIN and ACE/EPAM experiments, respectively. The HESPERIA REleASE Alert is a notification system based on the forecasts produced by the HESPERIA REleASE product and informs about the expected radiation impact in real-time using an illustration and a distribution system for registered users. We also present and discuss the Advance Warning Times derived for the compiled list of major SEP events successfully forecasted over the last 2.5 years during solar cycle 25 by HESPERIA REleASE and recent developments.

Keywords: Solar Energetic Particles · Space Weather · Human Exploration

1 Introduction

Solar Energetic Particles (SEPs) are transient injections into the heliosphere of protons, electrons and heavy ions, ranging in energy from tens of keV up to relativistic energies. As well as posing a threat to modern technology that heavily relies on spacecraft and posing a major radiation hazard to astronauts, they can also constitute a threat to avionics and commercial aircraft in extreme circumstances [1, 2]. A warning system is required to predict SEP occurrence to mitigate radiation exposure [3]. The SEP Real-Time Forecasting HESPERIA products have been developed under the HESPERIA H2020 project and since 2015 provide significant results concerning the prediction of SEP events. The real-time and highly accurate forecasts as well as the timely performance offered by the HESPERIA products have attracted the attention of various space organizations (e.g. NASA/CCMC, SRAG) and also led to the selection and their integration into the ESA Space Weather Service Network (SWESNET) (https://swe.ssa.esa.int/noa-hesperia-federated). HESPERIA REleASE (see top part of Table 2 in [4]), based on the Relativistic Electron Alert System for Exploration (REleASE) forecasting scheme [5], generates

© The Author(s) 2026
N. Gopalswamy et al. (Eds.): ISWI 2024, SPPHY 431, pp. 115–121, 2026.
https://doi.org/10.1007/978-981-95-1121-1_13

real-time predictions of the proton flux (30–50 meV) at L1, making use of relativistic and near-relativistic electron measurements by the SOHO/EPHIN and ACE/EPAM experiments, respectively [2]. In this work, we present and discuss the Advance Warning Times (AWTs) derived for the compiled list of major SEP events successfully forecasted over the last 2.5 years during solar cycle 25 by HESPERIA REleASE and recent developments.

2 HESPERIA REleASE Forecasting System

The greatest assets of the HESPERIA REleASE forecasting system are that there is no need for any prior solar flare (soft X-ray observations) to issue forecasts and that forecasts are also provided in the case of backside flares, which most other schemes cannot handle. We note that 25% of the Solar Proton Events (SPEs) events observed at Earth's orbit are due to backside solar events [6].

To further explore the capabilities of HESPERIA REleASE we have looked at the temporal characteristics of 20 significant SPEs that occurred between September 2021 and December 2023, namely events that the proton flux exceeds the 0.1 cm^{-2} s^{-1} sr^{-1} meV^{-1} threshold value in either the P3 (15.8–39.8 meV) or and P4 (28.2–50.1 meV) proton energy channel. The derived threshold crossing times as well as relevant proton alert times and onset times of the associated electron events are shown in Table 1.

Of particular interest for SEP forecasting is the estimation of the advance warning time, shown in the last column of Table 2, that denotes the advance warning time (AWT) in minutes from the first alert issue time (based either on ACE/EPAM or SOHO/EPHIN data) to the first onset of the significant SPE in either of the two energy channels. In Fig. 1 we show the distribution of AWT for the 20 considered SPEs. AWTs can go up to two hours with a mean value of ~ 70 min highlighting the unprecedented forecasting capabilities of HESPERIA REleASE.

The efficacy of the HESPERIA REleASE forecasting tool in successfully predicting the SPE events is demonstrated in Figs. 2 and 3, that show, respectively, the events of October 28, 2021, and March 28, 2022. For the first event, an electron event is detected at 15:54 UT, triggering the HESPERIA REleASE tool to issue a proton alert at 16:08 UT with the SPE event observed to occur at 17:55 UT on the same day. For the second event the respective times are at 11:53 UT, 11:59 UT and 13:44 UT. Both events confirm the accuracy of the HESPERIA REleASE forecasting system, providing AWTs of 107 and 105 min, respectively. The clear temporal correlation between the electron event, the subsequent proton alert, and the SPE event highlights the system's capability to deliver timely and reliable predictions based on electrons as precursors.

A considerable update of HESPERIA REleASE is now available, namely HESPERIA REleASE+ that improves forecasting with evidence of particle escape from the Sun, using and qualifying type III radio bursts, associated with electron beams accelerated in solar eruptive events and observed by STEREO A/SWAVES, as a timely and reliable proxy of particle escape from the Sun onto open field lines [7]. To this end, a radio module has been built that automatically identifies any present type III radio burst in real-time STEREO A/SWAVES beacon data. This pairing of HESPERIA REleASE with relevant radio information is expected to substantially eliminate false alarms.

Table 1. List of all 20 identified significant Solar proton Events (SPEs) during the period of September 2021 to December 2023, including relevant threshold crossing times, proton alert times and their associated electron event onset times. The last column shows the advance warning time. Adapted from [7].

	Electron event Onset Time [in UT]		Proton flux threshold crossing [in UT]		Proton Alert Time [in UT]		Advance Warning Time [in min]
Date	EPHIN	EPAM	P3	P4	EPHIN	EPAM	
28/10/2021	15:54	16:24	18:08	17:55	16:08	17:20	107
20/01/2022	06:06	–	08:07	07:47	06:19		88
28/03/2022	11:53	11:50	13:44	13:53	11:59	12:00	105
02/04/2022	–	13:39	14:38	14:39	14:03	13:39	35
09/07/2022	13:51	14:04	15:22	15:23	14:15	–	6
27/08/2022	03:20	–	10:32	13:52	10:30	–	2
13/01/2023	00:13	–	04:36	–	–	–	–
25/02/2023	–	19:50	21:26	21:29	19:58	20:00	88
13/03/2023	04:28	–	11:19	14:04	–	–	–
14/03/2023	–	08:30	10:44	15:13	09:48	10:24	56
23/04/2023	16:58	17:05	17:28	17:44	17:02	–	26
08/05/2023	–	09:05	13:12	07:23	–	–	–
09/05/2023	22:15	20:00	22:24	22:52	22:17	22:24	7
17/07/2023	18/07/2023 00:00	23:40	18/07/2023 01:06	18/07/2023 01:12	18/07/2023 00:06	23:54	60
28/07/2023	16:24	16:19	18:58	19:23	17:07	17:09	111
05/08/2023	07:42	07:40	10:03	10:14	–	09:59	4
05/08/2023	22:27	22:25	23:34	23:34	22:34	–	60
07/08/2023	21:32	21:00	22:09	22:19	21:33	–	36
01/09/2023	03:31	03:30	05:05	05:05	03:33	03:34	92
15/12/2023	–	–	15:09	15:09	11:07	–	242

Figure 4 shows a snapshot of the June 8, 2024 event as captured by HESPERIA REleASE+. The proton alert issue time (vertical blue dashed line) falls well within the forecasting window (FW, red rectangle) that denotes the maximum time interval (set by the radio module) within which a significant SPE is expected to occur after the onset of the qualified associated strong Type-III radio burst.

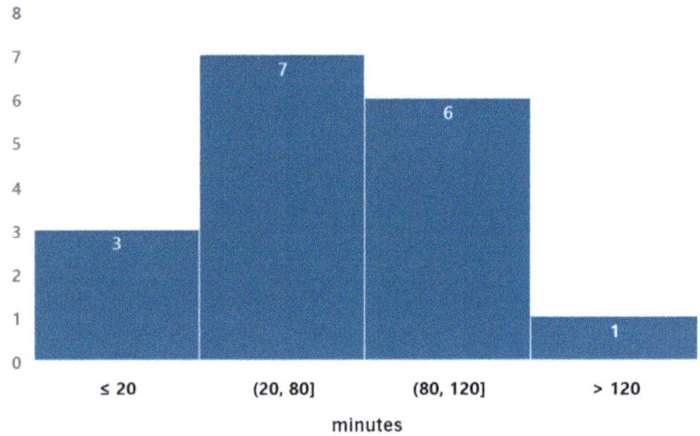

Fig. 1. Distribution of the Advance Warning Times (AWT), where applicable, for the SPEs presented in Table 1.

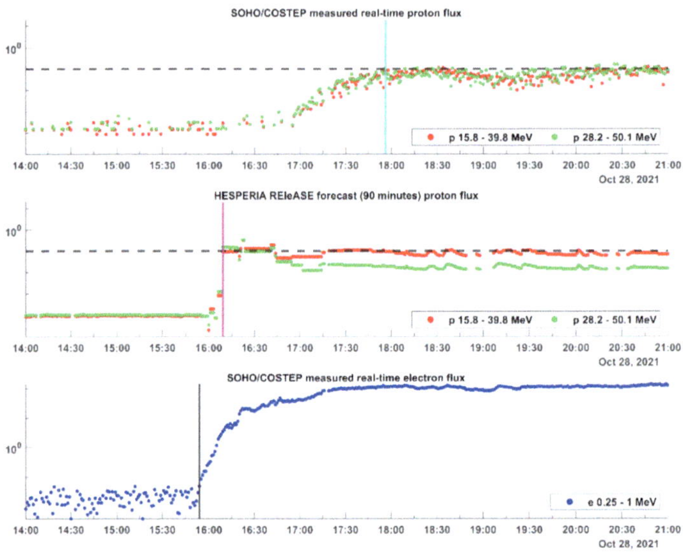

Fig. 2. The SPE event of October 28, 2021. The top panel presents the real-time proton fluxes also measured by SOHO/COSTEP, the mid panel the 60-min forecasted proton fluxes and bottom panel the real-time electron flux measured by SOHO/COSTEP. Black, magenta and cyan vertical lines indicate the onsets of the electron event, the proton alert issue time and the SPE occurrence time, respectively.

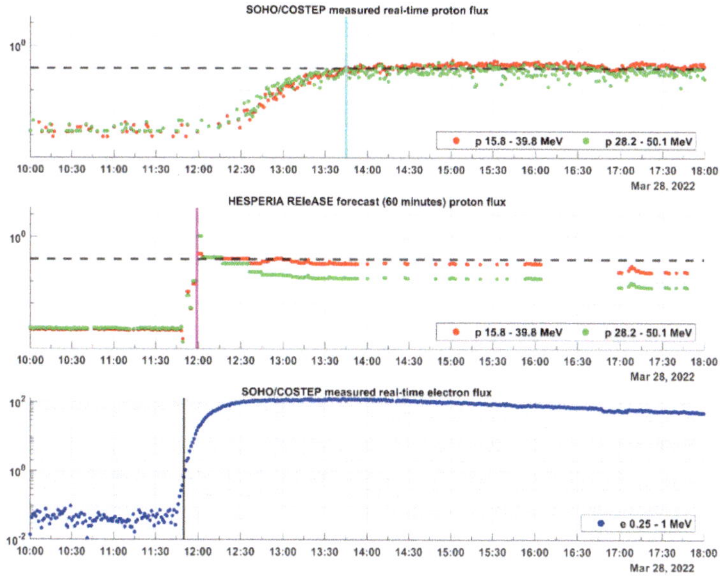

Fig. 3. The SPE event of March 28, 2022. See caption of Fig. 2 for description of panels.

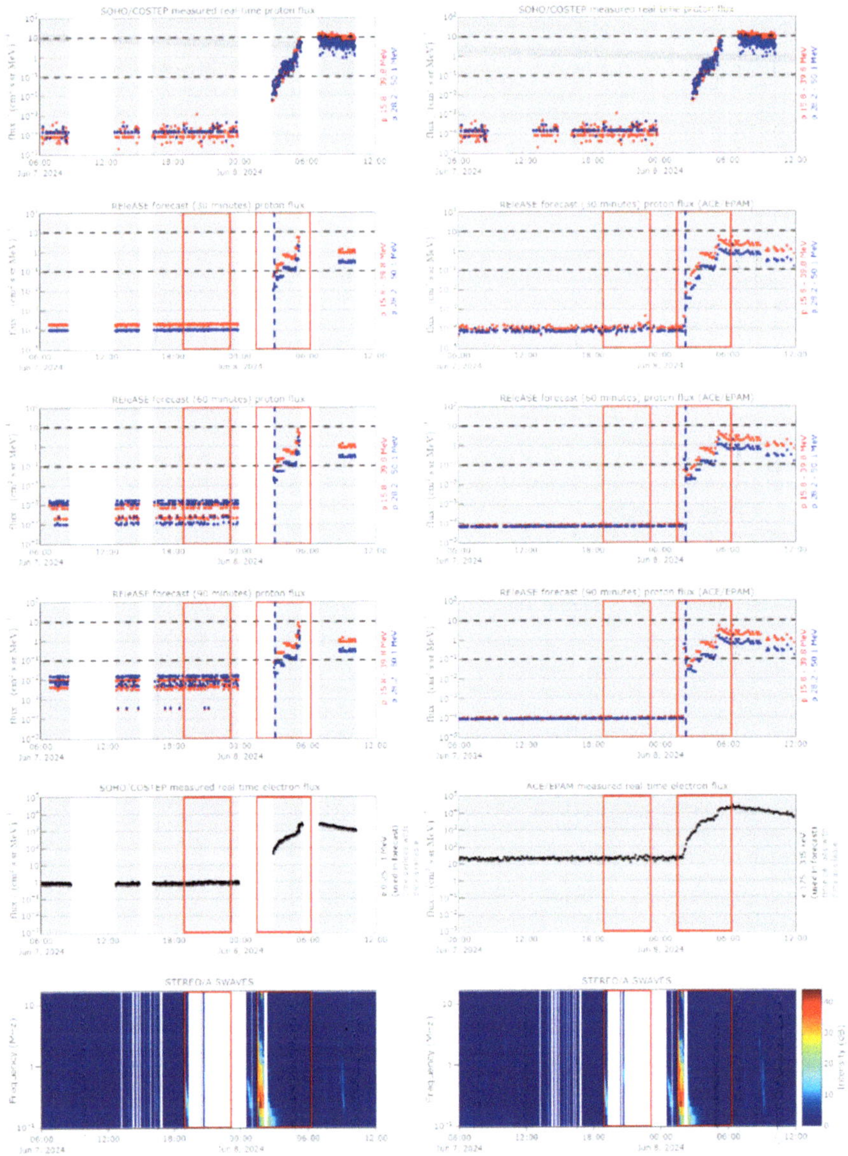

Fig. 4. The SPE event of June 8, 2024, as recorded by HESPERIA REleASE+. From top to bottom row panels show real-time proton fluxes, 30-min, 60-min and 90-min proton flux forecasts, real-time electron fluxes and real time STEREO/A SWAVES beacon radio fluxes. Vertical blue dashed lines indicate the time of the proton alert issue and red rectangles the forecasting window FW set by the radio module (see text).

3 Summary

We have further explored the capabilities of the HESPERIA REleASE forecasting system and carried out a study of 20 significant SPE events, successfully forecasted between September 2021 and December 2023. We have found that the AWT for the considered SPEs can go up to two hours with a mean value of ~ 70 min. Thus, HESPERIA REleASE significantly improves mitigation of adverse effects in space from a significant solar radiation storm, providing ample time of forewarning to space weather users, which is also valuable for astronaut protection in the context of human exploration.

References

1. Malandraki, O.E., Crosby, N.B.: Solar energetic particles and space weather: science and applications. In: Solar Particle Radiation Storms Forecasting and Analysis, Vol. 444, pp. 1–26. Springer International Publishing (2018)
2. Malandraki, O.E., Crosby, N.B.: Solar Particle Radiation Storms Forecasting and Analysis: The HESPERIA HORIZON 2020 Project and Beyond, vol. 444. Astrophysics and Space Science Library (2018)
3. Guo, J., Wang, B., Whitman, K., et al.: Particle radiation environment in the heliosphere: status, limitations, and recommendations. Adv. Space Res. (2024). https://doi.org/10.1016/j.asr.2024.03.070
4. Whitman, K., Egeland, R., Richardson, I.G., et al.: Review of solar energetic particle prediction models. Adv. Space Res. **72**, 5161–5242 (2022)
5. Posner, A.: Up to 1-hour forecasting of radiation hazards from solar energetic ion events with relativistic electrons. Space Weather **5**, S05001 (2007)
6. Richardson, I.G., von Rosenvinge, T.T., Cane, H.V., et al.: $>$ 25 MeV proton events observed by the high energy telescopes on the STEREO A and B spacecraft and/or at earth during the first $\sim$ seven years of the STEREO mission. Sol. Phys. **289**, 3059–3107 (2014)
7. Posner, A., Malandraki, O.E., et al.: HESPERIA REleASE+: improving solar proton event forecasting by means of automated recognition of type-iii radio bursts. Space Weather J., under review (2024)

Open Access This chapter is licensed under the terms of the Creative Commons Attribution 4.0 International License (http://creativecommons.org/licenses/by/4.0/), which permits use, sharing, adaptation, distribution and reproduction in any medium or format, as long as you give appropriate credit to the original author(s) and the source, provide a link to the Creative Commons license and indicate if changes were made.

The images or other third party material in this chapter are included in the chapter's Creative Commons license, unless indicated otherwise in a credit line to the material. If material is not included in the chapter's Creative Commons license and your intended use is not permitted by statutory regulation or exceeds the permitted use, you will need to obtain permission directly from the copyright holder.

The Correlation Between Spectral Shape of Energetic Protons and Cosmic Rays During Forbush Decrease Events

Mihailo Savić(✉), Nikola Veselinović, Aleksandar Dragić, Dejan Joković, Dimitrije Maletić, Radomir Banjanac, David Knežević, Miloš Travar, and Vladimir Udovičić

Institute of Physics Belgrade, Pregrevica 118, Belgrade, Serbia
msavic@ipb.ac.rs

Abstract. Some of the extreme events that occur on the Sun are solar flares and coronal mass ejections. These transient phenomena, driven by the realignment of the Sun's magnetic field, induce various secondary processes in the Sun's heliosphere and Earth's magnetosphere. The processes may include shock waves, acceleration, and consequent increases in the flux of charged particles, as well as interaction with and modulation of primary cosmic rays, among others. Concurrent increases in the flux of energetic protons measured near Earth and decreases in the flux of cosmic rays detected by Earth-based detectors could both be attributed to the passage of interplanetary shocks. To establish this relationship, connections between the parameters of induced Forbush decreases measured at Earth and the shape of event-integrated fluence spectra of energetic protons measured at L1 have been studied. Such analysis primarily focuses on investigating the correlation between power indices used to parameterize the differential fluence spectra and the magnitudes of associated Forbush decreases. In the presented work, the analysis is expanded to include additional models used to model the fluence spectra and compare them in order to find the best-performing one.

Keywords: Space Weather · Coronal Mass Ejections · Solar Energetic Particles · Forbush Decreases

1 Introduction

In case of events where both the increase in energetic protons and the decrease in cosmic ray (CR) fluxes (or Forbush decrease - FD) are likely induced by the same coronal mass ejection (CME), we should expect some connection between the properties of the CME and the properties of both induced phenomena, and thus, by extension, a correlation between the properties of the two otherwise unrelated phenomena. Indeed, we have previously observed such a correlation [1, 2]. In this work, we expand on our previous results by introducing an additional function to model the fluence spectra and further improve the reliability of the analysis.

© The Author(s) 2026
N. Gopalswamy et al. (Eds.): ISWI 2024, SPPHY 431, pp. 122–128, 2026.
https://doi.org/10.1007/978-981-95-1121-1_14

Based on the Izmiran Catalogue of Forbush effects and interplanetary disturbances provided by the Pushkov Institute of Terrestrial Magnetism, Ionosphere, and Radio Wave Propagation [3], we have selected a total of 20 solar events from Solar Cycles 23 and 24 that resulted in CMEs, where both the increase in energetic solar proton flux at L1 and the change in CR flux can likely be attributed to the passage of a CME associated interplanetary coronal mass ejection (ICME) and its accompanying shock. A detailed list of the selected events can be found in our previous work [1, 2].

The significant observed correlation between the shape of energetic proton fluence spectra and FD properties (such as FD magnitude for particles with 10 GV rigidity and FD magnitude for particles with 10 GV rigidity corrected for magnetospheric effects using the Dst index, both of which can be found in the Izmiran database) provides an alternative way to estimate these quantities, which are otherwise calculated using less straightforward procedures [4].

In the presented analysis, we employed three different functions to model the spectra and analyzed the connection between the model parameters and FD magnitudes. Additionally, we included several space weather parameters known to correlate significantly with FD parameters [5] for reference.

Data on energetic solar protons was measured by the SOHO/ERNE instrument [6] at L1, while the IZMIRAN catalog of Forbush decreases [3] was used as the source for FD and selected space weather parameters.

2 Methods

Energetic proton fluence spectra have a shape that may involve a characteristic "bend" or a "knee". A number of models have been proposed over the years to try and describe this shape. Here we are presenting three of these models and are comparing their effectiveness.

The first model was proposed by Ellison and Ramaty [7] to describe the particle acceleration by interplanetary shocks, and has the form of a power law with a exponential roll-over:

$$\frac{dJ}{dE} = AE^{\gamma} exp\left(-\frac{E}{E_0^{ER}}\right) \tag{1}$$

where J is the event-integrated fluence, E is the particle energy, γ is the spectral index, $E_0{}^{ER}$ is the rollover or cutoff parameter, while A is the free parameter. The second model has the form of a two-parameter stretched exponential, or the Weibull shape, which was successfully used to describe the SEP fluence spectra [8]:

$$\frac{dJ}{dE} = AE^{b-1} exp\left(-\frac{E}{E_0^{W}}\right)^{b} \tag{2}$$

where parameters of interest are spectral index b and the cutoff parameter $E_0{}^{W}$. The third model we used, proposed by Band et al. [9], was originally designed to describe gamma-ray burst spectra:

$$\frac{dJ}{dE} = \begin{cases} AE^{\alpha} exp\left(-\frac{E}{E_B}\right), & E \leq (\alpha - \beta)E_B \\ AE^{\beta}[(\alpha - \beta)E_B]^{\alpha-\beta} exp(\beta - \alpha), & E > (\alpha - \beta)E_B \end{cases} \tag{3}$$

where α and β are spectral indices and E_B is the "break" energy.

3 Results

3.1 Energetic Proton Fluence Spectra and Model Comparison

Overall, we found that the Band function provided the best fit for the spectra. However, the other two models did not perform significantly worse in most cases, as demonstrated for the November 2003 event in Fig. 1.

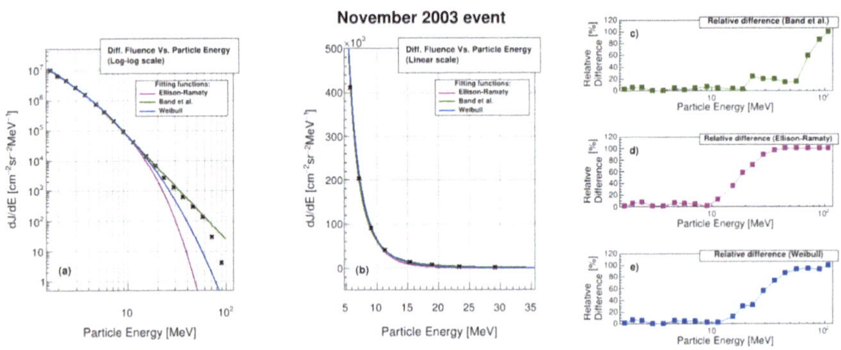

Fig. 1. Comparison of different models for the November 2003 event: (a) proton fluence spectrum in log-log scale, (b) portion of the spectrum around the 'knee' in linear scale, and (c), (d), and (e) relative difference between the model and the data for the Band, Ellison-Ramaty, and Weibull models, respectively.

The June 2015 event shown in Fig. 2 presents another characteristic case of difference in model performance.

Here, the Weibull function was somewhat less effective in modeling the region around the knee than the other two functions, while the Band function again performed the best overall. In case of the remaining events used in the analysis, the performance of the models was within the limits of the presented two cases.

3.2 Correlation Between Spectral Indices and Space Weather and FD Parameters

To study the assumed connection between the properties of CME induced enhanced flux of energetic protons and properties of concurrent Forbush decreases likely induced by the same CMEs, we looked into the correlation between the spectral indices from the three presented models and FD magnitudes, as well as the correlations between

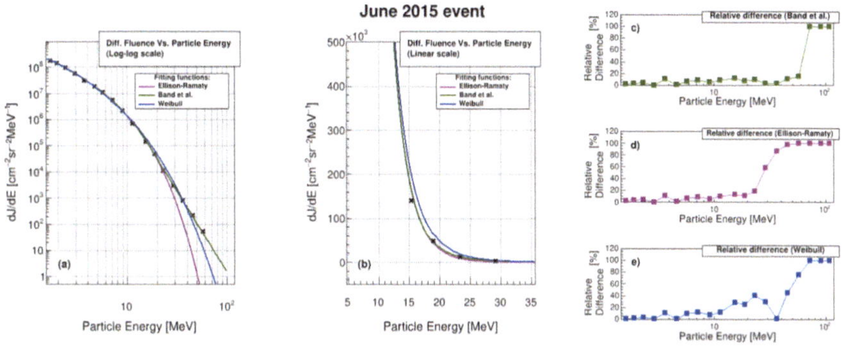

Fig. 2. Comparison of different models for the June 2015 event: (a) proton fluence spectrum in log-log scale, (b) portion of the spectrum around the 'knee' in linear scale, and (c), (d), and (e) relative difference between the model and the data for the Band, Ellison-Ramaty, and Weibull models, respectively.

these and selected space weather parameters. Two different values for FD magnitude were considered: one is the FD magnitude for 10 GV rigidity particles obtained from the measurements performed by the network of neutron monitors (M), and the other is the FD magnitude for 10 GV particles corrected for the magnetospheric effect (M_M). Both FD magnitudes and selected space weather parameters used in the analysis were obtained from the Izmiran catalog [3] where detailed description of the parameters can be found on the referenced page. The dependencies of FD magnitude on spectral indices α, γ and b are shown in Fig. 3.

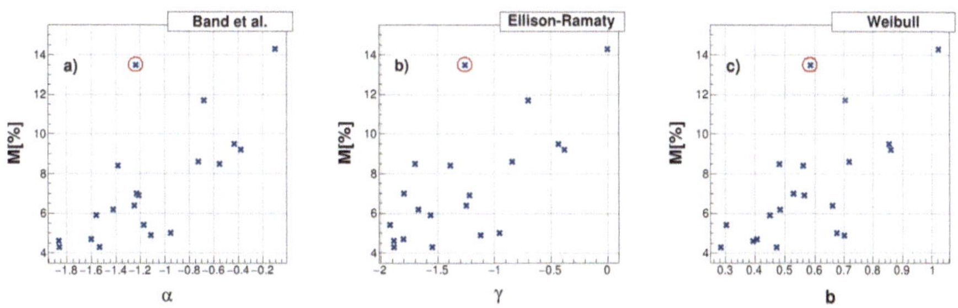

Fig. 3. Dependence of FD magnitude on spectral indices: (a) α (Band et al.), (b) γ (Ellison-Ramaty) and (c) b (Weibull). The outlier July 2004 event is indicated by the red circle.

Correlation coefficients between spectral indices and FD magnitudes as well as selected space weather parameters (detailed below) are shown in Table 1.

Other than spectral indices and rollover and cutoff parameters that appear in Eqs. 1, 2 and 3, additional parameters in Table 1 include $E_B{}^*$ equal to $(\alpha - \beta)E_B$, mean CME velocity (V_C), maximal hourly solar wind speed in the event (V_{max}) and maximal Kp index during the event (Kp_{max}). FD magnitudes (M) and FD magnitudes corrected for the magnetospheric effect (M_M) are presented in the last two columns.

Table 1. Correlation coefficients between spectral indices, FD magnitudes, and selected space weather parameters.

	α	β	E_B	E_B^*	γ	E_0^{ER}	b	E_0^W	V_C	V_{max}	Kp_{max}	M	M_M
α	1.00	−0.18	−0.46	0.04	0.83	−0.47	0.84	0.82	0.68	0.69	0.40	0.83	0.71
β	−0.18	1.00	0.21	−0.61	−0.44	0.47	−0.53	−0.27	−0.08	−0.10	−0.04	−0.07	−0.20
E_B	−0.46	0.21	1.00	0.52	−0.19	0.86	−0.37	−0.09	0.00	−0.05	−0.33	−0.23	−0.32
E_B^*	0.04	−0.61	0.52	1.00	0.47	0.18	0.37	0.42	0.37	0.28	−0.10	0.15	0.12
γ	0.83	−0.44	−0.19	0.47	1.00	−0.42	0.96	0.95	0.78	0.74	0.46	0.77	0.73
E_0^{ER}	−0.47	0.47	0.86	0.18	−0.42	1.00	−0.61	−0.27	−0.14	−0.14	−0.32	−0.33	−0.38
b	0.84	−0.53	−0.37	0.37	0.96	−0.61	1.00	0.90	0.72	0.70	0.45	0.76	0.70
E_0^W	0.82	−0.27	−0.09	0.42	0.95	−0.27	0.90	1.00	0.87	0.82	0.47	0.84	0.70
V_C	0.68	−0.08	0.00	0.37	0.78	−0.14	0.72	0.87	1.00	0.94	0.55	0.81	0.54
V_{max}	0.69	−0.10	−0.05	0.28	0.74	−0.14	0.70	0.82	0.94	1.00	0.48	0.80	0.52
Kp_{max}	0.40	−0.04	−0.33	−0.10	0.46	−0.32	0.45	0.47	0.55	0.48	1.00	0.47	0.66
M	0.83	−0.07	−0.23	0.15	0.77	−0.33	0.76	0.84	0.81	0.80	0.47	1.00	0.78
M_M	0.71	−0.20	−0.32	0.12	0.73	−0.38	0.70	0.70	0.54	0.52	0.66	0.78	1.00

The July 2004 event was a very complex one that included multiple successive FDs. The event was eventually excluded from the correlation analysis, as we suspect its relative magnitude related to specific proton flux enhancement to be smaller than the absolute value quoted in the IZMIRAN database.

The main focus of our study was on the correlation between spectral indices and FD magnitudes. We can see that these correlation coefficients for spectral indices are comparable with the ones for variables V_C and V_{max}, that typically show the largest correlation with FD magnitudes [5]. Even more interesting is that the correlation between spectral indices and the corrected FD magnitude M_M is noticeably larger than for any other parameter. Rollover and cutoff parameters, other than $E_0{}^W$, mostly show moderate to low correlation with spectral indices and other parameters.

All three functions used to model the energetic proton fluence spectra seem to be comparatively useful in this type of analysis, however, we would recommend the model proposed by Band et al. as it has proven to be the most adequate and reliable overall.

4 Discussion/Conclusion

In the presented work, we studied the potential connection between the enhancement of energetic proton flux and concurrent Forbush decreases. Our focus was on events where both phenomena are likely induced by the passage of a CME. It is already known that correlations exist between certain features of CMEs and the features of these induced phenomena. Thus, it is not unreasonable to assume that some relationship between the features of the induced phenomena should also reflect their common origin.

In our previous work, we established such a connection by identifying a correlation between the spectral indices, used to parameterize the shape of the event-integrated SEP fluence spectra, and the FD magnitude for particles with 10 GV rigidity. We also observed this correlation for FD magnitudes corrected for magnetospheric effects, with both parameters provided by the Izmiran catalog.

In the current analysis, we expand on our previous work by introducing another function to describe the energetic proton fluence spectra. The new results confirmed the previously observed significant correlation between the spectral indices and FD magnitudes. This correlation is comparable to, and in the case of the FD magnitudes corrected for magnetospheric effect even stronger than, the correlation between FD magnitudes and CME-related parameters.

This finding is somewhat unexpected, as the connection between these induced phenomena is one step further removed from the CME. One possible explanation for the observed correlation is that faster CMEs produce harder SEP proton spectra while also driving FDs of greater magnitude. However, it remains unclear why the correlation between the spectral indices and FD magnitudes is as strong, or stronger, than the correlation between FD magnitudes and CME parameters, such as CME speed. This may indicate that a different set of parameters or a redefinition of existing ones could provide more reliable predictors of FD magnitudes.

The practical implication of this analysis is that it offers an alternative method for estimating FD magnitudes found in the Izmiran catalog, which are otherwise calculated independently.

References

1. Savić, M., et al.: New insights from cross-correlation studies between solar activity indices and cosmic-ray flux during Forbush decrease events. ASR **71**(4), 2006–2016 (2023)
2. Savić, M., et al.: Further study of the relationship between transient effects in energetic proton and cosmic ray fluxes induced by coronal mass ejections. Universe **10**(7), 283 (2024)
3. Catalogue of the Forbush-effects and interplanetary disturbances. http://spaceweather.izmiran.ru/eng/dbs.html. last accessed 07 June 2024
4. Kolarski, A., et al.: Impacts of extreme space weather events on September 6th, 2017 on ionosphere and primary cosmic rays. Remote Sens. **15**(5), 1403 (2023)
5. Lingri, D., Mavromichalaki, H., Belov, A., et al.: Solar activity parameters and associated forbush decreases during the minimum between cycles 23 and 24 and the ascending phase of cycle 24. Sol. Phys. **291**, 1025–1041 (2016)
6. ERNE Homepage, https://srl.utu.fi/projects/erne/. last accessed 07 June 2024
7. Ellison, D.C., Ramaty, R.: Shock acceleration of electrons and ions in solar flares. ApJ **298**, 400–408 (1985)
8. Laurenza, M., Consolini, G., Storini, M., Damiani, A.: The Weibull functional form for SEP event spectra. J. Phys. Conf. Ser. **632**, 012066 (2015)
9. Band, D., Matteson, J., Ford, L., et al.: BATSE observations of gamma-ray burst spectra. I. Spectral diversity. ApJ **413**, 281 (1993)

Open Access This chapter is licensed under the terms of the Creative Commons Attribution 4.0 International License (http://creativecommons.org/licenses/by/4.0/), which permits use, sharing, adaptation, distribution and reproduction in any medium or format, as long as you give appropriate credit to the original author(s) and the source, provide a link to the Creative Commons license and indicate if changes were made.

The images or other third party material in this chapter are included in the chapter's Creative Commons license, unless indicated otherwise in a credit line to the material. If material is not included in the chapter's Creative Commons license and your intended use is not permitted by statutory regulation or exceeds the permitted use, you will need to obtain permission directly from the copyright holder.

Spacecraft Anomaly, Impact on GNSS, Ionospheric Irregularities

Ionospheric Irregularities, Space Weather Impact on GNSS, and Spacecraft Anomalies

Anthea J. Coster(✉)

MIT Haystack Observatory, 99 Millstone Road, Westford, MA 01886, USA
costera@mit.edu

Abstract. This paper provides an overview of space weather impacts in near-earth space. The primary focus is on those impacts related to radio wave propagation in the ionosphere. The ionosphere introduces range delays that can be increased during space weather events due to the formation of large scale horizontal gradients in electron density. The ionosphere can also impact radio signal reception due to the formation of small-scale irregularities that induce variations in the phase and amplitude of the received signal, in what is termed scintillation. Severe scintillation can cause receivers to stop tracking the signal and radars to lose the ability to coherently integrate. Separately, increased electromagnetic radiation due to solar flares and solar radio bursts can induce sudden jumps in the TEC (solar flares) or an increase in the background noise (solar radio bursts) causing some receivers to loose lock. Other space weather impacts described include the effect of increased radiation, including spacecraft charging, leading to satellite anomalies. In summary, electromagnetic radiation in the form of solar radio bursts and solar flares, radiation storms involving high speed ions and electrons, and geomagnetic storms related to coronal mass ejections and high speed solar storms are the primary agents of space weather. A brief discussion is also provided on the increasing importance of providing better atmospheric drag prediction during space weather event due to the exponential growth in the number of satellites in orbit.

Keywords: storm-induced ionospheric gradients · scintillation · spacecraft anomalies

1 Introduction

Space Weather as defined by the American Meteorological Society refers to the variable conditions on the Sun and in the space environment that can influence the performance and reliability of space-borne and ground-based technological systems, as well as endanger life or health. This aim of this paper is to provide a general overview of selected key technological impacts of space weather. This includes a discussion of disruptions in GNSS and satellite communication caused primarily by disturbances in radio-wave propagation.

All space weather effects discussed are caused by changes in the near-space environment due to events that begin on the Sun. The paper is divided into the following sections.

© The Author(s) 2026
N. Gopalswamy et al. (Eds.): ISWI 2024, SPPHY 431, pp. 131–141, 2026.
https://doi.org/10.1007/978-981-95-1121-1_15

The first concerns the causes of space weather disruption on satellite communication, including those on GNSS signals; what are the space weather effects on satellite electronics; and how are geomagnetic storms involved with satellite drag estimates. We begin with a brief discussion of radio wave propagation effects due to ionospheric gradients and ionospheric irregularities, followed by an outline of radiation effects, both internal and external, on satellite charging, and finally a discussion of the difficulties in estimating satellite drag during geomagnetic storms. The next section of this paper focuses on the solar conditions that induce the various effects, including solar flares and solar radio bursts, the increase in solar radiation, and finally the large geomagnetic storms due to coronal mass ejections and/or high speed solar winds. We conclude the paper with a brief description of the issues arising from the exponential growth of number of satellites in near Earth space, and what the space weather implications are.

2 Space Weather Effects in Near-Earth and Their Causes

2.1 Introduction

In this paper we will consider three categories of space weather effects. The first are those related to radio wave propagation, which includes effects on the GNSS and other satellite signals, such as HF-communication. The second are those related to satellite charging, both internal and external, which leads to satellite anomalies. Finally, we examine the changes in atmospheric density that occur during large geomagnetic storms, affecting our ability to correctly predict atmospheric drag.

2.2 Space Weather Effects: Radio Wave Propagation in the Ionosphere

All satellite communication must transverse the ionosphere, and it is this region that has the most impact on signal propagation. The ionosphere is the region of space that begins roughly at 60 km and extends to and beyond 1000 km, or higher if the region known as the plasmasphere is included. This region is defined by ionization, both ions and electrons, that are primarily produced by solar radiation. This ionization is embedded in a background of neutral atmospheric density. The Earth's magnetic field strongly affects the motion of the charged particles, although not the neutral particles. It is, for example, easier for a charged particle to propagate along the magnetic field than across it. Within the ionosphere there are distinct regions, the main ones being the D region(50–90 km), E region (90–150), F1region (150–220) and F2 region (220–1000) during the day, with D region decaying completely, the E region largely dissipating, and the F1 and F2 regions combining during the night. For space weather implications, what is important is that radio waves strongly interact with free electrons in space plasmas, and these interactions are responsible for the space weather impacts on communication and navigation systems.

Wave propagation in the ionosphere is governed by the index of refraction, n. In the ionosphere, n, is defined by Appleton Hartree equation which is (assuming negligible collisions):

$$n^2 = 1 - \frac{X(1-X)}{(1-X) - \frac{1}{2}Y_T^2 \pm \left[\frac{1}{4}Y_T^4 + (1-X)^2 Y_L^2\right]^{1/2}},$$

Where

$$X = \left(\frac{\omega_n}{\omega}\right)^2, Y = \frac{\omega_H}{\omega}, Y_L = Y\cos(\theta), \text{ and } Y_T = Y\sin(\theta),$$

with ω_n is the electron plasma frequency, equal to: $\omega_n = \sqrt{\frac{Ne^2}{\epsilon_o m_e}}$, where N is the electron density per m^3, e the electronic charge, ϵ_o the permittivity constant, and m_e, the mass of the electron.

ω_H is the electron gyrofrequency, equal to: $\omega_H = \frac{eB}{m_e}$, where e is the electronic charge, B the magnetic field strength, and m_e, the mass of the electron.

θ is the angle between the magnetic field vector, **B**, and the wave vector.

To first approximation, the index of refraction can be simplified to: $n^2 = 1 - X = 1 - \left(\frac{\omega_n}{\omega}\right)^2 = 1 - \left(\frac{Ne^2}{2\pi f \varepsilon_o m_e}\right)$; which can be further simplified to

$$n \sim 1 - \frac{Ne^2}{4\pi f \epsilon_o m_e}$$

These equations show that, in the ionosphere, the index of refraction is not equal to 1. Because of this, both a range delay and a phase advance are introduced. In addition, because the index of refraction changes as a function of altitude in the ionosphere due to changes in electron concentration, the angle of arrival is slightly changed leading to a correction to the signal's elevation. It is also important to also note that the index of refraction in the ionosphere is a function of the electron density and of frequency, meaning it is a dispersive medium.

The two ionospheric corrections are: the ionospheric range delay and the elevation bending angle. The ionospheric range delay, Δr_{ion}, in meters can be written as:

$$\Delta r_{ion} \sim \frac{40.3}{f^2} \int Ndl$$

The quantity within the integral is known as the total electron content, or TEC, in el/m^2. Average daytime ionospheric zenith delays at the GPS L1 frequency are 5–15 m. Zenith delays in the ionosphere are typically multiplied a mapping function to determine a line of sight delay. At 5 degrees elevation, the total delay is about 3 times the zenith delay.

Separate to the range delay, is the elevation bending angle correction. For the ionospheric correction term, this does not become significant until the elevation is 5 degrees or less. The ionospheric elevation bending is dependent on the geographical location and a simple equation to estimate its value does not exist for most radio frequencies, including L-band frequencies used by GNSS satellites. Typically for operational ground-based measurements a table is generated based on location using climatological data. In space, the bending angle can potentially be used to provide the estimate of ionospheric density in satellite occultation measurements. However, in practice, this method is not used as it is simpler and more straightforward to compute it from the delay difference between the two frequencies.

In general, with respect to GNSS applications, the primary space weather effects are the introduction of range and elevation errors and the loss of signal reception. Space weather driven effects that cause disruption to GNSS signals can be divided into these categories: 1) large horizontal gradients in ionospheric electron density that can introduce significant range errors and bending; 2) small-scale irregularities in the ionosphere that can cause loss of lock on the GNSS signals fluctuations (or scintillation); 3) solar radio bursts that can also introduce notable effects on GNSS signals by raising the background noise level, and 4) solar flares that can introduce a rapid change in the TEC which can also cause a temporary loss of lock.

2.3 Large-Scale Horizontal Ionospheric Density Gradients

Large horizontal ionospheric density gradients make it difficult to correctly predict the ionospheric range correction and the elevation correction term, and these terms are needed in high precision real-time positioning applications such as satellite tracking, precision farming, marine positioning. Large horizontal ionospheric gradients are observed during geomagnetic storms in features such as the storm enhanced density plumes, in polar cap patches, and the enhanced equatorial anomaly [1].

During large geomagnetic storms, there can be a change in the zenith range correction as much as 32 m in as little as 5 min as was observed in the October 2003 storm [2]. Large horizontal ionospheric gradients associated with geomagnetic storms also impact the density estimation retrieved from occultation measurements more difficult [3]. The SED plume is responsible for some of the largest space weather effects in the mid-latitudes [4]. Large horizontal gradients in the range of 50 TEC units per degree have been observed, and these gradients can persist for several hours, leading to large errors in the measured range, and in differential range.

2.4 Small-Scale Irregularities

Separate to the effects of large scale gradients on radio wave propagation, are those due to the signal propagating through a region of small-scale irregularities. The effects are observed both in the measured phase and amplitude of a signal due to the diffraction and refraction of the wave itself. These ionospheric irregularities form primarily in two distinct regions on the Earth: the equatorial region, primarily after sunset, and in the polar regions. Irregularities form due to different plasma instabilities, but the physics involved is strongly related to the magnetic field and its influence on the motion of electrons and ions. In standard GNSS receivers, severe scintillation can cause cycle slips or, in extreme cases, total loss of signal lock. During these times, even normal radio communication can be severely disturbed by scintillation. The occurrence of scintillation has large day-to-day variability. During periods of extreme space weather events scintillation is observed in the mid-latitudes.

The final two space weather effects on GNSS are solar radio bursts that can also introduce notable effects on GNSS signals by raising the background noise level, and solar flares that can introduce a rapid change in the TEC which can also cause a temporary loss of lock. We will discuss these two effects more in Sect. 3.1.

2.5 Satellite Anomalies

The exact amount of exposure to plasma that a satellite experiences is a function of their orbit and the background solar conditions. Space weather is recognized as the main driver in the control of the energetic particles and plasma that interact with the satellites. During solar storms, the radiation environment can change drastically during solar storms. Both low and high energy electrons play a role in spacecraft charging. Low energy electrons can cause surface charging whereas high energy relativistic electrons can penetrate the surface of the satellite and cause internal charging. In either case, a buildup of charge can lead to an electrostatic discharge, either externally or internally and result in damage to electronic components [1, 5]. Energetic protons and heavy ions from trapped radiation, solar particle events and cosmic rays can cause logic upsets in onboard satellite computers. Historically there are numerous examples of spacecraft anomalies associated with spacecraft charging, such as the Galaxy 15 satellite, which was a victim of spacecraft charging strong solar event during 3–5 April 2010. Eventually a workaround was found and the Galaxy 15 satellite was returned to operation [6].

2.6 Atmospheric Drag

Satellite drag is dependent on a number of factors, including the atmospheric density, the atmospheric winds, the ballistic coefficient, and the relative speed of the satellite within the atmosphere. For low earth orbiting satellites, atmospheric drag is the largest source of error. Geomagnetic storms produce variability that results in significant density uncertainty, and mesoscale structure. Changing atmospheric composition also plays an important role. With the gradual cooling and contraction of the thermosphere due to increasing carbon dioxide, orbiting debris have longer lifetimes, and continue to accumulate. Understanding thermospheric density and composition changes is crucial and motivates the development of improved atmospheric density and wind models, especially during geomagnetic storm periods. For low earth orbiting satellites, atmospheric drag is the largest source of error.

Space weather events can alter the thermospheric state very quickly and can produce variability that results in more than 4-sigma uncertainty in density. During solar storms, energetic particles deposit energy into the upper atmosphere thus heating the neutral atmosphere and altering the density and composition. Furthermore, most of the geospace energy input to the atmosphere during storms is deposited in the high latitudes and is associated with significant mesoscale structure. The energy associated with these storms can take several hours before the heating is globalized. The current understanding is that most of the energy input at high latitudes transfers to mid- and equatorial latitudes through waves, traveling atmospheric disturbances (TADs) and traveling ionospheric disturbances (TIDs). This contributes to variable drag for different satellites, depending on whether their orbit crosses mesoscale heating or not, especially during the main phase of the storm.

Current thermospheric models of neutral density need to be extended to exosphere heights to include higher altitudes. Modeling the ion upflow/outflow is important for changing high altitude densities in the exosphere. Better real-time estimation of the atmospheric density is critical to improving orbit prediction related products, and research on orbit determination, orbital uncertainty propagation and collision avoidance.

3 Three Agents of Space Weather

Space weather begins with the sun. Borrowing a concept from National Space Weather Service [7], there are three 'agents' of space weather, each generating different space weather effects on technology: 1) electromagnetic emission from the sun in the form of solar flares and radio bursts \, 2) charged particle radiation, and finally 3) magnetized plasma produced by coronal mass ejections or high speed solar streams.

3.1 Solar Flares and Radio Bursts

A solar flare is an intense burst of radiation coming from the release of magnetic energy associated with sunspots. A solar flare follows a violent explosion on the surface of the sun. It can affect the entire daylit side of the Earth. A solar flare is an intense burst of radiation coming from the release of magnetic energy associated with sunspots. They can be observed as bright areas on the sun and can last from minutes to hours. Solar flares are observed by the photons (or light) it releases, at most every wavelength of the spectrum. This visible light take 8 min to reach the Earth from the Sun. In Fig. 1, on the left hand side, the panel shows the increase in TEC (shown in red) over the entire daylight side of the Earth following a X9.3 solar flare that occurred at near 12:00 UT Sept 6, 2017. On the right hand side, the top plot shows all of the GPS zenith TEC measurements made in a 2 degree latitude bin between 20 and 22 degrees N. latitude and a 10 degree longitude bin between −80.0 and −70.0 degrees longitude. The bottom plot shows the GNSS zenith TEC measurements made in a 2 degree latitude bin between 50 and 52 degrees N. latitude, and 20 degree longitude bin between 0 and 10 degrees longitude. The jump in several of the TEC measurements slightly before noon when the solar flare effects are first observed are indications of loss of lock by various receivers.

Separately, GNSS signals degradation due to powerful solar radio bursts that significantly alter the distribution ionospheric density. Like solar flares, solar radio bursts take 8 min to reach the Earth. First reported in 2007, Cornell University researchers demonstrated that a solar radio burst in the L-band frequency could have a powerful impact on GNSS systems. A solar radio burst in the L-band frequency can raise the background noise level in the L-band channel (and thereby decrease the signal to noise (S/N) ratio of the incoming signal), causing GNSS receivers to loose lock [8, 9].

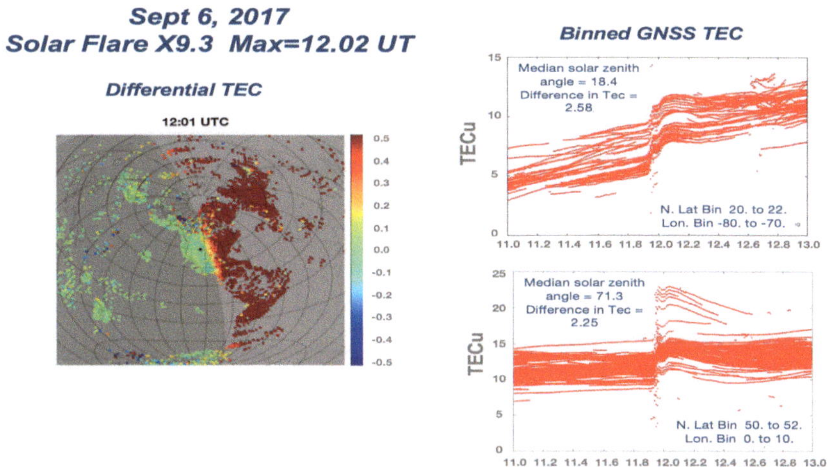

Fig. 1. TEC measurements on Sept 6, 2017 during an X9.3 solar flare. The figure on the left shows the increase in TEC (in red) over the entire daylit portion of the Earth. The two figures on the right show data in bins of 2 degrees of latitude and 10 degrees of longitude. The top plot is for 20–22 degrees latitude and −80.0 to −70.0° degrees longitude, while the bottom plot is for 50 to 52 degrees latitude and 0 to 10.0 degrees longitude.

3.2 Solar Radiation Storms

Solar radiation storms occur when large quantities of charged particles, protons and electrons, are accelerated by processes at or near the Sun, filling the near-Earth environment. Solar radiation storms (also known as a Solar Proton Events or SPEs) occur frequently after major eruptions on the Sun. During these events, protons can be launched at incredibly high speeds, up to several 10.000 km/s. These particles can take from 30 min to several hours to reach the near-Earth environment, and they can last for hours to days.

Associated with radiation storms are the risks associated with radiation, for example, radiation exposure for people on transpolar flights and for astronauts in space. Airlines frequently reroute planes as a consequence. In addition, HF-communication over the poles can suffer severe outages due to polar-cap absorption. This happens when fast moving protons are guided down along the open magnetic field lines into the polar cap where the fast-moving protons ionize the ionospheric D-layer. The increased absorption and ionization in this lower layer of the ionosphere prevents the HF waves from reaching the higher layers of the ionosphere where they can be reflected. This severely limits HF-communication channels. In addition, the increased radiation is responsible for satellite anomalies associated with the spacecraft charging events, degradation of solar panels, malfunctioning of electronics, and additional noise observed in star tracking systems.

During the recent May 10–12 2024 storm, it was reported that the GOES-16 satellite, the primary operational geostationary weather satellite in the GOES East position, stopped transmitting all data at 00:19 UT on May 13 for about 2 h. There was a second loss of transmission for 11 min at 3:19 UT [10]. It has not yet been ascertained that this occurred due to the storm, but the timing is worth looking into.

3.3 Geomagnetic Storms

A geomagnetic storm is a major disturbance of Earth's magnetosphere-ionosphere-thermosphere (MIT) system. Geomagnetic storms are typically associated with solar coronal mass ejections (CMEs) which consist primarily of electrons, protons, and other ionized particles with an embedded magnetic field. If the magnetic field conditions are right, there can be an efficient exchange of energy between the solar wind and the Earth's MIT system. The plasma in a CME moves much slower than the other two categories of solar disturbances and typically it takes several days for this plasma to arrive at Earth, although, for some of the most intense storms, it can arrive at the Earth in as little as 18 h.

Another solar wind disturbance that creates conditions favorable to geomagnetic storms is a high-speed solar wind stream (HSS). HSSs plow into the slower solar wind in front and can create co-rotating interaction regions (CIRs). These regions are often related to geomagnetic storms that are less intense than CME storms, yet can often deposit more energy in Earth's magnetosphere over a longer interval.

Geomagnetic storms produce variability in the ionosphere that results in significant density uncertainty, and mesoscale structure. During storms, perturbations in the solar wind, translate into sweeping changes throughout the geospace system. For example, energetic particles deposit energy into the upper atmosphere thus heating the atmosphere and altering the density and composition. Electric fields generated in the magnetosphere and/or by storm-time neutral winds are imposed on the ionosphere producing areas of extremely high total electron content, large electron density gradients, and regions of both large and small electron density irregularities. These space weather effects are responsible for degradation of GNSS positioning accuracy and satellite navigation, geomagnetic induced currents (GIC) affecting the electric grid, and loss of signal tracking (HF communication outages, loss of lock on GNSS signals). An example of some of these features is the storm enhanced density plume (SED) [11, 12]. Often associated with SED are large TEC gradients [4] and small-scale irregularities [13]. This feature in TEC is shown in Fig. 2 during the May 10, 2024 storm, along with measurements of the phase scintillation index. When the phase scintillation index is greater than 0.6, many receivers lose their capability to track the satellites.

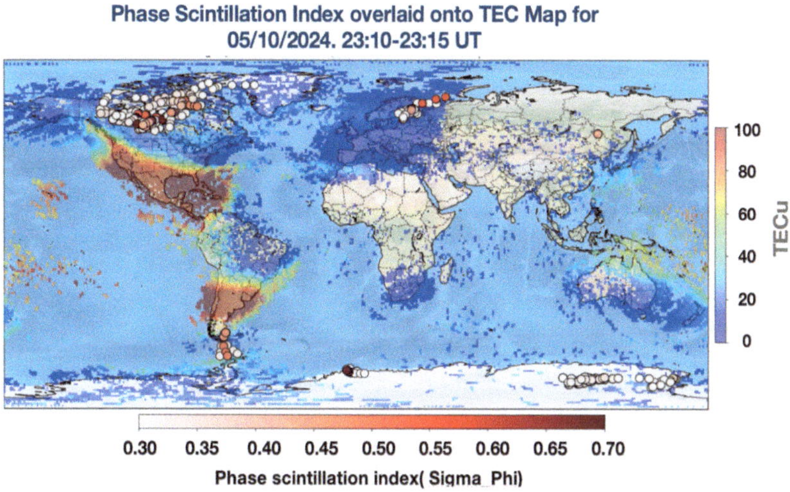

Fig. 2. TEC measurements during the recent May 10, 2024 storm. The higher TEC (in red) during the main phase of the geomagnetic storm, is likely due to a penetration electric field. The circles shown (white to red) indicate the level of phase scintillation. Red values are fairly significant. The primary scintillation visible is in the polar regions, likely due to electron precipitation associated with aurora or irregularities forming in the plasma moving across the pole.

4 Increasing Number of LEO Satellites

One of the main reasons why space weather will gain far more attention in the coming years is due to the increasing number of satellites in Low Earth Orbit (LEO) (<2,000 km altitude). In the next five years, an estimated 8,500 to 10,000 low Earth orbit (LEO) satellites are expected to be launched. This estimation is based on the annual projection of approximately 1,700 satellites being launched each year until 2030. Major contributors to these launches include companies like SpaceX with its Starlink constellation, Amazon's Project Kuiper, OneWeb, and other significant players in the satellite industry. LEO satellites are significantly impacted by atmospheric drag, and it is the largest source of error in orbit prediction. Inaccurate estimates of the thermospheric density $\boldsymbol{\rho}$, especially during storm times are the primary reason for this error.

Space weather events can alter the thermospheric state very quickly. Increases of the thermospheric density greater than 100% have been reported between 400 and 500 km during a minor geomagnetic storm [14]. During solar storms, energetic particles deposit energy into the upper atmosphere thus heating the neutral atmosphere and altering the density and composition. Furthermore, most of the geospace energy input to the atmosphere during storms comes through high latitudes with significant mesoscale structure and can take several hours before the heating is globalized. The current understanding is that most of the energy input at high latitudes transfers to mid- and equatorial latitudes through waves, traveling atmospheric disturbances (TADs) and traveling ionospheric disturbances (TIDs). This contributes to variable drag for different satellites, depending on whether their orbit crosses mesoscale heating or not, especially during the main phase of geomagnetic storms.

Finally, the current thermospheric models of neutral density need to be extended to exosphere heights to include higher altitudes. Modeling the ion upflow/outflow is important for changing high altitude densities in the exosphere. Better real-time estimation of the atmospheric density is critical to improving orbit prediction related products, and research on orbit determination, orbital uncertainty propagation and collision avoidance.

Acknowledgements.
We acknowledge NSF awards AGS-1952737, AGS-2033787, AGS-2149698, PHY-2028125, and NSFC41974184, NASA support 80NSSC22K0171, 80NSSC21K1310, 80NSSC21K1775, 80NSSC19K0834, 80NSSC20K1785, and 80GSFC22CA011, AFOSR MURI Project FA9559-16-1-0364, and ONR Grant N00014-24-1-2122 and N00014-23-1-2160. The research of JDH was supported by NASA grant 80NSSC23K1322 and AFOSR grant FA9550-22-C-0001. Data for TEC processing is provided from the following organizations: UNAVCO, SOPAC, IGN (France), IGS, CDDIS, NGS, IBGE (Brazil), RAMSAC (Argentina), CORS (Panama), Arecibo Observatory, LISN, Topcon, CHAIN (Canada), CRS (Italy), SONEL, RENAG (New Zealand), GNSS Reference Networks, Finnish Meteorological Institute, and SWEPOS.

References

1. Coster, A.J., Erickson, P.J., Lanzerotti, L.J. (eds.): Space Physics and Aeronomy, Space Weather Effects and Applications, vol. 5. John Wiley & Sons (2021)
2. Mannucci, A.J., et al.: Dayside global ionospheric response to the major interplanetary events of October 29–30, 2003 "Halloween Storms,". Geophys. Res. Lett. **32**, L12S02 (2005)
3. Habarulema, J.B., Carelse, S.A.: Long-term analysis between radio occultation and ionosonde peak electron density and height during geomagnetic storms. Geophys. Res. Lett. **43** (2016)
4. Coster, A., Foster, J.: Space weather impacts of the sub-auroral polarization stream. Radio Sci. Bull. **321**, 28–36 (2007)
5. Lohmeyer, W., Carlton, A., Wong, F., Bodeau, M., Kennedy, A., Cahoy, K.: Response of geostationary communications satellite solid-state power amplifiers to high-energy electron fluence. Space Weather **13**, 298–315 (2015)
6. Allen, J.: The galaxy 15 anomaly: another satellite in the wrong place at a critical time. Space Weather **8**, S06008 (2010)
7. Kunches, J.: Personal Communication (2012)
8. Cerruti, A.P., Kintner, P.M., Gary, D.E., Lanzerotti, L.J., de Paula, E.R., Vo, H.B.: Observed solar radio burst effects on GPS/wide area augmentation system carrier-to-noise ratio. Space Weather **4**, S10006 (2006)
9. Cerruti, A.P., et al.: Effect of intense December 2006 solar radio bursts on GPS receivers. Space Weather **6**, S10D07 (2008)
10. General Satellite Message, May 13, 2024, NOAA, Office of Satellite and Product Operations, www.ospo.noaa.gov. "Topic: GOES-16 All Products and derived products delivered to AWIPS, GRB, PDA", 13 May 2024
11. Foster, J.C.: Storm time plasma transport at middle and high latitudes. J. Geophys. Res. **98**(A2), 1695 (1993)
12. Foster, J.C., Erickson, P.J., Coster, A.J., Goldstein, J., Rich, F.J.: Ionospheric signatures of plasmaspheric tails. Geophys. Res. Lett. **29**(13), 1623 (2002)
13. Ledvina, B.M., Makela, J.J., Kintner, P.M.: First observations of intense GPS L1 amplitude scintillations at midlatitude. Geophys. Res. Lett. **29**(14), 1659 (2002)

14. He, J., et al.: Comparison of empirical and theoretical models of the thermospheric density enhancement during the 3–4 February 2022 geomagnetic storm. Space Weather **21**, e2023SW003521 (2023)

Open Access This chapter is licensed under the terms of the Creative Commons Attribution 4.0 International License (http://creativecommons.org/licenses/by/4.0/), which permits use, sharing, adaptation, distribution and reproduction in any medium or format, as long as you give appropriate credit to the original author(s) and the source, provide a link to the Creative Commons license and indicate if changes were made.

The images or other third party material in this chapter are included in the chapter's Creative Commons license, unless indicated otherwise in a credit line to the material. If material is not included in the chapter's Creative Commons license and your intended use is not permitted by statutory regulation or exceeds the permitted use, you will need to obtain permission directly from the copyright holder.

Sun Earth Connections: Impacts of Space Weather Events at Low Latitudes

Christine Amory-Mazaudier(✉)

Laboratoire de Physique Des Plasmas (LPP), Sorbonne Université, Ecole Polytechnique, Institut Polytechnique de Paris, Université Paris Saclay, Observatoire de Paris, CNRS, Paris, France
christine.amory@lpp.polytechnique.fr

Abstract. At the equator, the Earth's magnetic field is horizontal. The zonal ionospheric electric field and Earth's magnetic field produce ExB plasma vertical drift. On the daytime, at low altitudes between 90 and 200 km, ExB drift pushes upward the ionospheric plasma. Above 200km, under the influence of gravity and pressure gradients, the plasma diffuses along the Earth's magnetic field lines. These motions of the plasma create the Equatorial Ionization Anomaly (EIA/Equatorial Fountain). The EIA is composed by two density crests at ± 15° magnetic latitude and a density trough at the equator. The strengthening of the eastward electric field at the evening terminator at the magnetic equator (PRE) associated with the Rayleigh Taylor plasma instability generates the appearance of equatorial plasma bubbles (EPB). Another characteristic of the equatorial is the existence of the Equatorial Electrojet (EEJ) flowing along the magnetic equator. It is at low latitudes that the GNSS signal is the most disturbed by plasma bubbles (EPB). Space weather events affect the EIA, PRE, EPB and EEJ. In this paper, we will present the impact of various solar phenomena as Coronal Mass Ejections (CME), High Speed Solar Wind (HSSW), on middle and low latitudes and their effects on the GNSS signal and Earth's magnetic field.

Keywords: Space Weather · Ionosphere · Low Latitudes

1 Introduction

During the last decade, progress on the low latitude ionosphere has been made within the framework of the IHY (International Heliophysical Year 2007–2009) and ISWI (International Space Weather Initiative 2020–2012) scientific projects. The studies started within these two projects continue in the ISWI research network (www.iswi-secretariat.org since 2013). This article highlights some results obtained by the GNSS (Global Navigation Satellite System) receivers and magnetometers deployed during these projects. The first section presents the characteristics of the low latitude ionosphere EIA (Equatorial Ionization Anomaly), EEJ (Equatorial Electrojet), PRE (Pre-Reversal Enhancement), EPB (Equatorial Plasma Bubble) during magnetic quiet time. The second illustrates the new approach developed in Space Weather, the change from geophysics to Helio physics. The third section shows the changes of the low latitude ionosphere under the influence of the solar disturbance CME and HSSW flowing from solar coronal hole.

© The Author(s) 2026
N. Gopalswamy et al. (Eds.): ISWI 2024, SPPHY 431, pp. 142–153, 2026.
https://doi.org/10.1007/978-981-95-1121-1_16

2 Characteristics of the Low Latitude Ionosphere EIA, PRE, EEJ, EPB

During daytime, the ionospheric zonal electric field and the Earth's magnetic field produce the lift of plasma in E ionospheric region (altitude 90–170 km) by vertical **E X B drift**.

At higher altitudes, in F region (200 km up to 800 km), the plasma diffuses downward along the geomagnetic field lines into both hemispheres under the influence of gravity and pressure gradients, this produces the EIA characterized by an electron density trough at the magnetic equator, and two crests of enhanced electron density at about ± 15° magnetic latitudes (Fig. 1, left panel) [1, 2].

The right panel of Fig. 1 shows the ExB vertical drift measured by the Incoherent scatter sounder of Jicamarca [3] for different seasons and solar flux. Just post sunset, ~ 18.00 LT (right panel in Fig. 1), the upward vertical drift (eastward electric field) increases and then becomes downward (westward electric field). This is the Pre-Reversal Enhancement (PRE); a speed of 25m/s corresponds to an electric field of 1mV/m.

At the time of sunset, the electron density decreases very rapidly in the ionospheric E region. The strengthening of the eastward electric field (**PRE**) at the evening terminator associated with the Rayleigh Taylor plasma instability [4] creates gradients of electronic densisty forming the **E**quatorial **P**lasma **B**ubbles (**EPB**) [5]. The EPB appears after sunset during evening and nighttime.

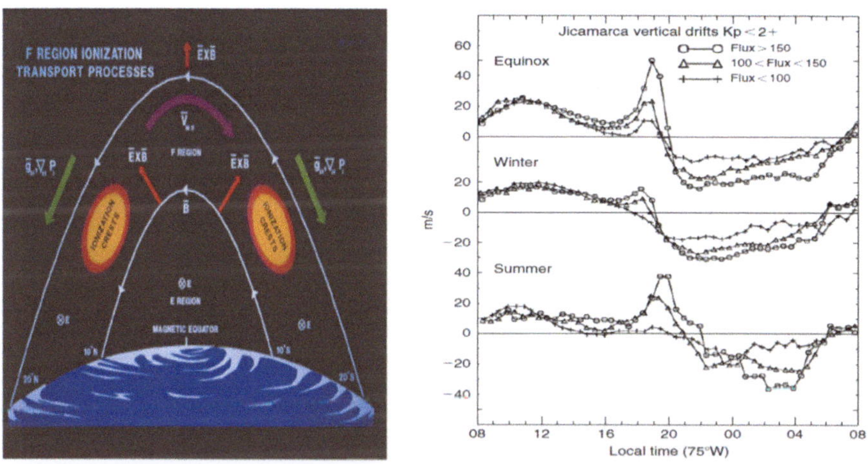

Fig. 1. EIA-Equatorial Fountain with the two crests and the trough (left panel), the Pre-Reversal Enhancement (right panel): increase of the vertical upward drift just before the sunset.

Along the magnetic equator, the electric current flowing in the ionospheric E region at altitudes between 100 km and 130 km (EEJ) is two and a half times more intense than the ionospheric electric currents circulating at mid-latitudes [6]. The EEJ flow on the daytime, 6.00–18.00 LT, when there are conductivities in the E region. The reinforcement of the conductivities is due to the geometry of the Earth's magnetic field, which is horizontal to the magnetic equator.

3 Systemic Approach of the Sun Earth System: Heliophysics

The IHY [7] and ISWI [8] projects made possible to develop the discipline of Space Weather with 2 essential points:

1) to understand the impact of all the solar disturbances on the Earth's electromagnetic environment and to define geophysical studies in a broader framework of Helio physics. During the IHY project, many scientific advances were possible by taking into account all the solar events and not just the spectacular ones, as had been done in the past. The aim was to understand the variability of the Earth's environment as a function of the variability of the Sun.
2) to plan and estimate how these solar events disrupt our new technologies such as GNSS. During the ISWI project, attention focused on the malfunctioning of our technologies due to solar phenomena. This has led to a link between the science developed during the IHY project and the applications of this science in everyday life.

Space weather needs the knowledge of all the connections between the Sun and the Earth. A recent review [9] presents the solar events that influence the terrestrial environment. It is necessary to link all the knowledge acquired on the ionosphere over more than 8 decades to the solar events, to establish variability of the ionosphere function of the solar variability.

For the development of space weather there is a need to break down the walls that separate the different disciplines and to understand how new technologies are affected by intense solar events. To succeed, scientists belonging to different disciplines covering sun-earth connections have trained numerous young people, all over the world, within the framework of multidisciplinary schools [10]. Thus, a new scientific community has emerged and the ISWI scientific network is coordinating its work [11]. It is in Africa that the emergence of new young space weather scientists has been most significant [12].

4 Some Scientific Results

4.1 Gnss/gps

GPS stations are the most widely used instruments on Earth and many data from these stations are accessible on the internet in a standard RINEX format. From these RINEX files we can extract the TEC, **T**otal **E**lectron **C**ontent (number of electrons along the path between the satellite and the earth) and deduce the ROTI index (Eq. 1) from the fluctuations of the TEC. The ROTI index reveals the existence of a type of irregularity. It is also possible for certain GPS to determine the S4 index (Eq. 2, I: signal intensity): which also provides information on the presence of irregularities in the ionospheric plasma [13, 14].

$$\mathrm{rot} = \frac{STEC_{k+1} - STEC_k}{time_{k+1} - time_k} * 60 \ \mathrm{roti} = \sqrt{\langle \mathrm{rot}^2 \rangle - \langle \mathrm{rot} \rangle^2} \quad (1)$$

$$s4 = \sqrt{\frac{\langle I^2 \rangle - \langle I \rangle^2}{\langle I \rangle^2}} \tag{2}$$

It is important to know the variation of a parameter on quiet magnetic days, when there are no solar disturbances, to assess the impact of the solar disturbance. A day was deemed quiet if Kp is less or equal to 3. Electron density gradients (irregularities) exist at low latitudes during magnetic quiet periods, very often after sunset, under the influence of several factors, in particular the PRE and the Rayleigh Taylor instability. Figure 2 shows the ROTI index (signature of irregularities) for the 2 crests and the equatorial trough, during 2013 [15]. This figure illustrates the seasonal variability of the ROTI and the asymmetry of the ROTI between the two crests of the EIA. '*The behavior of irregularities in summer and winter can also be explained in terms of the late time of reversal of PRE at the dip equator. The later (earlier) time of reversal of the PRE which occurs in the local summer (winter) at the magnetic equator favors (inhibits) the occurrence of irregularities*' [15].

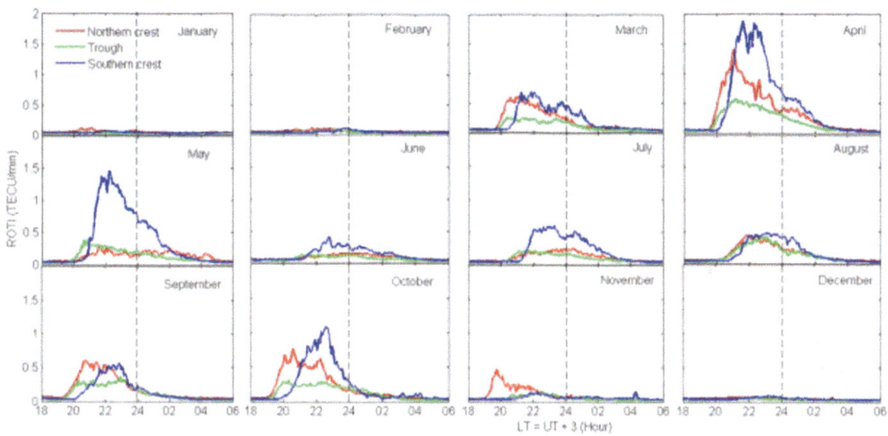

Fig. 2. Monthly mean variations of quiet time irregularities over the northern crest (red line), trough (green line), and southern crest (blue line) in 2013 l15]. There is an asymmetry between the two crests of the EIA.

Figure 3 shows the evolution of the VTEC during the Saint-Patrick storm on March 17, 2015. The Storm Sudden Commencement (SSC) occurs at 04.45UT (first vertical dotted line in the figure). Before the start of the storm, we can see the regular equatorial fountain with the two crests of density and the trough at the equator. The second vertical line indicates the minimum of the SYM-H magnetic index. In the Asian sector, the ionosphere disappears after this line, and a super-fountain develops in the American sector before it. The third vertical line indicates the arrival of a HSSW.

Figure 4 shows the variations in the ROTI and S4 indices measured in Vietnam at 2 stations, Phu Thuy (magnetic latitude: 14,89) and Hue (magnetic latitude: 9,58), for the same magnetic storm. The first vertical line is for the SSC, arrival of a CME, and the second vertical line the arrival of a HSSW.

The auroral zone is directly coupled to the interplanetary environment and the equatorial zone is connected to the interplanetary environment via the auroral region. The electrodynamic coupling between auroral and equatorial regions exists through different large-scale processes: 1) the transmission of the magnetospheric convection electric field at low latitudes, PPEF (Prompt Penetration of the Magnetospheric Convection Electric Field) [18], 2) the thermal expansion of the atmosphere due to Joule energy dissipated by auroral electric currents [19–21] and 3) the disturbance of the dynamo ionospheric, DDEF (Disturbance Dynamo Electric Field) [22].

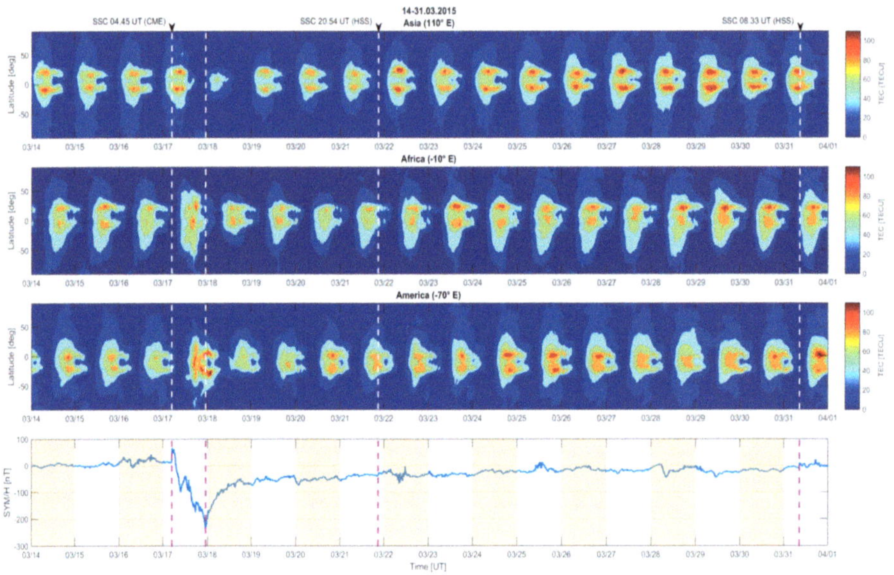

Fig. 3. Maps of vTEC from 14 to 31 March, Asian sector (first top panel), African sector (second), American sector (third panel, and SYM-H (last panel) [16]. The equatorial fountain disappears in Asia and a super fountain appears in America.

In Fig. 3 the super fountain in America observed at the beginning of the storm during the main phase is mainly due to the process of penetration of the electric field of the magnetospheric [18] and the disappearance of the Ionosphere in Asia mainly to the thermal expansion of the atmosphere [19–21] which changes pressure, temperature, motions and composition of the atmosphere. The disappearance of the ROTI on disturbed magnetic days (Fig. 4) is due to the ionospheric disturbed dynamo [22]. Ionospheric disturbance dynamo creates a westward electric field which inhibits the eastward electric field of the PRE and prevents the formation of irregularities [23].

Figure 5 shows the global TEC (ΔGEC) and regional TEC (ΔREC) variations in the top left and middle panels. The Bz component of the interplanetary magnetic field is shown in the lower left panel. The vertical dashed line indicate the arrival of the HSSW. During this episode of fast solar wind, the Bz component fluctuates rapidly. This variation is very different from the Bz component of a CME, which remains directed southwards for several hours and allows the magnetosphere to open. The panels on the right show

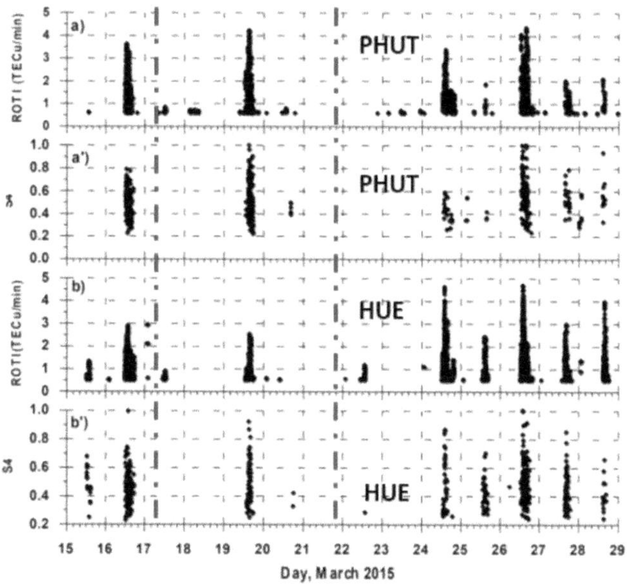

Fig. 4. ROTI and S4 indices in two stations of Vietnam a) and a') at PHUT (105.95° E 21.02°N) and b) and b') at HUES (107.59°E 16.45°N) from 15 to 28 March 2015 [17]

the variations in ⊗ VTEC for the 4 longitude sectors (Asia, Africa, America and the Pacific). Note that HSSW generally increases the amplitude of one or both crests of the equatorial fountain. Numerous positive ionospheric storms are observed throughout the period around the magnetic equator. There are very few negative ionospheric storms. This is contrary to the action of the CMEs (see Fig. 3), which cause a very strong asymmetry between the sectors of longitude.

4.2 Magnetometers

During the IHY [6] and ISWI [7] projects, in addition to the deployment of GPS receivers, magnetometers were installed, to study the impact of solar event on the Earth's magnetic field.

Solar disturbances amplify ionospheric and magnetospheric electrical currents [26] which induce telluric currents (Ground Induced Current, GIC), and GIC can damage electrical.

To process the magnetic data, we applied the law of Biot and Savart to the large scale electric current systems

$$DH = Sq + D_{iono} + D_{mag} \tag{3}$$

where ⊗ H is the variation of the H component of the Earth's magnetic field. Sq is the regular variation of the Earth's magnetic field during magnetic quiet days. D_{iono} is the magnetic disturbance due to the disturbed ionospheric electric currents and $D_{mag,}$ the

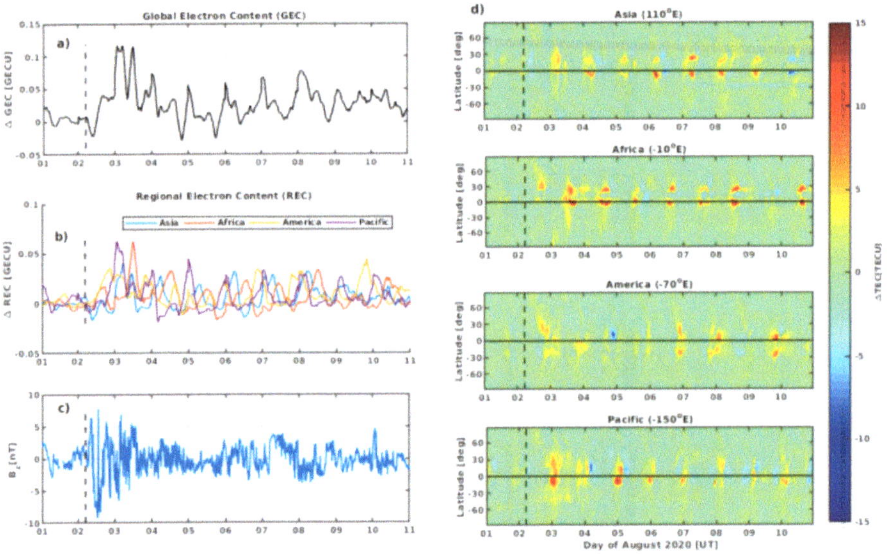

Fig. 5. A) ΔGEC (b) ΔREC in four longitudinal sectors Asia, Africa, America, and Pacific (c) Bz component of interplanetary magnetic field and (d) ΔvTEC at fixed longitudes, from top to bottom, Asia, Africa, America, and Pacific during 01 August–10 August 2020 [24]. EIA is strengthened.

magnetic disturbance due to the magnetospheric electric currents, estimated by SYM-H, ASYM-H.

We can estimate D_{iono}:

$$D_{iono} = DH - Sq - D_{mag} \tag{4}$$

At low latitudes, two large scale magnetic disturbances DP2 [26] and Ddyn [27] are the signature of PPEF [17] and DDEF [21].

$$D_{iono} = DP_2 + D_{dyn} \tag{5}$$

In Fig. 6, the top 3 panels illustrate ΔH, Sq and D_{iono} observed during the HSSW of April 2010 in three magnetic observatories near the magnetic equator (GUAM, MBO and KOU). ΔH is in blue, Sq in black and D_{iono} in red. The bottom 3 panels give the magnetic disturbance D_{dyn} due to the DDEF process. To extract D_{dyn}, we analyse the signal with wavelet + semblance technics. On the top three panels D_{iono} exhibits a variation composed by short fluctuations (DP_2) superimposed to an anti Sq variation (D_{dyn}). This anti-Sq variation is the signature of the ionospheric disturbed dynamo [21]. Joule heating in the auroral zone produces an atmospheric Hadley cell from the pole and the equator. This disturbed atmospheric circulation reverses the ionospheric electric current.

This processing of magnetic data was done for 19 geomagnetic storms [28] and we find that the magnetic disturbance D_{dyn} due to the disturbed ionospheric dynamo lasts longer for fast solar winds (HSSW) but is less strong than the disturbance D_{dyn} due to coronal mass ejections (CME).

During the IHY and ISWI projects we introduced Space Weather and sometimes ionosphere knowledge in different countries: Morocco [29], Algeria [30], Egypt [31], Vietnam [32, 33], Nepal [34], Senegal [35], Burkina Faso [36], Nigeria [37, 38]. In this document, we have presented some results and many other subjects have been covered: the study of travelling ionospheric disturbances [39], ground-induced currents [40], hemispheric asymmetry due to HSSWs [41], Ionospheric quasi-biennial oscillation of the TEC amplitude of the EIA [42], and the impacts of CMEs or HSSWs in Africa, Asia and America [43–47].

5 Conclusion

In this article we have highlighted a few examples. We have presented the ROTI index for East Africa because such a study can only be done in Africa, which is the continent with the widest latitudinal extent on either side of the magnetic equator. We chose the St Patrick's storm of March 17, 2015, the strongest storm of sunspot cycle 24. For this storm the scintillations of the GNSS system disappeared over the whole planet for two days. We presented the magnetic data observed during a fast solar wind in April 2010. For this event, an anti-Sq circulation of ionospheric electric currents lasted for five days. We compared studies which reveal that the impact of CME and HSSW is very different on EIA (Figs. 3 and 5) that on the other hand CME and HSSW suppress plasma irregularities highlighted by the ROTI index.

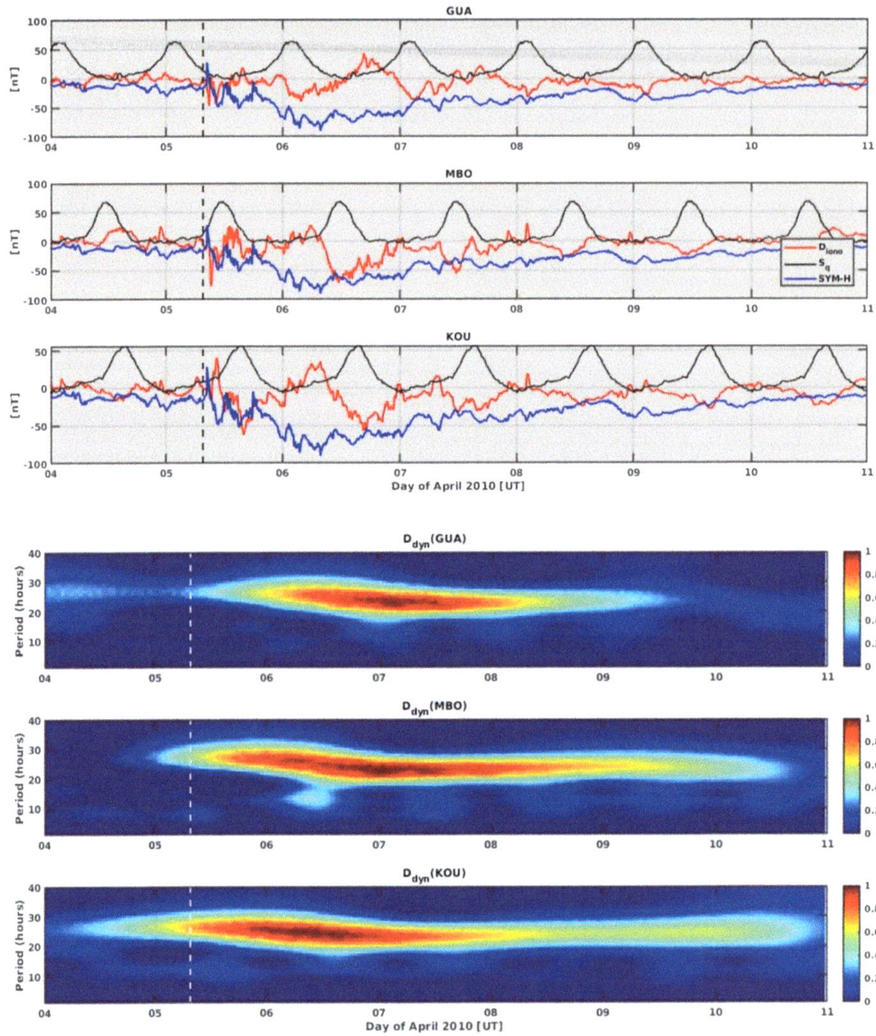

Fig. 6. On the top three panels: in black the Sq in blue the ⊗ H and in red D_{iono}, on the bottom three panels: D_{dyn} disturbance, for Asia, Africa and America. [28]. The D_{dyn} magnetic disturbance is observed first in the American sector, then in African and Asian sectors Its lasts longe in the Amrican than African and then Asian sectors.

References

1. Namba, S., Maeda, K.-I.: Radio Wave Propagation, p. 86. Corona Publishing, Tokyo (1939)
2. Appleton, E.: Two anomalies in the ionosphere. Nature **157**, 691–693 (1946)
3. Fejer, B.G., de Paula, E.R., Gonzalez, S.A., Woodman, R.F.: Average vertical and zonal F region drifts over Jicamarca. J. Geophys. Res. **96**(A8), 13901–13906 (1991)
4. Basu, B.: Characteristics of electromagnetic Rayleigh-Taylor modes in nighttime equatorial plasma. J. Geophys. Res. **110**(A2), A02303 (2005)
5. Bhattacharyya, A.: Equatorial plasma bubbles: a review. Atmosphere **13**(10), 1637 (2022)

6. Chapman, S.: Equatorial electrojet as detected from the abnormal electric current distribution above Huancayo, Peru, and elsewhere. Arch. Meteorol. Geophys. **4**, 368–390 (1951)
7. Davila, J., Goplaswamy, N., Thompson, B., Bogdan, T., Hapgood, M.: The international space weather initiative (ISWI). In: The Sun and Solar wind and the Heliosphere (2011)
8. Davila, J., Gopalswamy, N., Haubold, H.J., Thompson, B.: International heliophysical year 2007: basic space science initiatives. Space Policy **23**(2), 121–126 (2007)
9. Gopalswamy, N.: The sun and space weather. Atmosphere **13**(11), 1781 (2022)
10. Amory-Mazaudier, C., et al.: Development of research capacities in space weather: a successful international cooperation. J. Space Weather Space Clim. **11**, 28 (2021)
11. Home page; https://www.iswi-secretariat.org. last accessed 10 June 2024
12. Baki, P., et al.: The status of space weather infrastructure and research in Africa 2023. Atmosphere **14**(12), 1791 (2023)
13. Pi, X., Mannucci, A.J., Lindqwister, U.J., Ho, C.M.: Monitoring of global ionospheric irregularities using the worldwide GPS network. Geophys. Res. Lett. **24**(18), 2283–2286 (1997)
14. Amory-Mazaudier, C., Fleury, R., Gadimova, S., Touzani, A.: Training on daily global positioning system GPS data. Coordinates a Monthly Magazine on Positioning, Navigation and Beyond **13**(3) (2017)
15. Amaechi, P.A., et al.: Quiet-time ionospheric irregularities over the African Equatorial Ionization Anomaly (EIA) region. Radio Sci. **55**(8), e2020RS007077 (2020)
16. Nava, B., et al.: Middle and low latitude ionosphere response to 2015 St. Patrick's day geomagnetic storm. J. Geophys. Res. Space Phys. **121**(4), 3421–3438 (2016)
17. Le Huy, M., et al.: TEC variations and ionospheric disturbances during the magnetic storm on March 2015 observed from continuous GPS data in the Southeast Asian region. Vietnam J. Earth Sci. **38**(3), 287–305 (2016)
18. Vasyliunas, V.M.: Mathematical models of magnetospheric convection and its coupling to the ionosphere. In: Particles and Fields in the Magnetosphere. Springer, New York (1970)
19. Blanc, M., Richmond, A.D.: The Ionospheric disturbance dynamo. J. Geophys. Res. **85**(A4), 1669–1686 (1980)
20. Volland, H.: Magnetospheric electric fields and currents and their influence on large scale thermospheric circulation and composition. J. Atmos. Terr. Phys. **41**(7–8), 853–866 (1979)
21. Fuller-Rowell, T.J., Codrescu, M.V., Moffett, R.J., Quegan, S.: Response of the thermosphere and ionosphere to geomagnetic storms. J. Geophys. Res. **99**(A3), 3893–3914 (1994)
22. Fuller-Rowell, T.J., Codrescu, M.V., Rishbeth, H., Moffett, R.J., Quegan, S.: On the seasonal response of the thermosphere and ionosphere to geomagnetic storms. J. Geophys. Res. **101**(A2), 2343–2353 (1996)
23. Fejer, B.G., Jensen, J.W., Shin-YI, S.: Seasonal and longitudinal dependence of equatorial disturbance vertical plasma drifts. Geophys. Res. Lett. **35**(20), L20106 (2008)
24. Younas, W., Khan, M., Amory-Mazaudier, C., Amaechi, P.O.: Ionospheric response to the coronal hole activity of August 2020: a global multi-instrumental overview. Space Weather **20**(12), e2022SW003176 (2022)
25. Amory-Mazaudier, C.: Magnetic signatures of large-scale electric currents in the earth's environment at middle and low latitudes. Atmosphere **13**(10), 1699 (2022)
26. Nishida, A.: Geomagnetic DP2 fluctuations and associated magnetospheric phenomena. J. Geophys. Res. **73**(5), 1795–1803 (1968)
27. Le Huy, M., Amory-Mazaudier, C.: Magnetic signature of the Ionospheric disturbance dynamo at equatorial latitudes: "Ddyn." J. Geophys. Res. **110**(A10), A10301 (2005)
28. Younas, W., Amory-Mazaudier, C., Khan, M., Le Huy, M.: Magnetic signatures of ionospheric disturbance dynamo for CME and HSSWs generated storms. Earth Space Sci. **19**(9), e2021SW002825 (2021)

29. Azzouzi, I., Migoya-Orué, Y., Amory Mazaudier, C., Fleury, R., Radicella, S.M., Touzani, A.: Signatures of solar event at middle and low latitudes in the Europe-African sector, during geomagnetic storms, October 2013. Adv. Space Res. **56**(9), 2040–2055 (2015)
30. Hammou, A.O., Zaourar, N., Fleury, R., Amory-Mazaudier, C.: Transient variations of vertical total electron content at low latitude during (2013–2017). Adv. Space Res. **68**(12), 4857–4871 (2021)
31. Shimeis, A., et al.: Signature of the coronal hole on near the north crest equatorial anomaly over Egypt during the strong geomagnetic storm 5th April 2010. J. Geophys. Res. **117**, A07309 (2012)
32. Tran, T.L., Le Huy, M., Amory-Mazaudier, C., Fleury, R.: Climatology of ionospheric scintillation over the Vietnam low-latitude region for the period 2006–2014. Adv. Space Res. **60**(8), 1657–1669 (2017)
33. Nguyen Thanh, D., et al.: Characterization of ionospheric irregularities over Vietnam and adjacent region for the 2008–2018 period. Vietnam J. Earth Sci. **43**(4), 1–20 (2021)
34. Pandit, D., Amory-Mazaudier, C., Fleury, R., Chapagain, N.P., Adhikari, B.: VTEC observations of intense geomagnetic storms above Nepal: comparison with satellite data CODE and IGSG models. Indian J. Phys. **97**, 701–718 (2023)
35. Ndao, A., Gaye, I., Fleury, R., Amory-Mazaudier, C.: Effects of ionospheric plasma irregularities at the equatorial zone on GPS signal. J. Sci. Eng. Res. **9**(4), 109–117 (2011)
36. Zoundi, C., Ouattara, F., Fleury, R., Amory-Mazaudier, C., Lassudrie-Duchesne, P.: Seasonal TEC variability in West Africa equatorial anomaly region. Eur. J. Sci. Res. **77**(3), 309–319 (2012)
37. Amaechi, P.O., et al.: Ground-based GNSS and C/NOFS observations of ionospheric irregularities over Africa: a case study of the 2013 St. Patrick's day geomagnetic storm. Space Weather **19**(2), 631 (2021)
38. Amaechi, P.A., Oyeyemi, E.O., Akala, A.O., Amory-Mazaudier, C.: Geomagnetic activity control of irregularities occurrences over the crests of the African EIA. Earth Space Sci. **7**(7), e2020EA001183 (2020)
39. Shimeis, A., Borries, C., Amory-Mazaudier, C., Fleury, R., Mahrous, A.M., Hassan, A.F.: TEC variations along an East Euro-African chain during 5th April 2010 geomagnetic storm. Adv. Space Res. **55**(9), 2239–2247 (2015)
40. Doumbia, V., Boka, K., Grodji, O.D.F., Amory-Mazaudier, C., Menvielle, M.: Induction effects of geomagnetic disturbances in the geo-electric field variations at low-latitude. Ann. Geophys. **35**(1), 39–51 (2017)
41. Zaourar, N., Amory-Mazaudier, C., Fleury, R.: Hemispheric asymmetries in the ionosphere response observed during the high-speed solar wind streams of the 24–28 August 2010. Adv. Space Res. **59**(9), 2229–2247 (2017)
42. Nguyen Thanh, D., et al.: Ionospheric quasi-biennial oscillation of the TEC amplitude of the equatorial ionization anomaly crests from continuous GPS data in the Southeast Asian region. Vietnam J. Earth Sci. **45**(1), 1–18 (2023)
43. Mohamed, H.S., Amory-Mazaudier, C., Shalabiea, O.M., Mahrous, A.: Delayed response of low latitudes TEC during thirty-six geomagnetic storms from 2014 to 2017. J. Atmos. Solar Terr. Phys. **250**, 106109 (2023)
44. Vankadara, R.K., et al.: Signatures of equatorial plasma bubbles and ionospheric scintillations from magnetometer and GNSS observations in the Indian longitudes during the space weather events of early-September 2017. Remote Sens. **14**(3), 652 (2022)
45. Ranjan, A.K., et al.: Variability of ionosphere over indian longitudes to a variety of space weather events during December 2006. Space Weather **21**(11), e2023SW003595 (2023)
46. Tahir, A., et al.: Multi-instrument observation of the ionospheric irregularities and disturbances during the 23–24 March 2023 geomagnetic storm. Remote Sens. **16**(9), 1594 (2024)

47. Afolabi, O.O., Candido, C.M.N., Becker-Guedes, F., Amory-Mazaudier, C.: Study and modelling of the impact of June 2015 geomagnetic storms on the Brazilian ionosphere. Atmosphere **15**(5), 597 (2024)

Open Access This chapter is licensed under the terms of the Creative Commons Attribution 4.0 International License (http://creativecommons.org/licenses/by/4.0/), which permits use, sharing, adaptation, distribution and reproduction in any medium or format, as long as you give appropriate credit to the original author(s) and the source, provide a link to the Creative Commons license and indicate if changes were made.

The images or other third party material in this chapter are included in the chapter's Creative Commons license, unless indicated otherwise in a credit line to the material. If material is not included in the chapter's Creative Commons license and your intended use is not permitted by statutory regulation or exceeds the permitted use, you will need to obtain permission directly from the copyright holder.

Space Weather Radiation Impacts on Aviation

Rendani Nndanganeni(✉)

South African National Space Agency (SANSA), PO Box 32, Hermanus 7200, South Africa
rnndanganeni@sansa.org.za

Abstract. Measurements of radiation exposure within the African region are rare or non-existent and as such, different available models are utilised to evaluate the amount of radiation exposure at flight levels. This paper discusses the results computed using CARI7 model for selected periods 2–8 November 2021 and 8–13 May 2024. This is done for flight routes between Cape Town (CPT) (33.97° S, 18.60° E) and Johannesburg (JNB) (26.13° S, 28.24° E) in South Africa and between Johannesburg and Ethiopia (HAAB) (8.98° N, 38.803° E). The maximum computed effective dose for the CPT-JNB route during the period of 2- 8 November 2021 is ~ 5.9 μSv/h which was on the 4th of November 2021. However, during the period of 8–13 May 2024, the computed maximum effective dose is ~ 5.8 μSv/h and there was no distinction in the amount of dose computed for this period. On the other hand, the maximum computed effective dose for the JNB-HAAB route during the 2–8 November 2021 period is ~ 4.9 μSv/h which was on the 4th and for the period of 8–13 May 2024 the maximum computed dose is ~ 3.9 μSv/h for the entire storm period.

Keywords: Cosmic Radiation · Radiation Exposure · Effective Dose

1 Introduction

At aviation altitude, which is about 11 km, the passengers and crew are subjected to low levels of radiation exposure from the omnipresent galactic cosmic rays. This exposure to cosmic radiation is categorized as an existing exposure situation as the source exists before a protection decision can be made and this is occupational exposure [1] and as such the employers have a role to play in the protection of these employees. The risks from changes in cosmic radiation due to space weather and potential radiation exposure during flights are not currently fully understood or even acknowledged. Legislation and regulation for such enterprises is also in its infancy with little or no guidance for commercial entities or potential passengers [2]. Exposure of aircrew to cosmic radiation has been recognized globally as an occupational health risk for several decades [3].

The intensity of the radiation field at the cruising altitudes of civil aviation is usually still about one to two orders of magnitude higher than at sea level. Although not as harsh as the radiation environment in space, the radiation exposure for people such as aircrew and frequent flyers, who spend many hours at these altitudes, is significant. Aircrew are typically exposed to 1.5–5 mSv per year from GCRs [4]. The radiation field at aviation

© The Author(s) 2026
N. Gopalswamy et al. (Eds.): ISWI 2024, SPPHY 431, pp. 154–163, 2026.
https://doi.org/10.1007/978-981-95-1121-1_17

altitudes consists of mainly neutrons between 1 and 150 MeV [5]. The contribution that solar proton events (SPEs) and other space weather events make to the dose rate is not well known. It is estimated to increase the effective dose anywhere between several μSv and 20 mSv, depending on the severity of the SPE and latitude of the aircraft [6].

The main source of radiation is galactic cosmic rays (GCRs) [7] which is modulated by the solar cycle (GCRs are more important at a solar minimum than at a solar maximum). In addition to GCRs, sudden changes due to space weather i.e., SEPs events may produce short-term enhanced radiation levels at aircraft altitudes and even at ground level during so-called ground-level enhancements (GLEs) [8]. Geomagnetic cutoff rigidity describes the Earth's magnetic field shielding against the arrival of charged cosmic ray particles from outside the magnetosphere [9]. The influence of space weather such as geomagnetic storms can change the geomagnetic cutoff rigidity at a given location, and this can lead to a small modification in cosmic ray intensities that can alter the atmospheric radiation environment. The dose rates in mid-latitude regions are sensitive to small changes in cutoff rigidity and as such the changes in the radiation environment might be significant. During commercial flights, primary and secondary high-energy particles interact with aircraft components to further produce more high-energy secondary particles [10]. The importance of having routine airborne radiation measurements for model validation and improvement was emphasized by [11]. Furthermore, there is still much more work to be done in terms of understanding the influence of solar storms such as CMEs, SEPs and solar flares on how they influence the changes in the Earth's atmospheric radiation environment which is part of efforts to better understand space weather [12].

This study shows the results computed using CARI7 model for the following selected period 2–8 November 2021 and 8–13 May 2024 respectively. This is done for flight routes between Cape Town (CPT) (33.97° S, 18.60° E) and Johannesburg (JNB) (26.13° S, 28.24° E) in South Africa and between Johannesburg and Ethiopia (HAAB) (8.9838° N, 38.7963° E). The model results show the dose dependence on the latitude during the selected period. The results show that the dose at mid-latitude is much higher than the equatorial latitude as expected.

2 Methodology

Enhancement of radiation doses at aviation altitude is of concern, especially to aircrew and frequent flyers. To investigate radiation dose exposure at aviation altitude, flight data was obtained from FlightAware website [13]. This data gives information about flight time, latitude, longitude and altitude, which is then used as an input to the models to compute the dose. CARI 7 radiation models were used to compute effective dose rate received in-flight [14].

The CARI-7 computer program, developed at the FAA's Civil Aerospace Medical Institute, calculates different types of doses e.g., effective dose, absorbed dose, etc. of galactic cosmic radiation received by an individual or a detector (based on an anthropomorphic phantom) on an aircraft. The program accommodates both waypoint defined and shortest route (geodesics) flight paths. Particle specific doses can also be calculated for route or at specific locations. This model starts from the particle spectrum and then models the propagation of the particles through the Earth's magnetosphere and atmosphere using a collection of previously developed physics models [15].

The model uses modern GCRs models (local interstellar spectrum from the ISO 2004 GCRs model modulated by the heliocentric potential method [16], combined with atmospheric shower data calculated using Monte Carlo particle transport soft-ware MCNPX 2.7.0. Calculation of flight doses is done by integrating single location doses along the flight path calculated from user input flight data. CARI 7 assumes that a geodesic route between the origin and destination airports describes flight routes. The user must specify the start time of the flight [16]. Geodesic flight route information is calculated using a computer program called "FORWARD and INVERSE" [17]. Speed is assumed to be constant, and the output radiation dose is calculated for each minute of the flight and summed for the total radiation dose [15].

Figure 1 (top panel) shows the Dst and Kp index values for the period 2–8 November 2021 obtained from the World Data Center for Geomagnetism, Kyoto [17] and kp.gfz-potsdam.de/en [18], respectively. The Kp index is illustrated as a blue bar graph with the values ranging between $1 >= \text{Kp} < 2$ indicating the quiet, and $\text{Kp} = 3$ active, $\text{Kp} = 4$ unsettled and $\text{Kp} > 4$ stormy periods, respectively, and the Dst index is shown as a black line. The minimum Dst (−125 nT) occurred on 4th of November at 13:00 UT and Kp reached a maximum of 7+ between 09:00 and 12:00 UT. The recovery phase of the geomagnetic storm started after 4th of November at 13:00 UT. On 5 November the Kp index was mainly unsettled throughout the day and on 6 and 7 November the values of Kp were just ranging between 1 and 4, indicating quiet to unsettled levels with two active intervals. The bottom panel shows the Dst and Kp index but for the period 8–13 May 2024. The minimum Dst (−410 nT) occurred on the 11 May at 2:00 UT and Kp reached a maximum of 9 around the same time at 2:00 UT on the 11 May. On the 11 May the recovery phase of the geomagnetic storm started after 12:00 UT. From the 12 May to the 13 May the Kp index was between 7 and 3 with one interval of $\text{Kp} = 7$ observed at 3:00 UT, two intervals of $\text{Kp} = 6$ observed between 21:00 UT on 12 May and 6:00 UT on the 13 May. The recovery phase of the geomagnetic storm started on 12 May at 3:00 UT.

3 Results and Discussion

Figure 2 shows Dst and Kp indices for the 2–8 Nov 2021 (top left panel) and 8–13 May 2024 (top right panel), respectively. The Kp index is illustrated as a blue bar graph with the values ranging between $1 >= \text{Kp} < 2$ indicating the quiet, and $\text{Kp} = 3$ active, $\text{Kp} = 4$ unsettled and $\text{Kp} > 4$ stormy periods, respectively, and the Dst index as a black line (see Sect. 2 under methodology). The middle panels show the computed effective dose rate as a function of latitude and the bottom panel is the effective dose as a function of altitude for the computed for the route between JNB-HAAB for the period of 2–8 Nov 2021 and 8–13 May 2024 respectively. The effective dose rate is computed using CARI 7 model.

The trend is that during takeoff the dose rapidly increases as a function of increasing altitude indicating the decrease in atmospheric shielding to reach a maximum peak and then becomes constant thereafter. The does peaks at the highest latitude of 24°S. As the

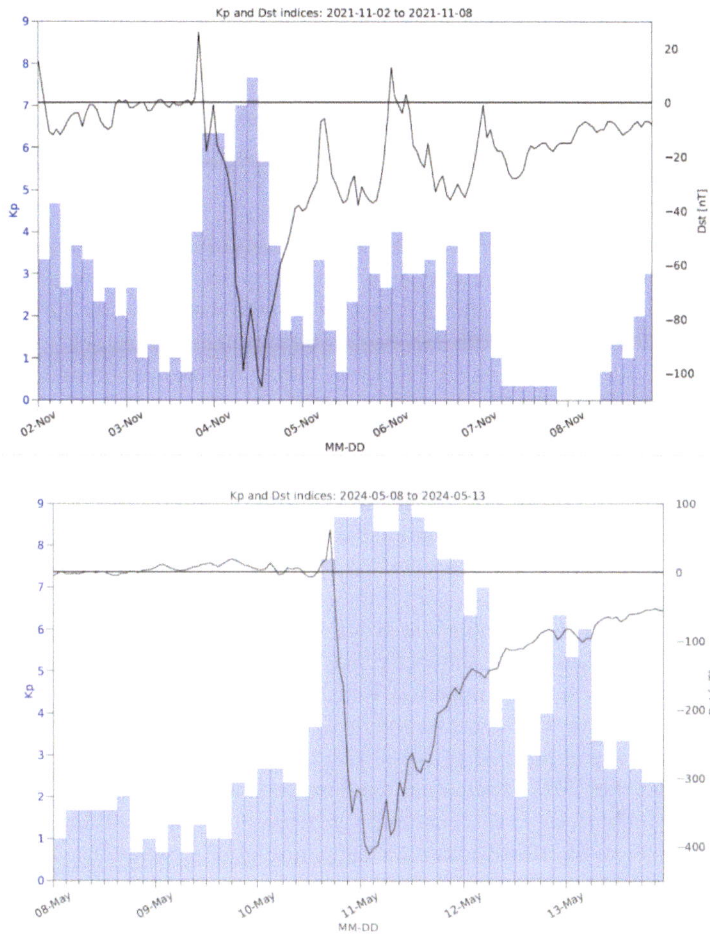

Fig. 1. Dst and Kp indices for the 2–8 Nov 2021 and 8–13 May 2024. The Kp index is illustrated as a blue bar graph with the values ranging between 1 > = Kp < 2 indicating the quiet, and Kp = 3 active, Kp = 4 unsettled and Kp > 4 stormy periods, respectively, and the Dst index as a black line.

plane moves from 26°S to 10°N the effective dose decreases due to changes in the cutoff rigidity which is highest at the equatorial regions and there is maximum geomagnetic shielding as seen from the top panel of Fig. 2. This is attributed to the fact that in terms of the amount of radiation, the geographic equator receives the maximum solar radiation whereas the at the geomagnetic equator the cosmic radiation is more deflected as can be seen from the result presented here.

From the middle-left panel, there is a clear change in terms of the dose computed for these different days with the maximum being seen for the 4th of November 2021 with the effective dose rate ranging between ~ 3.2 μSv/h and 4.8 μSv/h and the minimum dose on the 3rd of November ranging between ~ 1.3 and ~ 3.2 μSv/h. With reference to Fig. 1 top panel the 3rd of November was a quiet day and there was no space weather storm that can lead to changes in the radiation environment at aviation altitude [8]. Effective dose rate first increased reaching its highest peak at 24°S, then constantly decreasing as a function of changing latitude with the lowest value obtained between ~ 5°S and 4°N. This is because of the magnetic shielding which is stronger along the equatorial region than elsewhere [10]. On the other hand, the results shown at the middle right panel show no effect of the space weather events on May 2024. With reference to the top left panel which shows the storm time disturbance index, both the 10th and 11th were disturbed days, and one expects the dose rate of the 10th and the 11th to be variable from the quiet days. This is however not the case; this may be due to that the contribution or changes in terms of enhancing the radiation at aviation altitude that is mostly attributed to the occurrence of the GLE events with energies > 500 meV that they are able to be observed by neutron monitors on the ground. However, during the two period under discussion there was no significant GLE events observed but there was a forbush decrease (See Fig. 4) which will potentially lead to decrease in dose rate when strong enough, but as seen from the results in Fig. 2 middle right panel and bottom right panel there were no changes or variations in terms of the dose during the entire period of 8- 13 May 2024. During large solar eruptions, large fluxes of solar energetic particles (SEPs) can be created via acceleration processes at the Sun. SEPs can pose a serious space weather risk to aircrews and airline passengers if the eruption is directed towards the Earth by enhancing the radiation environment at aviation altitudes [19].

The bottom panels of Fig. 2 show effective dose rate as a function of altitude during ascent and descent for both storm period. Results for both selected periods show similar trends, where effective dose increases with altitude until maximum effective dose is reached at a cruising altitude of 11.5 km with the highest effective dose computed on the 4th of November 2021. Thereafter, the dose decreases as a function of decreasing altitude as the flight starts to descend until it reaches the ground at 0 km. Similar trend of variation in terms of effective dose for the period of 2- 8 November 2021 is evident here. For the 8–13 May 2024 there is no variation as seen in both top right and bottom right panels.

Figure 3 (top panels) shows the computed effective dose rate as a function of latitude for the same period as in Fig. 2 and the bottom panel shows computed effective dose as a function of altitude, but for a route between CPT-JNB. Looking at the top left panel, the computed effective dose rate is much higher on the 4th of November 2021 as compared to that of 2 November 2021.

The 3rd of November was mostly quite with the minimum dst value of −20 nT as compared to the 4th of November which was disturbed with the minimum dst value of −125 nT as can be seen from the top panel of Fig. 1. The effect of the storm is evident on the computed dose rate for the 4th of November 2021 with the effective dose rate ranging between ~ 3.5 μSv/h and 4.8 μSv/h as compared to the 3rd of November 2021 where the dose was ranging between ~ 1.3 and ~ 2.5 μSv/h.

Geomagnetic disturbances have the potential to change the levels of radiation at around ~ 11 km altitude [20]. The computed results for other days within this period are almost the same depicting minimal changes in terms of the radiation environment at this selected route (3, 5, 6, 7 and 8 November 2021). This agrees with what is shown in

Fig. 1 (top panel) which shows the storm disturbance index for these days which varies from quiet to unsettled, or during the recovery phase of the storm.

The top right panel shows the computed effective dose for 8–13 May 2024. From this plot it is evident that the results show no effect of the space weather events for this period.

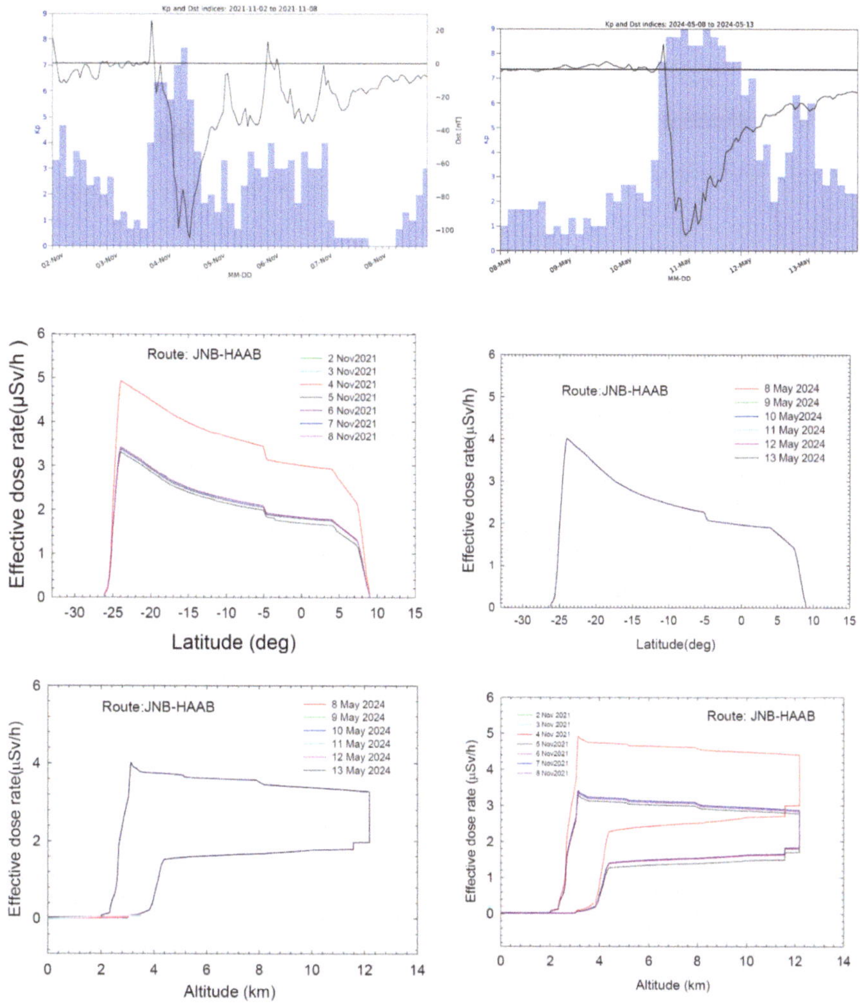

Fig. 2. Subplots showing Dst and Kp indices for the 2–8 Nov 2021(top left panel) and 8–13 May 2024 top right panel. The Kp index is illustrated as a blue bar graph with the values ranging between 1 > = Kp < 2 indicating the quiet, and Kp = 3 active, Kp = 4 unsettled and Kp > 4 stormy periods, respectively, and the Dst index as a black line. The middle panels show the computed effective dose rate as a function of latitude computed for the route between JNB-HAAB and bottom panel is the effective dose as a function of altitude for the same route for the period of 2–8 Nov 2021 and 8–13 May 2024 respectively. The results are computed by CARI7 model.

The bottom panels of Fig. 3 show the effective dose rate as a function of altitude during ascent and descent for both storm periods. Results for both selected periods show similar trends, where effective dose increases with altitude until maximum effective dose is reached at a cruising altitude of 11.5 km with the highest effective dose computed on the 4th of November 2021. Thereafter, the dose decreases as a function of decreasing altitude as the flight starts to descend until it reaches the ground at 0 km. Similar trend of the effective dose for the period of 2- 8 November 2021 is evident here and for the 8–13 May 2024 there is no variation in effective dose for these different dates as seen in both top right and bottom right panels.

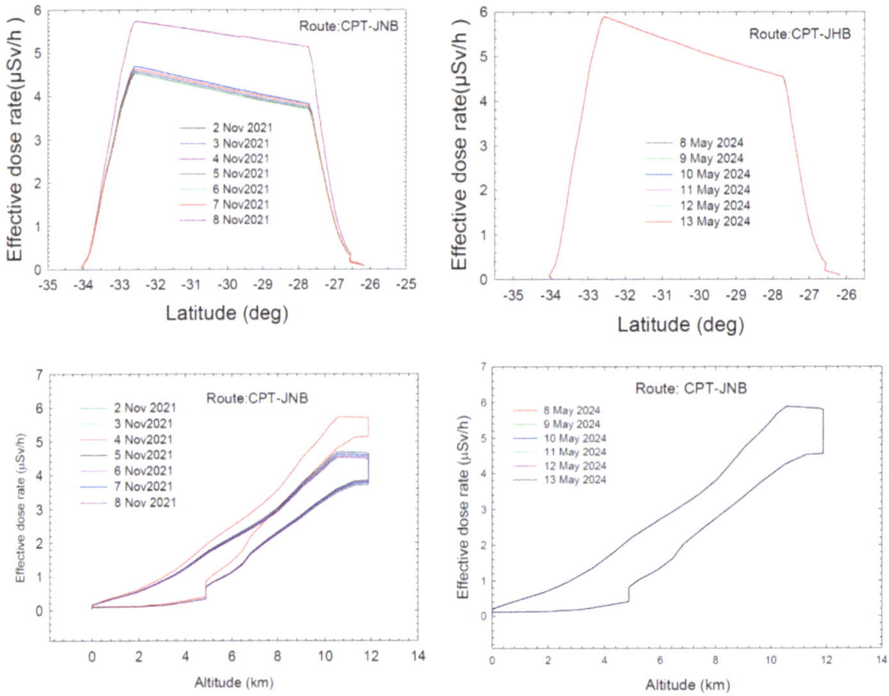

Fig. 3. Subplots showing effective dose rate as a function of latitude (top panels) computed for the route between (CPT-JNB) and the bottom is the effective dose as a function of altitude for the same route for the period of 2–8 Nov 2021 and 8–13 May 2024 respectively. The results are computed by CARI7 model.

4 Conclusion

In this study the latitude dependence of radiation dose was computed based on the CARI 7 model. This was done for the period of 2–6 November 2021 and 8–13 May 2024 for the route between CPT-JNB and JNB-HAAB. We find that the effective dose rate is much more pronounced at mid-latitude flight than at equatorial latitude which is emphasizing the known fact about the strongly present geomagnetic shielding at the equator.

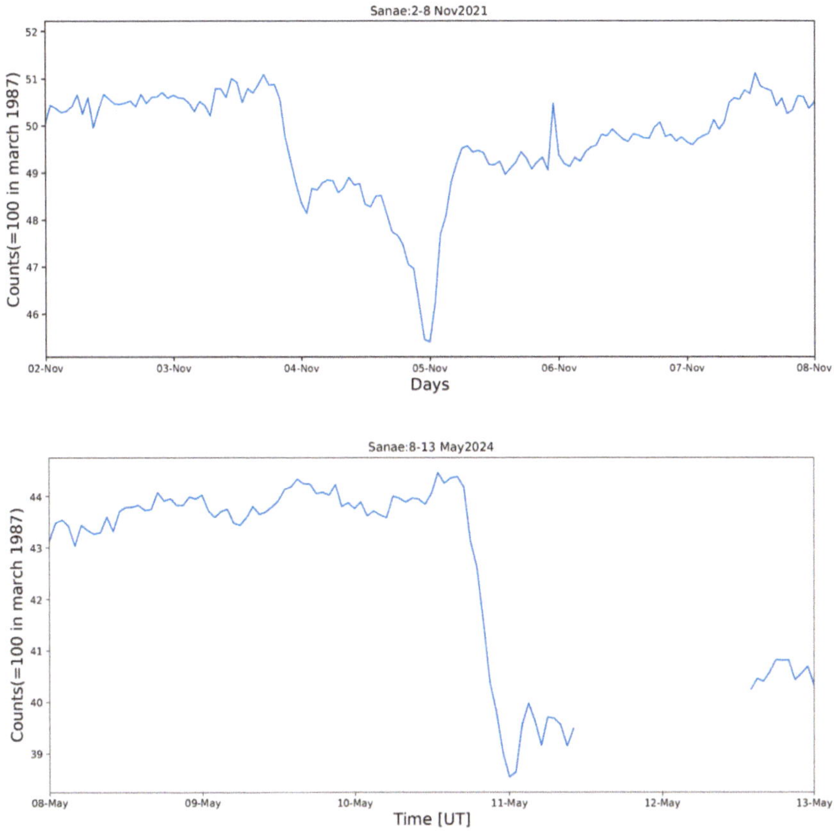

Fig. 4. Sanae neutron monitor count rate for 2–8 November 2021(top panel) and 8–13 May 2024. Both the storm period indicates a Forbush decrease in the observed counts rates. Data was obtained from: https://natural-sciences.nwu.ac.za/neutron-monitor-data.

The results show that the dose is higher at midlatitude flight (CPT-JNB) than at equatorial latitude flight (JNB-HAAB) during storm time. In general, for the route between CPT-JNB, the computed effective dose rate is estimated to be between ~ 4.5 and ~ 5.8 μSV/h whereas for the JNB-HAAB the estimated effective dose rate is between ~ 3.5 and ~ 4.8 μSV/h. Further analysis using different available models is necessary for the same study period to make any solid conclusions. If other models give similar results, then one can with confidence conclude the doses on these specific routes. There is still more that needs to be considered to fully understand the changes due to space weather events that contribute to changes in the radiation environment, especially the changes that occur due to SEPs events. The importance of real-time onboard measurements especially during the GLE event will improve the knowledge significantly in terms of understanding the changes that are brought about by this space weather phenomena and indicate what the expected value of the radiation dose are at these latitudes. Additionally, the effects of different solar storms contribute to making changes in the radiation environment, especially at the aviation altitude. Complementary to these the inflight radiation data will play a crucial role in terms of giving the true picture of what these values are supposed to be especially during the (GLE) events and what long term health implications will these have on the crew members, passengers and avionics.

Acknowledgements. The author would like to acknowledge the United Nations Office for Outer Space Affairs (UNOOSA) and ISWI for their financial support to enable me to participate in this workshop. The Dst index data used in this study were obtained from http://wdc.kugi.kyoto-u.ac.jp/index.html operated by Data Analysis Center for Geomagnetism and Space Magnetism. The Kp index data was obtained from the GFZ German Research Centre for Geosciences, at https://www.gfz-potsdam.de/en/kp-index/. The radiation dose data was generated using the FAA CARI7 model https://www.faa.gov/data_research/research/med_humanfacs/aeromedical/radiobiology/cari7. I thank Dr. Matamba and Ms. Murovhi for their contribution.

References

1. Harrison, J.: Overview of ICRP Committee 2 'doses from radiation exposure.' Ann. ICRP **44**(1_suppl), 15–23 (2015)
2. Rees, C.T., Ryden, K.A., Hands, A.D.P., Clewer, B.: Radiation risk assessment for varying space weather conditions for very high altitude 'near space' tourism balloon flights. J. Space Saf. Eng. **10**(2), 197–207 (2023)
3. Goldfinch, E.P.: Publication of ICRP 60 and 61. Int. J. Radioact. Mater. Transp. **1**(3), 143–144 (1990)
4. Meier, M.M., et al.: Radiation in the atmosphere—A hazard to aviation safety? Atmosphere **11**(12), 1358 (2020)
5. Goldhagen, P., Clem, J.M., Wilson, J.W.: The energy spectrum of cosmic-ray induced neutrons measured on an airplane over a wide range of altitude and latitude. Radiat. Prot. Dosim. **110**(1–4), 387–392 (2004)
6. Tobiska, W.K., et al.: Advances in atmospheric radiation measurements and modeling needed to improve air safety. Space Weather **13**(4), 202–210 (2015)
7. Matthiä, D., Berger, T., Mrigakshi, A.I., Reitz, G.: A ready-to-use galactic cosmic ray model. Adv. Space Res. **51**(3), 329–338 (2013)
8. Saito, S., Wickramasinghe, N.K., Sato, T., Shiota, D.: Estimate of economic impact of atmospheric radiation storm associated with solar energetic particle events on aircraft operations. Earth Planets Space **73**, 1–10 (2021)
9. Smart, D.F., Shea, M.A., Tylka, A.J., Boberg, P.R.: A geomagnetic cutoff rigidity interpolation tool: accuracy verification and application to space weather. Adv. Space Res. **37**(6), 1206–1217 (2006)
10. O'Sullivan, D., et al.: Investigation of radiation fields at aircraft altitudes. Radiat. Prot. Dosim. **92**(1–3), 195–197 (2000)
11. Bain, H.M., et al.: Improved space weather observations and modeling for aviation radiation. Front. Astron. Space Sci. **10**, 1149014 (2023)
12. https://flightaware.com/, last accessed 16 June 2024
13. CARI-7 and CARI-7A | Federal Aviation Administration. https://www.faa.gov/data_research/research/med_humanfacs/aeromedical/radiobiology/cari7
14. Copeland, K.: CARI-7A: development and validation. Radiat. Prot. Dosim. **175**(4), 419–431 (2017)
15. Copeland, K.: CARI-7 Documentation: Particle Spectra (No. DOT/FAA/AM-21/04). United States. Department of Transportation. Federal Aviation Administration. Office of Aviation. Civil Aerospace Medical Institute (2021)

16. Copeland, K.: Recent and planned developments in the CARI Program (No. DOT/FAA/AM-13/6). United States, Office of Aerospace Medicine (2013)
17. https://wdc.kugi.kyoto-u.ac.jp/dstae/wwwtmp/WWW_dstae02069206.dat
18. https://kp.gfz-potsdam.de/en. last accessed 16 June 2024
19. Larsen, N., Mishev, U.: Investigating the relationship between GLE neutron monitor data and radiation dose at flight altitudes. In: 38th International Cosmic Ray Conference (ICRC2023), 1–8, Proceedings of Science, Nagoya, Japan (2023)
20. Mishev, A., Tuohino, S., Usoskin, I.: Neutron monitor count rate increase as a proxy for dose rate assessment at aviation altitudes during GLEs. J. Space Weather Space Clim. **8**, A46 (2018)

Open Access This chapter is licensed under the terms of the Creative Commons Attribution 4.0 International License (http://creativecommons.org/licenses/by/4.0/), which permits use, sharing, adaptation, distribution and reproduction in any medium or format, as long as you give appropriate credit to the original author(s) and the source, provide a link to the Creative Commons license and indicate if changes were made.

The images or other third party material in this chapter are included in the chapter's Creative Commons license, unless indicated otherwise in a credit line to the material. If material is not included in the chapter's Creative Commons license and your intended use is not permitted by statutory regulation or exceeds the permitted use, you will need to obtain permission directly from the copyright holder.

On the Diurnal Variation of Doppler Frequency Shift of Trans-Ionospheric Radio Waves Within Equatorial Ionospheric Anomaly Region in Africa

Babatunde Rabiu[1,2,3](✉), Aderonke Akerele[1,3,4], Daniel Okoh[1,2,4], Anton Kascheyev[5], Bruno Nava[6], Elijah Oyeyemi[7], and Busola Olugbon[7]

[1] National Space Research and Development Agency, Abuja, Nigeria
tunderabiu2@gmail.com
[2] Institute of Space Science and Engineering, African University of Science andTechnology, Abuja, Nigeria
[3] Department of Physics and Solar Energy, Bowen University, Iwo, Nigeria
[4] United Nations African Regional Centre for Space Science and Technology Education - English, Ile Ife, Nigeria
[5] Physics Department, University of New Brunswick, Fredericton, Canada
[6] The Abdus Salam International Centre for Theoretical Physics, Trieste, Italy
[7] Department of Physics, University of Lagos, Lagos, Nigeria

Abstract. This paper examined the diurnal variation of the doppler frequency shift (DFS) of a trans-ionospheric high frequency (HF) radio wave signal within the equatorial ionospheric anomaly region. Within the framework of the International Space Weather Initiative, a multi-institutional cooperation set up a HF transmitter and receiver systems at the Space Environment Research Laboratory, Abuja, Nigeria (Geographic: 7.39°E, 8.99°N; dip latitude – 1.37°) and the Department of Physics, University of Lagos, Nigeria (Geographic: 3.27°E, 6.48°N; dip latitude – 1.72°) respectively. Medium-term digital recording of waveforms of trans-ionospheric HF radio signals were obtained from the experimental observation to investigate the diurnal variation of DFS along Abuja – Lagos radio path between 2019 and 2022. The DFS exhibited diurnal variation with maximum at sunrise (5:00–7:00 UT) and sunset (17:00–18:00 UT) periods, as well as day-to-day variability which was attributed to day-to-day variability in the dynamics of flow of plasma in the Equatorial Ionization Anomaly region, meridional neutral winds and local atmospheric conditions in the thermosphere. A large spread of amplitudinal variations of DFS was observed in post Sunset and post-midnight period, while the observed small variations of DFS in midday and afternoon hours is attributed to the displacement of reflection point downward.

Keywords: equatorial ionosphere · Doppler Frequency Shift · High Frequency

© The Author(s) 2026
N. Gopalswamy et al. (Eds.): ISWI 2024, SPPHY 431, pp. 164–170, 2026.
https://doi.org/10.1007/978-981-95-1121-1_18

1 Introduction

The ionosphere is a very important medium due to its role in radio communication. It has continued to attract interest due to increasing patronage of modern space-dependent technologies that dominate today's global market. Some notable effects of the dynamics of the ionosphere on trans-ionospheric radio signals include attenuation, refraction, polarization, or even scintillation [1]. Diverse experimental techniques are being applied to probe the ionosphere and its impact of trans-ionospheric radio waves.

Doppler frequency shift (DFS) measurements have been an age long technique for monitoring the Ionosphere. [2] noted that the measurements of Doppler frequency shifts reveal short-term variations in ionization of the ionosphere rather than time-averaged values. These short-term variations have critical implications in propagation of High Frequency (HF) radio waves through the medium. DFS observations are particularly useful for detecting/monitoring travelling ionospheric disturbances/short-period gravity waves (GWs) in the ionosphere. A number of studies, such as [3] and [4] have investigated the variability of the ionosphere at middle latitudes using doppler measurements. [5] studied the Doppler frequency shift (DFS) variations of middle-latitude radio line signals during different seasons and for different geomagnetic conditions.

This present work investigated the diurnal variation of the DFS of a trans-ionospheric radio signal along Abuja-Lagos radio path, within African equatorial region, using a radio propagation technique whose equipment were deployed within the framework of the International Space Weather Initiative.

2 Methodology

The HF Doppler system engaged in this research consists of a transmitter located at Abuja (ABU; geographic coordinates: 7.39° E, 8.99°N; dip latitude: −1.37°) and a receiver stationed in Lagos (LAG; geographic coordinates: 3.27°E, 6.48°N; dip latitude: −1.72°). The system operates at a frequency of 6.957 MHz. The transceiver in Abuja consists of an ICOM 718, a stable reference oscillator (OCXO 32 MHz), a terminated folded dipole antenna (Diamond WD330), a Raspberry Pi 3B (Rpi), and a PC monitor, while the receiver in Lagos consists of a digital receiver (WiNRADiO WR-G313i), a stable external reference oscillator, an active loop antenna, and a personal computer. More details about the experimental setup and derivation of DFS can be found in [6] and [1] respectively.

The daily spectrogram showing the frequency as against time of the day and the corresponding values of the DFS were obtained for each of the days of the years under study. The hourly values of DFS were obtained by averaging the values of DFS at every 15 min interval within a particular hour. Thus, daily hourly values were obtained for all the days of observation within the period of study (2019–2022). Appropriate code was developed using Matlab® programming language, to obtain the contour plots of the daily hourly variations with days of the year for each of the years 2019, 2020, 2021 and 2022.

3 Results and Discussion

Figure 1 presents typical daily spectrogram and the corresponding values of the DFS for the selected days of 28th September 2019 (A slight magnetically disturbed day with Kp ≤ 5 and Ap of 12). And 26th February 2022 (A quiet day with Kp ≤ 1 and Ap of 1). The Ap and Kp indices were obtained from the "Geomagnetic Data Service" of the World Data Centre, Kyoto.

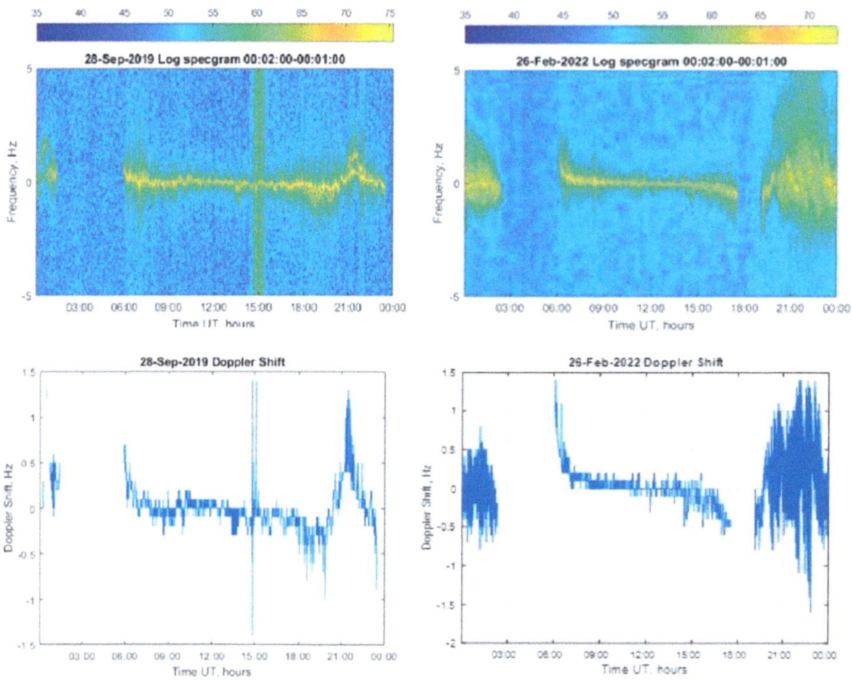

(a) 28 September 2019 (b) 26 February 2022

Fig. 1. Spectrogram with Doppler shift frequency on selected days

Doppler Frequency shift (DFS) undergoes consistent diurnal variation across the days of the year as seen in Fig. 1 for the selected days on 28 September 2019 and 26 February 2022. It is observed that on daily basis, the DFS of ionospheric signals on quiet days in daytime is of the order of 0.4 Hz between 8:00 UT and 18:00 UT. On the individual quiet days, large amplitudinal variations of DFS are observed during sunrise (5:00–7:00 UT corresponds to 6:00–8:00 LT) and sunset (17:00–18:00 UT corresponding to 18:00–19:00 LT). This is observed on all the regular quiet days exemplified. These sunrise and sunset effects are also observed in Fig. 2 which shows the contour plots of the hourly DFS for each of the years of 2019–2022. It is also observed that the DFS variation during the sunrise appear to be larger than that of the Sunset period. This may be due to the fact that the ionospheric processes responsible for DFS are stronger at the sunrise time than the sunset time, obviously due to the photo-ionization effect that seems to decline as the Sun sets. [7] reported a sudden increase in ionization during sunrise due to increasing thermospheric temperatures.

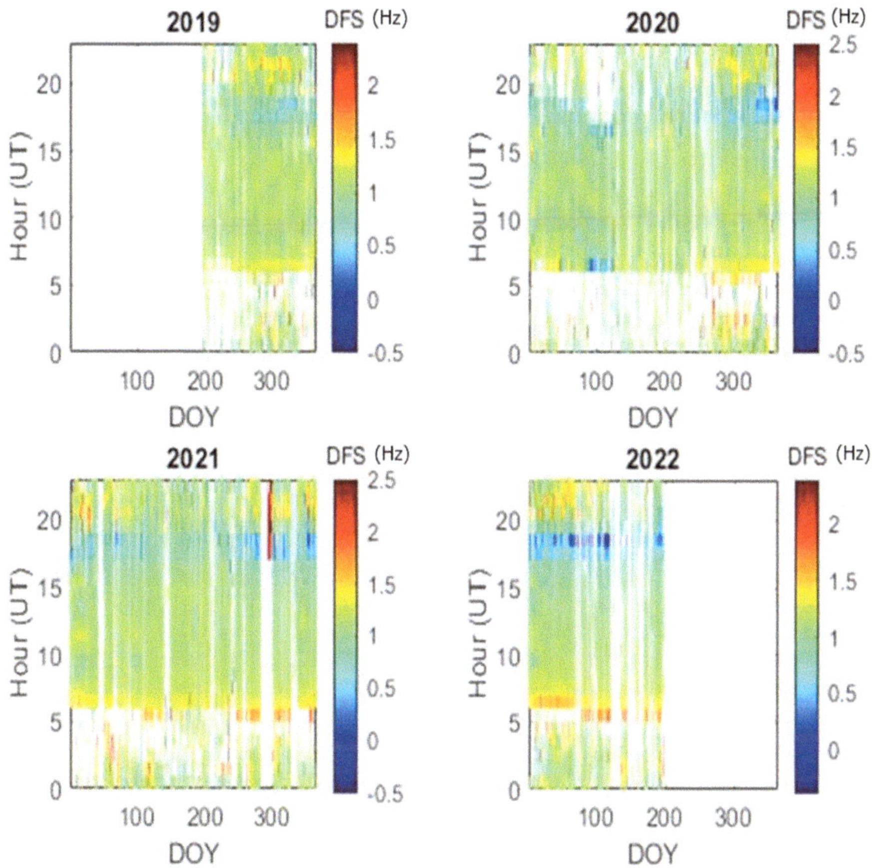

Fig. 2. Contour plots of the hourly values of DFS for years 2019, 2020, 2021, 2022 (White patches show where data were not available)

The rapid variation in DFS observed at the sunrise and sunset period as monitored in Lagos has also been reported by [5] along the Moscow-Kazan radio path. This effect is attributed to the rapid changes in the atmospheric current distribution during the sunrise and sunset periods. The redistribution of ionospheric concentration has direct influence on the high frequency radio wave propagation.

The large amplitudinal variation observed in DFS during sunrise and sunset periods is explicable by the electrodynamics of the equatorial ionosphere during these periods. At the magnetic equatorial region, the sunset period has been observed to be characterized by an enhancement of the vertical E × B drift due to the eastward electric field at the evening terminator, a phenomenon known as 'Pre-reversal enhancement (PRE)' [8–13]. The enhancement in vertical drift during the post sunset period is explained in terms of the polarization fields produced by the large westward gradient of the ionospheric conductivity across the evening terminator [12].

Also, [12] explained what happened around sunrise terminator in equatorial ionosphere, when the plasma polarizes to produce an electric field that is downward, then upward, on each side of sunrise terminator. 'These electric fields, when mapped to the E-region, produce Hall currents convergent at the terminator boundary. The plasma polarizes, preventing a build-up of positive charge at sunrise terminator, and these polarization fields produce downward, then upward drifts on each side of sunrise terminator. The observed drift profile, downward enhancement followed by upward reversal, agrees with these suggestions pointing to the existence of a morning phenomenon complementary to PRE'. Thus, the equatorial ionosphere during sunrise period experiences downward enhancement of vertical plasma drift, during which, according to [12], many physical processes like photo-ionization, chemical loss, diffusion, electrodynamic drift, coupling between E- and F-regions of the ionosphere among others come together into action.

Furthermore, a large spread of amplitudinal variations of DFS is obviously observed in post Sunset and post-midnight according to Figs. 1 and 2. This observed large spread can be attributed to the complete absence of D region and the modification of the ionospheric structure in such a manner that it affects the propagating radio wave signal. [14] observed an increase in amplitude of DFS variation during sunrise and near the evening terminator, together with a low value of DFS in daytime hours. They explained the observed diurnal variation in DFS along the path of observation using the following statements:

1. The large amplitudinal variation observed during sunrise and sunset could be due to the approach of the maximum usable frequency of propagation or probably due to the increase in the amplitude of disturbances during the passage of the evening terminator.
2. The small variations of DFS in midday and afternoon hours (daytime) is attributed to the displacement of reflection point downward towards the earth's surface caused by an increase in the electron concentration in the ionosphere.

Figures 1 and 2 clearly show that the DFS experiences more amplitudinal variation in the nighttime than daytime. Earlier works by [15] and [16] have attributed the daytime variability in the ionospheric condition to electric field strength that induces vertical drift and subsequent ionization anomalies; whereas nighttime variability may result from the Rayleigh–Taylor plasma instability, alongside variabilities in thermospheric neutral winds and temperature.

4 Conclusion

The analysis of the variability of the Doppler frequency shift of ionospheric radio signal within the equatorial anomaly region of Nigeria using HF Doppler sounder installed in Abuja with the receiver in Lagos has been carried out. The DFS exhibits diurnal variation with greater amplitudes at sunrise (5:00–7:00 UT) and sunset (17:00–18:00 UT) periods. A large spread of amplitudinal variations of DFS was observed in post

Sunset and post-midnight period. The small variations of DFS in midday and afternoon hours (daytime) is attributed to the displacement of reflection point downward; while the large amplitudinal variation observed during sunrise and sunset could either be due to the approach of the maximum usable frequency of propagation or the increase in the amplitude of disturbances during the passage of the evening terminator.

Acknowledgements. The HF radio data used in this work were obtained from the HF Doppler instrument jointly deployed and operated by the Space Environment Research Laboratory, Nigeria; University of Lagos, Nigeria; Institute for Radio Astronomy, National Academy of Science, Ukraine; and Telecommunication and Information and Communication Technology for Development (T/ICT4D) Lab of the Abdus Salam International Centre for Theoretical Physics (ICTP), Italy. The Ap and kp indices used in this paper/presentation was provided by the WDC for Geomagnetism, Kyoto (http://wdc.kugi.kyoto-u.ac.jp/wdc/Sec3.html). The authors appreciate UNOOSA and International Space Weather Initiative for the support to participate in the 2024 UN/German Workshop on ISWI and to publish this paper. BR would like to acknowledge support from the ICTP through the Associates Programme (2023–2028).

References

1. Akerele, A., et al.: Complexity and nonlinear dependence of ionospheric electron content and doppler frequency shifts in propagating HF radio signals within equatorial regions. Atmosphere **15**(6), 654 (2024)
2. Boldovskaya, I.: Doppler frequency shifts of radio waves reflected by parabolic and quasi-parabolic ionospheric layers. J. Atmos. Terr. Phys. **44**(4), 305–311 (1982)
3. Liu, J., Chiu, C., Lin, C.: The solar flare radiation responsible for sudden frequency deviation and geomagnetic fluctuation. J. Geophys. Res. Space Physics **101**(A5), 10855–10862 (1996)
4. Petrova, I.R., Bochkarev, V.V., Latipov, R.R.: Application of HF Doppler measurements for the investigation of internal atmospheric waves in the ionosphere. Adv. Space Res. **44**(6), 685–692 (2009)
5. Petrova, I.R., Bochkarev, V.V., Teplov, V., Sherstyukov, O.N.: The daily variations of Doppler frequency shift of ionospheric signal on middle-latitude radio lines. Adv. Space Res. **40**(6), 825–834 (2007)
6. Olugbon, B., et al.: Daytime equatorial spread f-like irregularities detected by HF doppler receiver and digisonde. Space Weather **19**(4), e2020SW002676 (2021). https://doi.org/10.1029/2020SW002676
7. Oron, S., D'ujanga, F.M., Ssenyonga, T.J.: Ionospheric TEC variations during the ascending solar activity phase at an equatorial station, Uganda. IJRSP **42**(1) (2013)
8. Farley, D.T., Bonelli, E., Fejer, B.G., Larsen, M.F.: The prereversal enhancement of the zonal electric field in the equatorial ionosphere. J. Geophys. Res. **91**(A12), 13723–13728 (1986)
9. Kelley, M.C.: The Earth's Ionosphere. Plasma Physics and Electrodynamics. International Geophysics Series, Academic Press, San Diego, CA (1989)
10. Fejer, B.G., de Paula, E.R., Gonzales, C.A., Kelley, M.C., Woodman, R.F.: Average vertical and zonal F-region plasma drifts over Jicamarca. J. Geophys. Res. **96**, 13901–13906 (1991)
11. Nayar, S., Prabhakaran, R., Sreehari, C. V.: Investigation of height gradient in vertical plasma drift at equatorial ionosphere using multi-frequency HF Doppler radar. J. Geophys. Res., 109(A12308) (2004)
12. Nayar, S.R.P., et al.: Electrodynamics of the equatorial F-region ionosphere during pre-sunrise period. Ann. Geophys. **27**, 107–111 (2009)

13. Adebesin, B.O., Rabiu, A.B., Adeniyi, J.O., Amory-Mazaudier, C.: Nighttime morphology of vertical plasma drifts at Ouagadougou during different seasons and phases of sunspot cycles 20–22. J. Geophys. Res. Space Physics **120**(11), 10020–10038 (2015)
14. Reznychenko, A.I., Koloskov, A.V., Sopin, A.O., Yampolski, Y.M.: Statistic of seasonal and diurnal variations of doppler frequency shift of HF signals at mid-latitude radio path. Radio Phys. Radio Astron. **25**(2), 118–135 (2020)
15. Rishbeth, H., Mendillo, M., Wroten, J., Roble, R.G.: Day-by-day modelling of the ionospheric F2-layer for year 2002. J. Atmos. Solar Terr. Phys. **71**(8), 848–856 (2009)
16. Rastogi, R.G., Chandra, H.: Response of equatorial ionosphere during the super geomagnetic storm of April 2000. Indian J. Radio & Space Phys. **45**(2), 67–78 (2016)

Open Access This chapter is licensed under the terms of the Creative Commons Attribution 4.0 International License (http://creativecommons.org/licenses/by/4.0/), which permits use, sharing, adaptation, distribution and reproduction in any medium or format, as long as you give appropriate credit to the original author(s) and the source, provide a link to the Creative Commons license and indicate if changes were made.

The images or other third party material in this chapter are included in the chapter's Creative Commons license, unless indicated otherwise in a credit line to the material. If material is not included in the chapter's Creative Commons license and your intended use is not permitted by statutory regulation or exceeds the permitted use, you will need to obtain permission directly from the copyright holder.

A Glance on the Monitoring Network for Evil Waveform and Ionospheric Characterization H037-MoNEWIC

David Wenzel(✉), Youssef Tagargouste, Martin Kriegel, and Volker Wilken

Deutsches Zentrum für Luft- und Raumfahrt e.V. (German Aerospace Center – DLR), Kalkhorstweg 53, 17235 Neustrelitz, Germany
david.wenzel@dlr.de

Abstract. This paper introduces the Monitoring Network for Evil Waveform and Ionospheric Characterization (MoNEWIC). Detailing the design of the testbed focusses on the stations for ionospheric recording via GNSS observations. Established RINEX and SCINTEX files provide a snapshot of the conditions in the ionosphere. But for reliably recreating a certain event, high-rate binaries are in need. The respective bitgrabber device requires target-oriented configuration for not quickly exceeding limits of permanent storage. First estimates on the effects of a trigger mechanism working on a volatile buffer are presented.

Keywords: GNSS · Scintillation · Bitgrabber

1 Project Overview

The first section, after a short look back on the origins, embarks on objectives and state of the Monitoring Network for Evil Waveform and Ionospheric Characterization (MoNEWIC), and details its testbed. Section 2 then looks in more detail on one key component, the stations for recording the ionosphere. The last section is devoted to challenges in capturing the antenna signal, preferably for extreme events only, to be able to reproduce sceneries of interest for receiver development.

1.1 The Historical Background Leading to the Current Project

MONITOR. In 2010, ESA launched the project MONITOR (MONitoring of Ionosphere by innovative Techniques coordinated Observations and Resources) to improve knowledge of ionospheric variability and its impacts on the performance of GNSS signals, cp. [1]. Receivers were distributed globally to analyze total electron content (TEC) and scintillation data. The major task was to design, develop, deploy, and operate a system for ionospheric data collection, processing, and archiving. Some specific topics

© The Author(s) 2026
N. Gopalswamy et al. (Eds.): ISWI 2024, SPPHY 431, pp. 171–179, 2026.
https://doi.org/10.1007/978-981-95-1121-1_19

were temporal and spatial electron density gradients, ionospheric scintillations, better characterization of short-term events and traceability to physical causes, long-term statistics, and validation of existing models. It was experimented with different receivers and settings, but most stations provided one-minute indices S4 and SigmaPhi.

MONITOR 2. With MONITOR 2, the second phase ran from 2014 to 2016, cp. [2]. The network was united with the network of CNES in collaboration with the Agency for Aerial Navigation Safety in Africa and Madagascar (L'Agence pour la Sécurité de la Navigation aérienne en Afrique et à Madagascar – ASECNA) in the West African sector and supplemented with new receivers. Special focus was also on improving central data management and processing. There were also first tests on recording the raw antenna signal. Several partners of the project consortium from phase one were also involved in its second phase. A handful of the receivers is still active.

1.2 An Introduction to MoNEWIC

Originally proposed under the title eMONITOR, the research started in the MONITOR projects continues since late 2019 in the Horizon 2020 project MoNEWIC (Monitoring Network for Evil Waveform and Ionospheric Characterization), funded by the European Union and administrated by ESA. In the ongoing phase of the project, the emphasis is laid on technical distortions of GNSS signals that are of interest for safety-of-life applications.

Three new stations designed to monitor so-called Evil Wave Forms (EWF) are thus installed in central Europe. Moreover, there are five new extended Ionospheric Recording Stations (IRS) distributed between low and high latitudes, three of them in the equatorial region of Africa and two in northern Scandinavia. These eight stations are connected to a Central Processing Facility (CPF) and constitute the GNSS experimental receiver network within the EGNOS extension area. Heritage data gathered in the previous projects shall be transferred to the CPF. Ongoing measurements from stations deployed in earlier phases of the project, and which are still in operation, must be similarly continuously integrated. Three of the IRS are equipped with a bitgrabber, enabling the system to record the antenna signal directly. Utilizing this, one particular aim is to build up a library of example scenarios, e.g., to test receiver robustness. The testbed components are described in more detail in the following section. An impression of the evolution of the network throughout the three phases is given in the interactive map [3].

The consortium working on MoNEWIC involves several European institutions: DLR leads the project and, with support from the International Centre for Theoretical Physics (ICTP), is responsible for the development, deployment, and evaluation of the IRS. Airbus Defence and Space (ADS) designs and deploys the EWF stations, while the Netherlands' National Aerospace Laboratory (NLR) builds the CPF server system. The Polytechnic University of Catalonia (UPC) and Informatique, Electromagnetisme, Electronique, Analyse Numerique (IEEA) provide dedicated scientific data processors.

1.3 The Testbed Description

The main task of the project is to establish the testbed consisting of eight new GNSS stations connected to a central computing facility, and to integrate the surviving older stations in the capacity these allow. Currently, the last new station is in testing and shall be deployed soon. After installation and verification by DLR, the testbed shall be operated for several more years under EU control, enabling long-term scientific evaluation of the data. The integral components are as follows. Figure 1 illustrates their interplay.

CPF. The Central Processing Facility is a hardware system designed and tested by NLR, consisting of the primary multi-core server and a secondary server acting as a mirror. Data integrity and completeness was a major requirement, hence the secondary CPF acts as backup and is capable to step in if the primary should be malfunctioning. The synchronization is done automatically, while a failover switch requires some manual clearance. After operation tests at NLR, the hardware is set up at its destined location at the European Space Agency (Noordwijk/Netherlands).

The system software automatically pulls the data from the stations and schedules the MoNEWIC UPC-IonSAT Ionospheric Processors (MUIIP) for computing rate of TEC (ROTI), along-arch-TEC-rate (AATR), TEC-based I-scale, electron density gradients, and TEC patterns. Similarly, the scintillation tools of IEEA to SCORE the occurrence of events and get statistics on FADES and inter fades durations, as well as the EWF processor of ADS is invoked daily with the recent measurements.

EWF Stations. Evil waveforms are non-nominal signal distortions at the satellite that were first detected after the 1993 signal anomaly event on a GPS Block II satellite (SVN19), resulting in range errors of several meters at some airborne receivers but could not be detected by monitoring pseudo ranges at stations alone. The causes and the types of these distortions are manifold: high-power amplifier distortions, frequency selectivity of the transmission medium, satellite internal multipath. So, though not 'evil' per se, EWF are likely to happen in the presence of a satellite hardware fault. During the project, the partner ADS has set up the following three EWF stations:

- Galileo Reference Centre (Noordwijk/Netherlands, repurposed MONITOR site)
- German Aerospace Center (Oberpfaffenhofen/Germany)
- International Center for Theoretical Physics (Trieste/Italy)

The stations use a NovAtel GNSS-850 antenna connected to a frontend for the IFEN SX3 software receiver ([4]), running on a cooled in-rack computer. Continuously processing the signals of the satellites in view, logs of, e.g., complex correlator values, and code and carrier phase measurements are transmitted to the CPF for further analysis.

Ionospheric Recording Stations/Bitgrabber. Technical infrastructure like the IRS is the key in the space weather community's efforts to investigate the ionosphere via well-established indices, as the collected data is valuable for the analysis of space weather events and scintillation. Within the overall project scope, data and knowledge has been continuously growing from the early MONITOR phase. Five new IRS are installed at:

- Andøya Space Center (Andenes/Norway)
- Université Félix Houphouët-Boigny (San Pedro/Côte d'Ivoire)
- Tromsø Geophysical Observatory (Tromsø/Norway)
- Center for Atmospheric Research (Abuja/Nigeria)
- ICTP-East African Institute for Fundamental Research (Kigali/Rwanda)

of which the last-listed three are extended with a controllable bitgrabber. The configurations of the stations are described in full detail in the next sections.

The data is stored in known formats like RINEX3, and the local hosts may directly access and analyze it together with DLR and ICTP. While the GNSS measurements and indices are fully integrated into automated CPF transfer, recordings from the bitgrabber are way too large and must be transported via parcel services to DLR, where they will be checked and manually uploaded to the CPF.

Heritage Stations. A few of the previously installed MONITOR stations are still working. They are the ancestors of MoNEWIC's IRS and provide similar data, but, depending on their varying receiver specifications, with lower resolution or accuracy, and not covering the full GNSS range. The data is gathered at the original CNES FTP in France set up in MONITOR2. Acquired by the CPF, it is further processed in a limited fashion.

2 Ionospheric Recording Station

2.1 Configuration

DLR has been deploying and operating GNSS stations within another network called EVNet ([5]). Its measurements are used to produce, e.g., scintillation indices, provided in our Ionosphere Monitoring and Prediction Center ([6]). Although the IRS is not meant to be part of EVnet, it has greatly benefited from that experience. The IRS is similarly constructed with a specially fitted station laptop communicating with the receiver hardware. The core of the MoNEWIC stations is a Septentrio PolaRx5S high rate GNSS

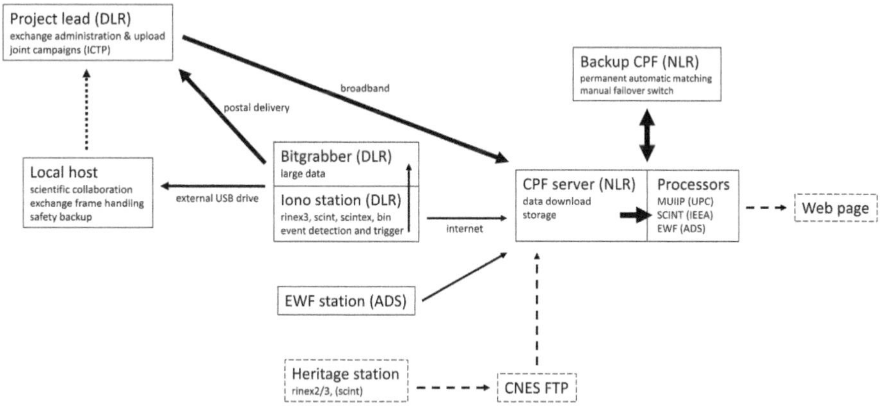

Fig. 1. The testbed of the MoNEWIC project comprises several stations connected to a central server, on which several partner tools are evaluating the measurements regularly. Most raw data will be transferred automatically via internet, but the data of the bitgrabbers turn out to be too large for this and requires manual treatment. Dashed boxes are outside the current phase.

receiver capable of generating 50 Hz raw data and computing S4 and SigmaPhi, ionospheric indices for amplitude and phase scintillation. These are stored in RINEX and SCINTEX files, collected and further processed by the CPF server. The operation of the IRS is monitored continuously internally, and automatic alerts will be raised in case of malfunctioning. The major difference is that EVnet stations use Javad receivers and the IRS's connection to the CPF is designed with an expanded range of processors. Particular attention is paid to extreme scintillation events that are determined directly in the IRS system in real time for controlling the optional bitgrabber.

2.2 Installation

The deployment of the IRS is done by DLR in close cooperation with ICTP. The equipment was configured and tested at DLR in Neustrelitz, assembled and further prepared by ICTP in Trieste for deployment, and installed by the local partner institutions at the sites with remote support. As some connections using cables of the same type could be misplaced (doing harm to the controllability of the components), we employ a combined color and naming scheme for labelling the respective cables and ports. For the simple chart implementing this and a quick look on the real hardware see Fig. 2.

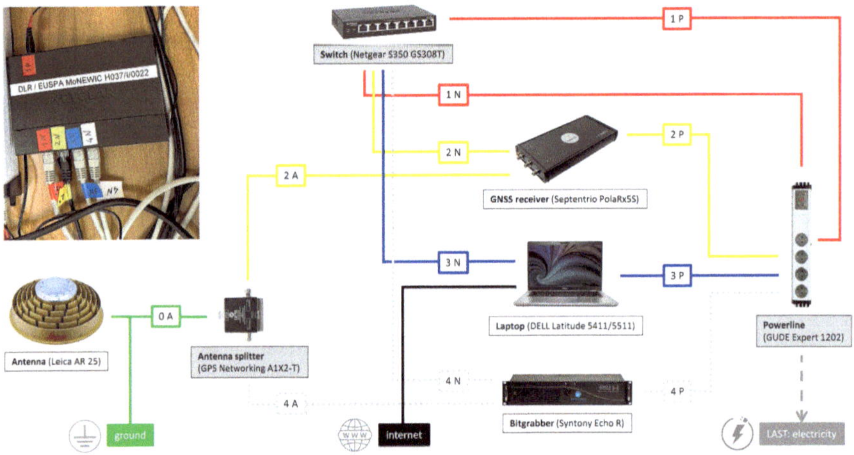

Fig. 2. The connection plan for setting up the IRS at the host locations ensures proper installation. Components and cables are prepared with colorized/named labels. For instance, antenna cables have the reference 'A' and the receiver is numbered component '2' and double coded in yellow. The bitgrabber (and splitter) is available in three selected sites: Tromsø, Abuja, and Kigali.

3 Bitgrabber Extension

3.1 Properties

A bitgrabber can record and store raw radio data in a dedicated frequency band. L band communications generate an enormous amount of data, whence the antenna signal is down-converted to a lower frequency band (e.g. baseband) to save space, and then digitized and stored to file with settable precision. Conversely, the original signal can be reconstructed from these records via up-conversion with only slight deviations. One can replay the same signal to different receivers and evaluate correction algorithms, enabling to test their behavior during solar/ionospheric events. The aim is to develop a standardized verification and certification process based on an event catalog. In addition, the data can be analyzed for investigating scintillation phenomena and their impact on signal quality, or manipulated for simulating unrecorded critical events. Such a feature is desirable in general, but only three stations could be selected under the budget.

The requirements of the project were to simultaneously observe the frequencies L1 and L5 in an optional fashion to not disturb the IRS's other duties. The basic component is a Syntony ECHO-R ([7]), with configuration possibilities and choices in Table 1. Major advantage is that there are separate devices for recording and replaying, and the latter is not required on site. Most important, the records are in an open format on exchangeable drives, easing postal data circulation. Due to the huge amounts of data produced in uninterrupted operation, the bitgrabber activity should be triggered based on the current S4 levels. A challenge was that such devices are usually GUI based, whence scripts were created for remote control (like idle/monitor switching, event trigger, disk change). Moreover, when detecting events, they are indeed already in progress. As their formation is of high interest, the system was modified to work with a RAM-based ring buffer for saving the recent past, hence being able to capture events as a whole.

Table 1. Key values of the minimal and maximal configurations of the bitgrabber system, and the chosen setting's properties in between (bold). Configurable parameters are italic. The higher the settings, the better certain effects can be replicated, but also the more increases data size. The RAM can buffer 24 GB for saving recent measurements. The disk has an overall capacity of 2 TB for permanently storing recorded data for a couple of separated intervals in time.

Channels	*Quantization*	*Bandwidth*	Data rate	Buffer (24 GB)	Record (2 TB)
1	4 bit	10 MHz	10 MB/s	40 min	55.5 h
2	**8 bit**	**10 MHz**	**40 MB/s**	**10 min**	**13.8 h**
3	16 bit	100 MHz	1200 MB/s	0.3 min	0.4 h

3.2 Parametrization of the Trigger

As Table 1 illustrates, mindless recording would exhaust the available disk capacity after just some hours. The parameters are a tradeoff between desired accuracy and the (scientifically motivated) buffer time. With the above configuration, it is possible to record 10 min back in time when high scintillation is noticed as such. However, we are limited to the actual storage space, bounding the number of recordable events. Determining whether a record is suitable for the library requires a deep analysis and cannot be done in real time or at the IRS without disturbing it. Consequently, we have to find a way for being selective from the very beginning, storing only 'occasionally' when it is promising. Otherwise the local hosts would have to switch disks too often, and there are not enough disks for circulating them every day considering transportation times.

To trigger the bitgrabber and control disk space, an accessible indicator is required. The index S4 obtained from the IRS naturally seems a good choice (or SigmaPhi when searching for phase scintillations). In order to find an adequate level, before being fully activated, the bitgrabber is run in a pretend-to-record mode for estimating consequences on the storage. First results from the Abuja station are illustrated in Fig. 3. The observed distribution of high S4 levels at any satellite shows that limiting the daily monitoring period to daytime or nighttime campaigns is an option. For judging on possible events, the more advanced scintillation index SI is even more restrictive. Nevertheless, in a single month we would register 32 event candidates, distributed unevenly over the days. With such settings, the disk would be full already in three weeks. Doing this once at the project's start would be okay, but we clearly must be even more selective afterwards. One sees that most events have just 10 min in buffer plus the configured 10 min after detection. These short events could be posed to direct deletion in favor of long events or series of events within short time. As the goal is to save a variety of events, the assumption of the frequent short events to be similar shall be tested on the first high-rate records only, justifying the future waiving of many in-depth analyses.

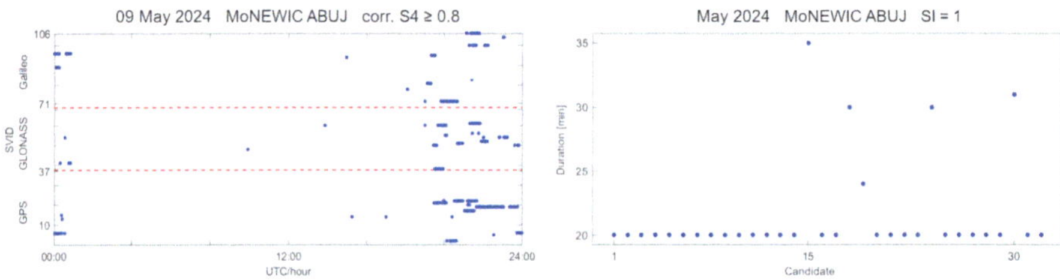

Fig. 3. Left: High S4 is rare and mostly at nights, so we can limit the active observation time. Right: May 2024 had 32 event candidates with SI = 1, but only 5 of them lasted long.

Acknowledgements. This paper is based on joint work with Bruno Nava (ICTP), Hein Zelle, Erik Baalbergen (NLR), Miguel Odriozola Olavarria (ADS), Manuel Hernández-Pajares (UPC), Yannick Béniguel (IEEA), as well as their colleagues from the working groups.

References

1. Prieto Cerdeira, R., Béniguel, Y.: The MONITOR project: architecture, data and products. In: Ionospheric Effects Symposium, Alexandria VA (2011)
2. Béniguel, Y., et al.: MONITOR Ionospheric Network: two case studies on scintillation and electron content variability. Ann. Geophys. **35**, 377–391 (2017)
3. MoNEWIC Map: https://umap.openstreetmap.fr/en/map/monewic-sites_957516. Last accessed 4 Nov 2024
4. IFEN Homepage: https://www.ifen.com/receivers/sx3-gnss-software-receiver-for-rd. Last accessed 4 Nov 2024
5. Kriegel, M., et al.: An experimentation and verification network to analyze small scale ionospheric disturbances. In: VII Brazilian Symposium on Space Geophysics and Aeronomy, Santa Maria (2018)

6. IMPC Homepage: https://impc.dlr.de/products. Last accessed 4 Nov 2024
7. Syntony Homepage: https://syntony-gnss.com/products/gnss-recorder-and-player-echo. Last accessed 4 Nov 2024

Open Access This chapter is licensed under the terms of the Creative Commons Attribution 4.0 International License (http://creativecommons.org/licenses/by/4.0/), which permits use, sharing, adaptation, distribution and reproduction in any medium or format, as long as you give appropriate credit to the original author(s) and the source, provide a link to the Creative Commons license and indicate if changes were made.

The images or other third party material in this chapter are included in the chapter's Creative Commons license, unless indicated otherwise in a credit line to the material. If material is not included in the chapter's Creative Commons license and your intended use is not permitted by statutory regulation or exceeds the permitted use, you will need to obtain permission directly from the copyright holder.

Different Plasma Bubble Dynamics over the EIA: a Comparative Study of Geomagnetic Storm Responses in Two Distinct Regions of South America

Gilda González[1(✉)], Maria Graciela Molina[2,3,4], Marcos Paz[3], and Yamila Melendi[2,3,5]

[1] Space Sciences Laboratory, University of California Berkeley, Berkeley, CA, USA
g.gonzalez@berkeley.edu

[2] Tucumán Space Weather Center (TSWC), Facultad de Ciencias Exactas y Tecnología (FACET-UNT), Tucumán 4000, Argentina

[3] Consejo Nacional de Investigaciones Científicas y Técnicas (CONICET), Tucumán, Argentina

[4] Instituto Nazionale di Geofísica e Vulcanología (INGV) Roma, Rome, Italy

[5] Departamento de Física - UNS, Bahía Blanca, Argentina

Abstract. Plasma bubbles have the potential to disrupt radio communications and satellite navigation systems. The influence of geomagnetic storms on these irregularities remains not fully understood, particularly over the Equatorial Ionization Anomaly (EIA). This research analyzes the behavior of post-sunset plasma bubbles over two regions located under the south crest of the EIA: East of Brazil and Northwest of Argentina, during an intense geomagnetic storm on November 3, 2021. We used ionosonde data, Global-scale Observations of the Limb and Disk (GOLD), and the Ionospheric Connection Explorer (ICON) Michelson Interferometer for Global High-resolution Thermospheric Imaging (MIGHTI) data. Our observations indicate significant differences in post-sunset effects between these two EIA regions. Eastern Brazil exhibited plasma bubbles before and after the storm, but these were seemingly inhibited during the storm's recovery phase when observed using space-based instrumentation. However, these structures were not observed in ionosonde data, suggesting that the irregularities were located on the ionospheric top side. In contrast, post-sunset plasma bubbles were not observed at any time during the storm over Northwest Argentina. The potential factors responsible for these differing responses in the post-sunset ionosphere between the two regions will be further investigated.

Keywords: Plasma bubbles · Geomagnetic storms · Spread-F

1 Introduction

The ionosphere is critical for several applications, including communication and navigation systems. Electron density irregularities in the F region of the

© The Author(s) 2026
N. Gopalswamy et al. (Eds.): ISWI 2024, SPPHY 431, pp. 180–187, 2026.
https://doi.org/10.1007/978-981-95-1121-1_20

ionosphere pose significant threats to sub-ionospheric and trans-ionospheric radio signals, adversely affecting telecommunications and navigation. These irregularities can degrade and disrupt GNSS signals, reducing, for instance, the number of available GNSS satellites for positioning. Thus, evaluating and predicting the risk of their occurrence is essential for developing applications and space systems. However, the ionosphere's dynamic nature makes it susceptible to various environmental and space weather factors, resulting in significant changes.

Ionospheric irregularities, or plasma bubbles, are temporal and spatial variations in electron density, ranging from a few minutes to several hours. Their occurrences depend on the season, solar cycle, latitude, and longitude. It is widely accepted that these plasma irregularities develop through the Rayleigh-Taylor instability process operating at the bottom side of the F-region [7].

2 Data and Methods

2.1 Geomagnetic Data

To analyze the **solar wind-magnetosphere coupling conditions**, we used the geomagnetic indices SYM-H, Kp, and SME. The SYM-H index was obtained from the World Data Center (WDC) for Geomagnetism, Kyoto, Japan (http://wdc.kugi.kyoto-u.ac.jp), the SME index from the SuperMAG website (https://supermag.jhuapl.edu/), and the Kp index from the OMNIWeb database (https://omniweb.gsfc.nasa.gov).

2.2 ACE Solar Wind Data

We used the North-South component (Bz) of the interplanetary magnetic field (IMF) and the zonal component (Ey) of the interplanetary electric field (IEF) to characterize the solar wind. Bz data were obtained from the Advanced Composition Explorer (ACE) mission (https://science.nasa.gov/mission/ace/), and Ey data from OMNIWeb (https://omniweb.gsfc.nasa.gov).

2.3 Global-scale Observations of the Limb and Disk (GOLD) Airglow Imaging Instrument

GOLD takes far-ultraviolet images of the Earth from geostationary orbit and measures densities and temperatures in Earth's thermosphere and ionosphere [2]. We used GOLD Dark Limb Observation Type L1D channel A version 05. These are 135.6 nm nightglow images that can be used to detect plasma bubble signatures. The data is available at https://gold.cs.ucf.edu/.

2.4 Ionospheric Connection Explorer (ICON)

The Ionospheric Connection Explorer (ICON) was launched in October 2019. It has a circular low inclination orbit (27°) and an altitude of about 575 km [5]. We used the Michelson Interferometer for Global High-Resolution Thermospheric Imaging (MIGHTI) instrument onboard ICON to measure meridional and zonal neutral wind velocity [4].

2.5 Ionosondes

We used two ionosondes: Tucumán (GEO -26.9N 294.6R) and Cachoeira Paulista (GEO-22.7N 315.0E). Each ionogram was manually examined for the presence of spread-F (Range Spread-F and Frequency Spread-F). Spread F observed on ionograms is caused by irregularities of 10 s of meters. It is called Range Spread-F (RSF) if the broadening is in height and Frequency Spread-F (FSF) if the broadening is in frequencies [6]. The ionosonde sounding time resolution was 10 min. More information about the ionosonde can be found here https://spaceweather.facet.unt.edu.ar/. For Cachoeira Paulista we used data from GIRO observatory (https://giro.uml.edu/)

3 Results

A cannibal Coronal Mass Ejection (CME) hit Earth's magnetic field on November 3, 2021, triggering an intense geomagnetic storm. This phenomenon occurs when a faster CME overtakes and engulfs slower CMEs in front of it, creating a single, massive wave of plasma. The event was well-documented in the Space Weather Prediction Center (SWPC). Figure 1 shows the Kp index, SYM-H, SME, B_z, and E_y from November 2 to 6, 2021. SYM-H recorded two minimums: -106 nT on November 4 at 8:30 UT and -111 nT at 12:30 UT. The Kp index reached a maximum of 7 on November 4 at 9:00 UT. SME increased from the beginning of the storm, peaking on November 3 at 21:48 UT (2149 nT) and November 4 at 9:14 UT (3197 nT). B_z exhibited irregular behavior, with northward and southward turnings between November 3 at 20:00 UT and November 4 at 3:00 UT, remaining southward on November 4 between 4:00 UT and 12:00 UT. E_y showed a similar pattern, peaking at 10.9 mV/m on November 4 at 8:00 UT.

Figure 2 shows that topside (approximately 500 km) postsunset plasma bubbles were inhibited over eastern Brazil on November 4 between 22:00 and 23:00 UT. Ionospheric irregularities were present on November 2, 3, and 5, but were absent on November 4 around post-sunset. This inhibition occurred during the recovery phase of the geomagnetic storm, when B_z was north and SME showed quiet time values [1,3,8].

Blanc and Richmond [9] developed the theory of the ionospheric disturbance dynamo due to Joule heating of the thermosphere in the auroral zone. Joule heating generates a circulation of thermospheric winds from the pole towards the equator and a westward circulation associated with the first meridional circulation due to the Coriolis force. This disturbance dynamo electric field (DDEF) has been shown to inhibit the pre-reversal enhancement (PRE) in equatorial electric fields, which usually leads to the rapid rise of the ionospheric F layer just before sunset, thus preventing the development of the Rayleigh-Taylor instability responsible for the development of equatorial plasma bubbles. Fejer et al. [10] produced an empirical model of the DDEF based on ROCSAT satellite data, confirming this inhibitory effect on PRE and equatorial plasma bubble formation. Recently, Kashcheyev et al. [11] used the ROTI index to demonstrate that

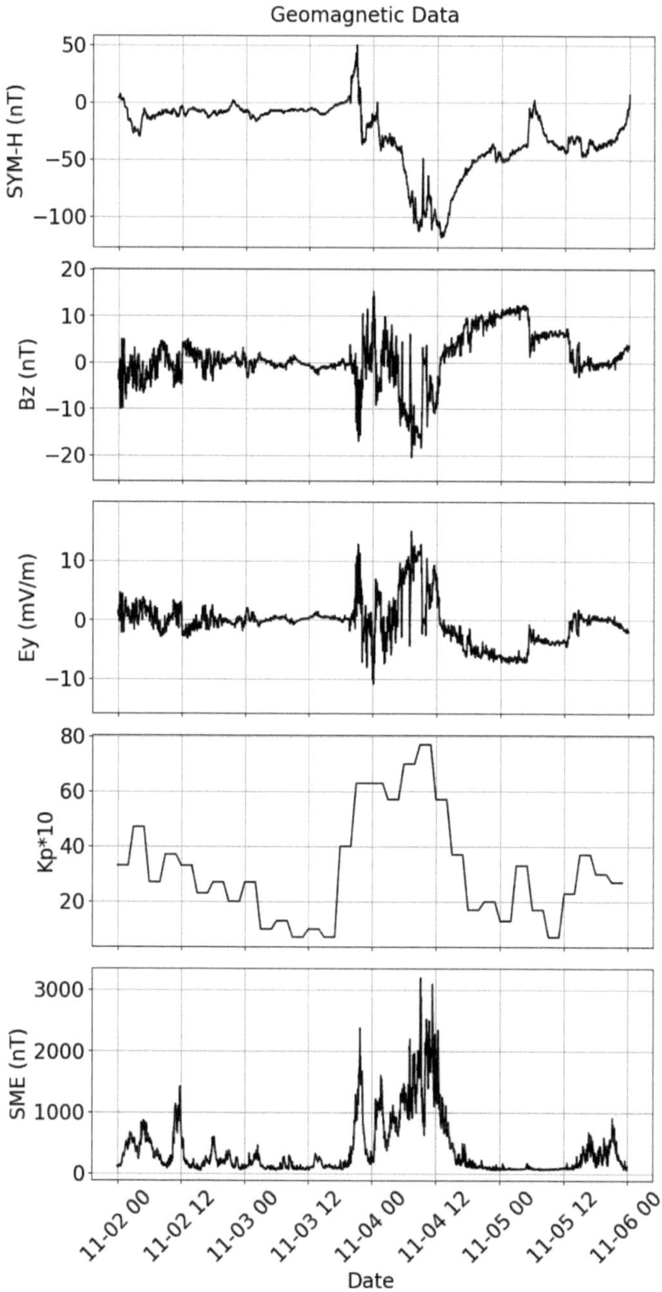

Fig. 1. SYM-H, Bz, Ey, Kp and SME for November 2 to November 6, 2021

plasma irregularities disappeared during the recovery phase of the St. Patrick's Day storm due to the DDEF effect.

The inhibition of postsunset plasma bubbles during the recovery phase could be attributed to a westward DDEF, as shown in Fig. 3. On November 4, the zonal wind at an altitude of 200-300 km turned westward, which may cause a downward $\mathbf{E} \times \mathbf{B}$ drift during postsunset. This drift could inhibit the development of the Rayleigh-Taylor instability. Additionally, no plasma bubble signatures were observed over western South America during any phase of the storm, suggesting that the response of topside plasma bubbles to geomagnetic storms differs between the eastern and western regions of South America.

To further analyze the different signatures of plasma bubbles, we used ionosonde data from two stations at the southern crest of the Equatorial Ionization Anomaly: one in the west of South America (Tucumán) 4 and another in the east (Cachoeira Paulista) 5. Both stations show no spread-F in the ionograms, indicating that plasma bubbles were not present over these ionosondes (below 350 km approx).

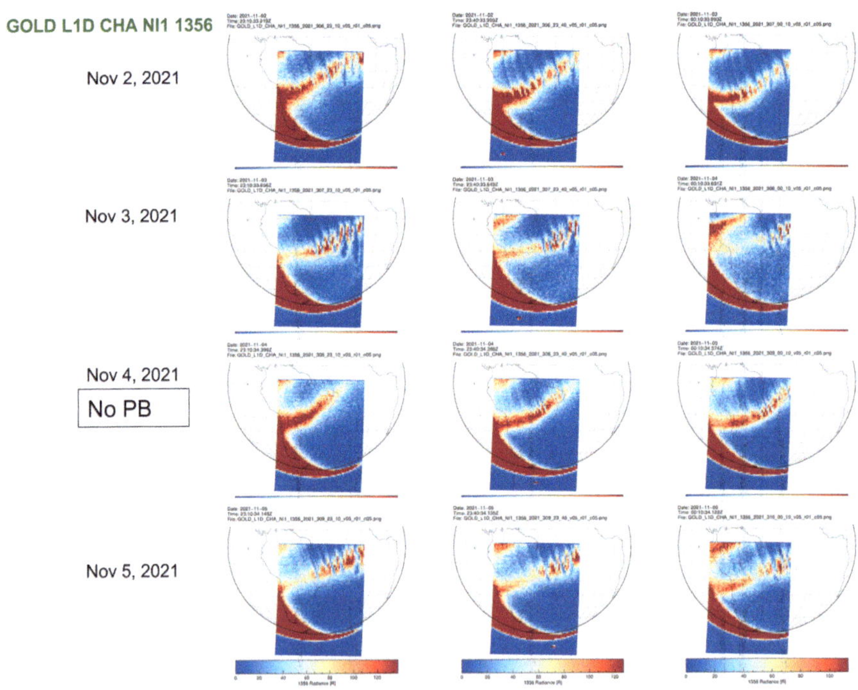

Fig. 2. GOLD LID CHA NI1 1356 for November 2 to November 5. Plasma bubbles are present on Nov 2, 3 and 5

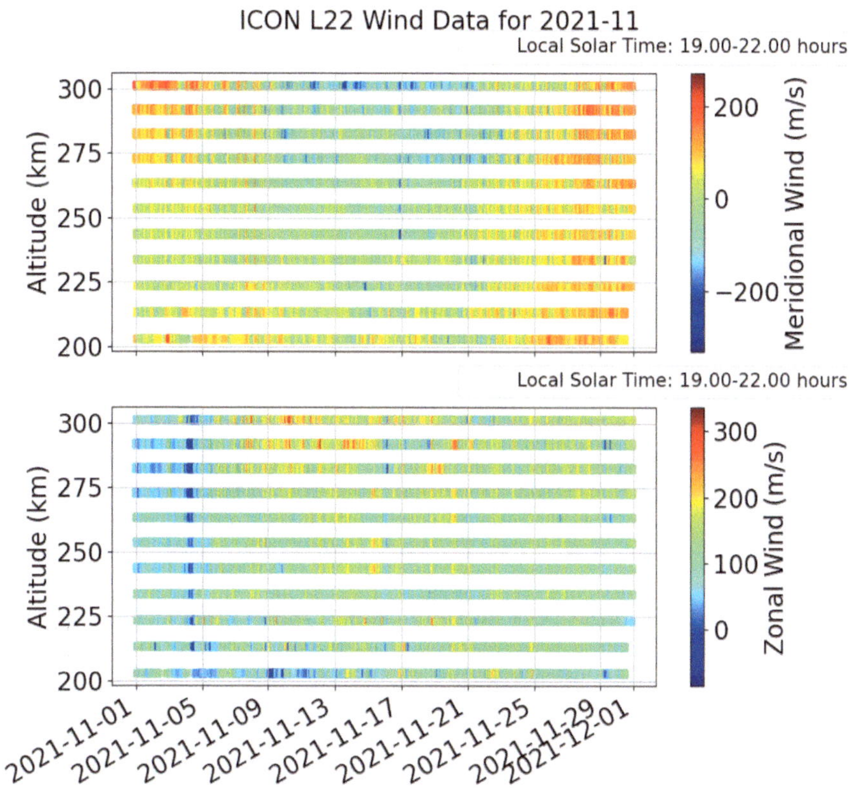

Fig. 3. ICON MIGHTI zonal and meridional Winds between 200 and 300 km altitude at postsunset during November 2021

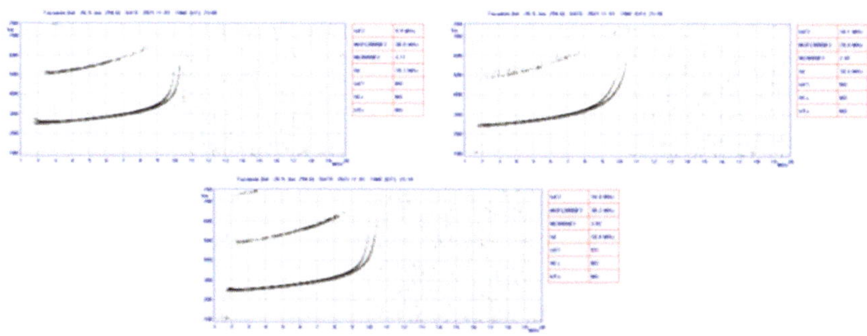

Fig. 4. Ionograms from Tucumán for November 3, 2021, around postsunset

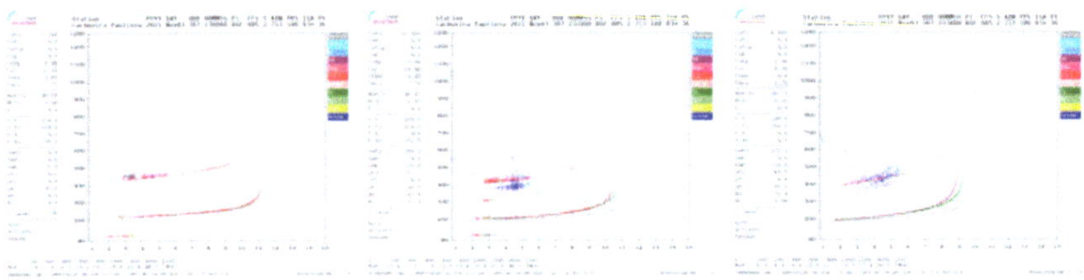

Fig. 5. Ionograms from Cachoeira Paulista for November 3, 2021, around postsunset

4 Conclusions

1 This study demonstrates the critical importance of using multiple instruments to comprehensively analyze various aspects of ionospheric irregularities.
2 Ionospheric irregularities were detected at higher altitudes (approximately 500 km, topside) but were absent at lower altitudes (around 300 km).
3 Westward disturbance winds may inhibit topside irregularities over eastern Brazil during the storm recovery phase.
4 The response of topside postsunset plasma bubbles to geomagnetic storms appears to vary between the western and eastern regions of South America.
5 Further investigation, incorporating additional instruments, is essential to fully understand this phenomenon.

Acknowledgements. The OMNI data were obtained from the GSFC/SPDF OmniWeb. GOLD data is courtesy of NASA/GOLD and the mission science team. ICON is supported by NASA's Explorers Program through contracts NNG12FA45C and NNG12FA42I. Data from the Brazilian Ionosonde Cachoeira Paulista is made available through the EMBRACE program from the National Institute for Space Research (INPE).

References

1. Abdu, M.A., Batista, I.S., Brum, C.G.M., Sobral, J.H.A., de Paula, E.R.: Planetary wave signatures in the equatorial atmosphere–ionosphere system, and mesosphere–stratosphere–troposphere coupling. J. Atmos. Solar Terr. Phys. **68**(3–5), 509–522 (2006). https://doi.org/10.1016/j.jastp.2005.02.020
2. Eastes, R.W., et al.: The global-scale observations of the limb and disk (gold) mission. Space Sci. Rev. **212**, 2383–2408 (2017). https://doi.org/10.1007/s11214-017-0459-2
3. Fejer, B.G., Scherliess, L., de Paula, E.R.: Equatorial ionospheric electric fields during the november 24, 1996 magnetic storm. Geophys. Res. Lett. **26**(16), 2601–2604 (1999). https://doi.org/10.1029/1999GL900563
4. Harding, B.J., et al.: The MIGHTI wind retrieval algorithm: description and verification. Space Sci. Rev. , pp.585–600 (2017). https://doi.org/10.1007/s11214-017-0359-3

5. Immel, T.J., et al.: The ionospheric connection explorer mission: mission goals and design. Space Sci. Rev. **214**(1), 1–36 (2017). https://doi.org/10.1007/s11214-017-0449-2
6. Piggott, W. R., Rawer, K.: URSI handbook of ionogram interpretation and reduction. Revision of chapters 1-4. Elsevier (1978)
7. Sultan, P.J.: Linear theory and modeling of the Rayleigh-Taylor instability leading to the occurrence of equatorial spread F. J. Geophys. Res. Space Physics **101**(A12), 26875–26891 (1996). https://doi.org/10.1029/96JA00682
8. Huang, C. S., Foster, J. C.: Long-duration penetration of the interplanetary electric field to the low-latitude ionosphere during the main phase of magnetic storms. J. Geophys. Res. Space Phys. 110(A11) (2005). https://doi.org/10.1029/2004JA010782
9. Blanc, M., Richmond, A.D.: The ionospheric disturbance dynamo. J. Geophys. Res. **85**(A4), 1669–1686 (1980). https://doi.org/10.1029/JA85iA04p01669
10. Fejer, B.G., et al.: Seasonal and longitudinal dependence of equatorial disturbance vertical plasma drifts. Geophys. Res. Lett. **35**, L20106 (2008). https://doi.org/10.1029/2008GL035584
11. Kashcheyev, A., et al.: Multi-variable comprehensive analysis of two great geomagnetic storms of 2015. J. Geophys. Res. Space Physics **123**,(2018). https://doi.org/10.1029/2017JA024900

Open Access This chapter is licensed under the terms of the Creative Commons Attribution 4.0 International License (http://creativecommons.org/licenses/by/4.0/), which permits use, sharing, adaptation, distribution and reproduction in any medium or format, as long as you give appropriate credit to the original author(s) and the source, provide a link to the Creative Commons license and indicate if changes were made.

The images or other third party material in this chapter are included in the chapter's Creative Commons license, unless indicated otherwise in a credit line to the material. If material is not included in the chapter's Creative Commons license and your intended use is not permitted by statutory regulation or exceeds the permitted use, you will need to obtain permission directly from the copyright holder.

Intense Spread-F Event at a Low Latitude Station and Its Possible Relationship with a Geomagnetic Storm and Substorms

M. Graciela Molina[1,2,3](✉), Yamila Melendi[1,4,5], Gallardo-Lacourt Bea[6,7], Marcos Paz[1], and Lourdes Ruesjas[1]

[1] Tucumán Space Weather Center (TSWC), Facultad de Ciencias Exactas y Tecnología (FACET), Universidad Nacional de Tucumán (UNT), Tucumán, Argentina
gmolina@herrera.unt.edu.ar

[2] Consejo Nacional de Investigaciones Científicas y Técnicas (CONICET), Buenos Aires, Argentina

[3] Istituto Nazionale Di Geofisica E Vulcanologia, Rome, Italy

[4] Departamento de Física - UNS, Bahía Blanca, Argentina

[5] Instituto de Física del Sur (CONICET-UNS), Bahía Blanca, Argentina

[6] NASA Goddard Space Flight Center, Greenbelt, MA, USA

[7] Catholic University of America, Washington, DC, USA

Abstract. In this study, we analyze an intense spread-F event observed in Tucumán (27°S; 65,5°W), a low latitude station located under the southern crest of the Equatorial Ionization Anomaly. Typically, post-sunset and post-midnight spread-F phenomena can be observed in this region. However, in this particular case, spread-F was observed throughout the entire night and into the daytime hours. In this case study, we investigate the possible linkage between the spread-F observed during March 24th, 2023, in Tucumán with a severe geomagnetic storm (Kp = 8) and several intense substorms. Continuous spread-F was observed in manually corrected ionosonde data from 01:20 UT to 06:50 UT, from 07:40 to 07:50 UT, from 08:10 to 10:50 UT, and also at daytime from 13:30 to 14:30 and from 17:10 to 17:20 UT (LT = UT-3). The onset of a geomagnetic storm (March 23rd at 08:00 UT) was observed one day prior to the spread-F reported here, the main phase peak occurred on March 24th at 02:00 UT (Dst = -163 nT), followed by a recovery phase lasting about 5 days. The spread-F event primarily occurred during both the main and the recovery phases of the geomagnetic storm, including instances of daytime spread-F. SML, SMU and SME geomagnetic indices showed several successive intense substorms on March 24th, with SME reaching up to 2100 nT. Notably, the occurrence of the intense spread-F was predominantly aligned with the expansion phase of these substorms. This study represents the first analysis using data from the aforementioned low-latitude ionospheric station. We intend to incorporate additional analysis, data, and instruments in future studies, as solar activity increases.

Keywords: Low latitude ionosphere · Substorm · Geomagnetic Storm

© The Author(s) 2026
N. Gopalswamy et al. (Eds.): ISWI 2024, SPPHY 431, pp. 188–194, 2026.
https://doi.org/10.1007/978-981-95-1121-1_21

1 Introduction

Spread F can be observed in ionograms as radio pulse echo traces spread at F2 layer heights and it occurs due to the presence of ionospheric irregularities originating at the equator. Equatorial plasma bubbles are geomagnetic field-aligned ionospheric plasma structures or irregularities occurring in spatial scales from few centimeters to hundreds of kilometers as well as in different time scales (Zolesi & Cander, 2014). Time scales can be observed as: a) long-term variability, linked mainly to the solar cycle activity, b) medium term (seasonal) variability and, c) short-term variability as day-to-day variability and other transient related to space weather events. The latter, short-term variability (from minutes to hours and several days), can be originated from magnetospheric forcing (at high latitude) and they are the most challenging ones to predict and often more interesting to forecast since many technologies (e.g. space base technologies and telecommunications) rely in the ionospheric conditions.

We investigate the possible linkage between the spread-F observed in Tucumán with a severe geomagnetic storm (Kp = 8) starting on March 23th, 2023 and several intense substorms. Many ionospheric disturbances are reported during the different phases of geomagnetic storms involving several mechanisms due to the complex coupling between the solar wind, the magnetosphere, and the high latitude ionosphere. Within the ionosphere, changes in composition, electric fields, and temperature can be observed (Molina et al., 2020). The ionosphere can also be disturbed during the occurrence of substorms. Substorms correspond to the phenomenon where energy extracted from the solar wind is intermittently stored in the magnetosphere. This stored energy is periodically released through explosive instabilities, affecting both the magnetosphere and the ionosphere (McPherron et al., 1973, 1979; Baker et al., 1997). This process corresponds to the three substorm phases: growth, expansion, and recovery.

2 Data and Methodology

We analyze data acquired by an AIS-INGV ionosonde deployed at Tucumán low latitude ionospheric station (27° S; 65,5° W) from the Tucumán Space Weather Center at FACET-UNT. This HF radar performs a vertical sounding to obtain a graphical representation (ionogram) of the virtual height of the ionospheric layers as a function of the frequency (1–20 MHz). One of the main parameters derived from ionograms is the critical frequency of the F2 layer (foF2). We manually corrected the ionograms. In addition we tagged, manually, spread F cases into three cases: a) Range spread F, where the echoes traces are spread in heigh, b) frequency spread F, where it can be observed, often, as a thick trace in the frequency axis, and c) mixed spread F (both range and frequency spread F at the same time). Examples of each type of spread F can be observed in Fig. 1. We used ionograms at a time resolution of 10 min to catalog Spread F events.

We used the interplanetary magnetic field (IMF), Bz and Bt, as well as Dst to analyze the geomagnetic storm's phases. Similarly, SML, SMU, and SME from the SuperMAG database were used to study the conditions of the auroral electrojet and substorm phases (see Fig. 2).

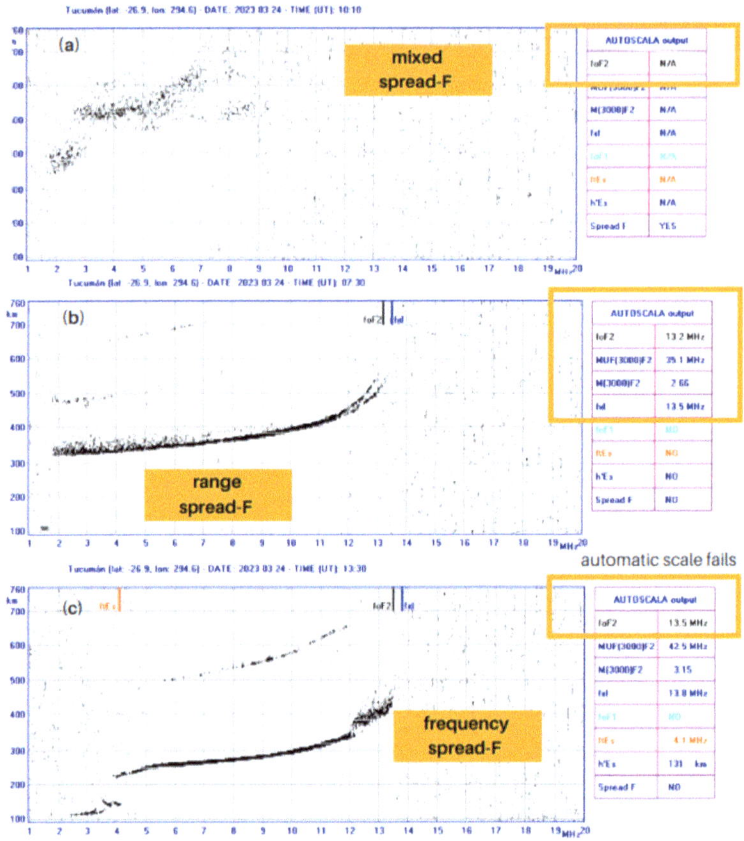

Fig. 1. Ionograms showing each type of spread F cases: (a) mixed spread F, (b) range spread F and (c) frequency spread F. Ionograms correspond to different hours on March 24th, 2023 at Tucumán (LT = UT - 3).

3 Results

The onset of a geomagnetic storm, or sudden storm commencement (SSC) (see Fig. 2), was observed one day prior to the spread-F reported here, occurring on March 23rd at 08:00 UT. During the main phase of the storm, the Dst index reached a minimum value of -163 nT (see Fig. 2, second parallel dashed orange line) on March 24th at 02:00 UT, followed by a prolonged recovery phase lasting about 5 days. In ghFig. 2 panel (b) Dst shows the development of the geomagnetic storm. SML, SMU and SME geomagnetic indices (Fig. 2 c, d, e respectively) show several successive intense substorms on March 24th, with SME reaching up to 2100 nT.

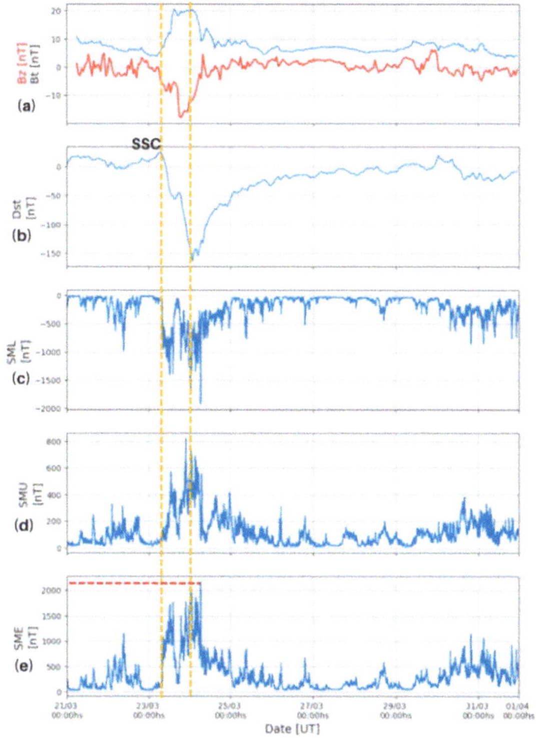

Fig. 2. Panels show: a) Bt (blue) and Bz (red); b) Dst index; c) SML index; d) SMU index; and e) SME index. The left vertical dashed line marks the SSC and the second vertical dashed line marks the peak of the main phase of the geomagnetic storm.

The spread-F event reported here occurred during March 24th, 2023, at Tucumán subequatorial ionospheric station. Typically, post-sunset and post-midnight spread-F phenomena can be observed in this region. However, in this particular case, spread-F was observed throughout the entire night and into the daytime hours. Continuous spread-F was observed in manually corrected ionosonde data from 01:20 UT to 06:50 UT (LT = UT-3). Three different types of spread-F were identified: frequency, range, and frequency + range (Fig. 1). A similar situation was observed from 07:40 to 07:50 UT, from 08:10 to 10:50 UT. Also at daytime from 13:30 to 14:30 and from 17:10 to 17:20 UT spread F is observed but only frequency spread F in these cases. Figure 3 shows the SME index during March 24th, 2023 plus the three types of spread F observed in Tucumán. It is worth mentioning that on March 23rd the ionosphere showed a quiet day with no registered perturbations.

Regarding the types of spread F, from 01:20 UT to 06:50 UT the majority of observed spread F were range spread F but also several cases of mixed spread F were identified. Most of the cases were detected at the beginning of the studied time interval (red cases in Fig. 3). From 07:40 to 07:50 UT, the majority of the spread F cases are range spread F while from 13:30 to 14:30 and from 17:10 to 17:20 UT almost all the cases belong to frequency spread F.

It is worth mentioning that, typically, ionograms are automatically scaled and when spread F events occur, the scaling may fail as in Fig. 1c where foF2 is wrongly scaled. Figure 1a and 1b show cases where the automatic scaling is correct. In the case of Fig. 1a, it is not possible to obtain the parameters automatically because the occurrence of spread F near the peak of the F2 layer meaning actually the foF2 is can not be obtained, so the interpretation is correct. While in Fig. 1b, the parameters are actually correct because the range spread F is happening near the lower frequencies not affecting the interpretation in the F layer. As a bottom line, when monitoring and analyzing ionograms it is essential both the automatic scaling but also the manual validation/correction.

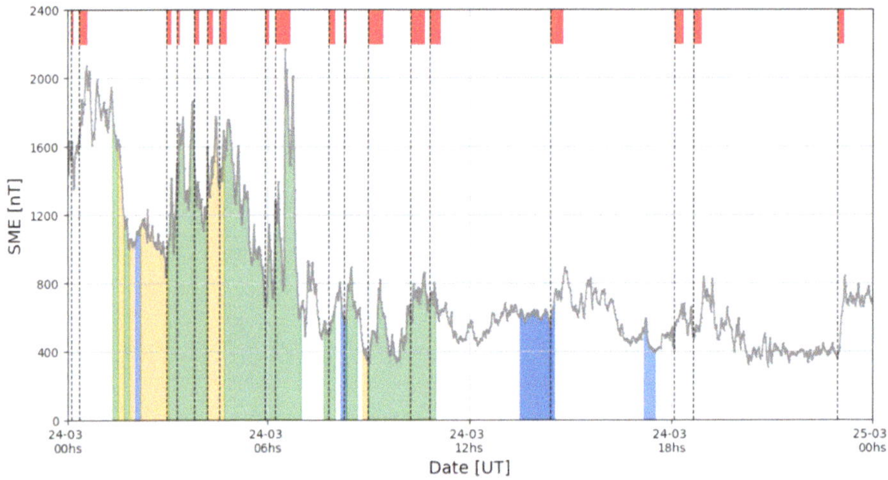

Fig. 3. SME index showing very intense substorms reaching 2100 nT on March 24th 2023. Dotted black vertical line shows a substorm onset reported by SuperMAG (Newell and Gjerloev, 2011) and has been derived from the geomagnetic indices. The pink segments represent the substorm expansion phase determined using a similar methodology as in Partamies et al. (2013). Spread F cases are also shown divided in 3 types of spread F: range spread F (green), frequency spread F (blue) and mixed (range + frequency) spread F (yellow).

In Fig. 3, the substorm onset at 02:57 UT (indicated by the black vertical dotted line) coincides with the beginning of a range spread-F event lasting until 04:10 UT. During this interval, multiple substorms occur. At 04:10 UT, a new substorm is onset, and its expansion phase aligns with another range spread-F event. The next substorm expansion phase starts at 04:32 UT, coinciding with the onset of a range spread-F that continues until approximately 07:00 UT. This ionospheric condition persists until the substorm begins its recovery phase, at which point the spread-F ends. Another substorm expansion phase is onset at 08:16 UT coinciding with a new range spread-F event. Finally, at 09:00 UT, another substorm expansion phase begins and, again, another range Spread-F event is onset. Daytime spread-F events do not coincide with observed substorms.

4 Conclusions

Typically during a storm/substorm development phase, electric fields promptly penetrate from auroral to equatorial latitudes with eastward polarity in the sunset sector, causing the growth of instability and the development of equatorial spread-F (Fejer et. al,1990; Pezzopane et al., 2013; among others). The auroral electrojet recovery associated with Bz turning north, in addition with large westward electric fields, foster stability of the F region.We analyzed the occurrence of intense spread F events at Tucumán low latitude ionospheric station during the development of a geomagnetic storm and several intense substorms.

Here, we report intense spread F events occurring at Tucumán low latitude station during nighttime (after midnight), during early morning hours on May 24th and daytime spread F which are extremely rare (Jing et al., 2016, among others) specially at lower heights where the ionosonde is able to observe them.

The spread-F event primarily occurred during both the main and the recovery phases of the geomagnetic storm, including instances of daytime spread-F (with majority of frequency spread-F).

Notably, we observed that the occurrence of the intense spread-F was predominantly aligned with the expansion phase of these substorms. It is worth mentioning that more in depth analysis is needed and is planned in future work, as solar activity increases. Consequently, we anticipate having more events to study in the future.

Acknowledgements. We acknowledge Dr. Dibyendu Chakrabarty for the fruitful and useful discussions during the United Nations/Germany Workshop on the International Space Weather Initiative (ISWI), Neustrelitz, Germany. BGL was supported by the NASA Mesoscale Magnetospheric Dynamics Heliophysics Internal Science Funding Model (HISFM).

References

Baker, D.N., Pulkkinen, T.I., Hesse, M., McPherron, R.L.: A quantitative assessment of energy storage and release in the Earth's magnetotail. J. Geophys. Res. **102**(A4), 7159–7168 (1997)

Fejer, B.G., Scherliess, L., dePaula, E.R.: Effects of the vertical plasma drift velocity on the generation and evolution of equatorial spread F. J. Geophys. Res. **104**(A9), 19859–19869 (1999)

Jiang, C., et al.: Ionosonde observations of daytime spread F at low latitudes. J. Geophys. Res. Space Physics **121**(12), 12093–12103 (2016)

McPherron, R.L.: Magnetospheric substorms. Rev. Geophys. **17**(4), 657–681 (1979)

McPherron, R.L., Russell, C.T., Aubry, M.P.: Satellite studies of magnetospheric substorms on August 15, 1968: 9. Phenomenological model for substorms. J. Geophys. Res. **78**(16), 3131–3149 (1973). https://doi.org/10.1029/JA078i016p03131

Molina, M. G., Dasso, S., Mansilla, G. et al.: Consequences of a Solar Wind Stream Interaction Region on the Low Latitude Ionosphere: Event of 7 October 2015. Sol. Phys. **295**(173) (2020).

Newell, P. T., Gjerloev, J. W.: Evaluation of SuperMAG auroral electrojet indices as indicators of substorms and auroral power, J. Geophys. Res. **116**(A12211) (2011).

Partamies, N., Juusola, L., Tanskanen, E., Kauristie, K.: Statistical properties of substorms during different storm and solar cycle phases. Ann. Geophys. **31**(2), 349–358 (2013)

Pezzopane, M., et al.: Low-latitude equinoctial spread-F occurrence at different longitude sectors under low solar activity. Ann. Geophys. **31**(2), 153–162 (2013)
Zolesi, B., Cander, J.: Ionospheric Prediction and Forecasting. Springer, Berlin (2014)

Open Access This chapter is licensed under the terms of the Creative Commons Attribution 4.0 International License (http://creativecommons.org/licenses/by/4.0/), which permits use, sharing, adaptation, distribution and reproduction in any medium or format, as long as you give appropriate credit to the original author(s) and the source, provide a link to the Creative Commons license and indicate if changes were made.

The images or other third party material in this chapter are included in the chapter's Creative Commons license, unless indicated otherwise in a credit line to the material. If material is not included in the chapter's Creative Commons license and your intended use is not permitted by statutory regulation or exceeds the permitted use, you will need to obtain permission directly from the copyright holder.

Climate Governance State of Art: A Snapshot on Current Discussions on Solar Radiation Management

Yvette Ramos(✉)

ICS, Doctoral Program On Climate Change and Sustainable Development Policies, Lisbon University, Av. Prof. Aníbal Bettencourt 9, 1600-189 Lisboa, Portugal
yvette@edu.ulisboa.pt

Abstract. Governance in the climate field involves a range of actions, processes, traditions, and institutions through which authority is exercised and decisions are implemented. This includes social norms, rules, institutions, and processes, involving both legal and non-legal measures like codes of conduct and standards, collectively termed "transnational regulation." Key purposes include organizing and coordinating technologies, conducting risk analysis, ensuring sustainable and responsible research, and defining mandates to influence technology development via policy and economic incentives. Effective governance strategies must be flexible and inclusive, engaging diverse stakeholders to manage the impacts and risks of various climate technologies. The International Space Weather Initiative (ISWI) is a global project that studies the effects of solar activity on Earth's environment and technology.

Solar Radiation Management (SRM) aims to mitigate global warming by reflecting sunlight back into space. Both ISWI and SRM rely on scientific research and data to evaluate their potential benefits and risks. ISWI can provide critical information on solar activity that may influence SRM methods such as like Stratospheric Aerosol Injection (SAI). Conversely, SRM may impact space weather and the climate system, which ISWI can monitor and evaluate.

The paper discusses the current state of climate governance, focusing on SRM and the potential contributions of ISWI to the governance process.

Keywords: Solar Radiation Management (SRM) · climate governance · risk assessment

1 Climate Governance

1.1 Challenges and Opportunities

The primary question raised here "How should we use technology for the benefit of the planet?" is fundamentally about governance. As scientists predict, we are deviating from the Paris Agreement's 1.5 °C target. What, then, is the current state of climate governance? The Napkin scheme presented below illustrates (Sheperd, 2009) that technologies like solar geoengineering have been considered, might have been patented, and even

© The Author(s) 2026
N. Gopalswamy et al. (Eds.): ISWI 2024, SPPHY 431, pp. 195–202, 2026.
https://doi.org/10.1007/978-981-95-1121-1_22

tested in various parts of the world. These efforts compare Solar Radiation Management (SRM) to historical volcanic eruptions as natural antilogues, highlighting both potential benefits and adverse impacts on the Earth, including damage to biodiversity and human health. While these technologies offer opportunities to prevent climate tipping points, significant uncertainties remain regarding their long-term effects on socioeconomic and ecological systems (Tilmes et al., 2009) (Fig. 1).

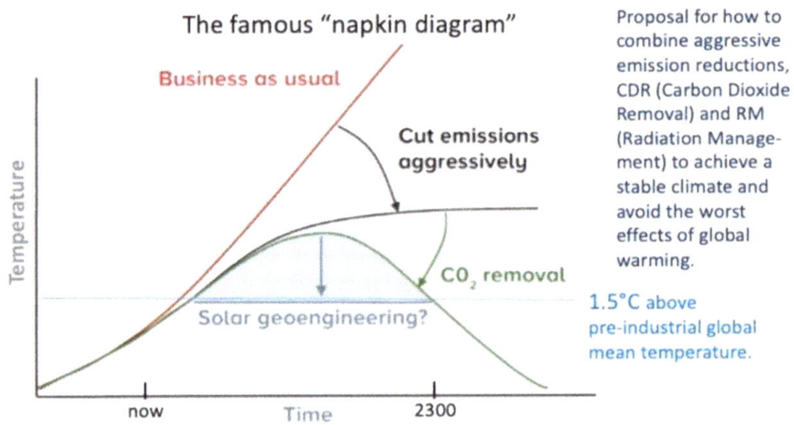

Fig. 1. The Napkin diagram: Future climate change scenarios based on different strategies Shepherd (2009)

Geoengineering is defined as the "deliberate large-scale manipulation of the planetary environment to counteract anthropogenic climate change". (Sheperd, 2009).

Governance involves a broad range of actions, processes, traditions, and institutions through which authority is exercised and decisions are implemented to achieve explicit goals. It includes legal instruments at international and national levels and non-legal measures such as codes of conduct and standards, collectively termed "transnational regulation." In the climate field, governance aims to coordinate technologies, provide safeguards for sustainable and responsible research and development, and define clear mandates, encouraging or discouraging technologies through policy and economic incentives. Effective governance strategies must be flexible and inclusive, engaging a wide range of stakeholders to address the diverse impacts and risks of different climate technologies.

At the transnational level, the role of non-state actors and the private sector in driving climate action is crucial, along with the contributions of cities and regional governments, which often pioneer innovative approaches to mitigation and adaptation. Other important factors include the development and implementation of new ideas and technologies for reducing emissions and adapting to climate impacts, leadership that inspires change and mobilizes action across sectors and governance levels, and the economic aspects of climate policy, including market mechanisms, carbon pricing, and the financial implications of climate action. Internationally, the influence of agreements and organizations such as the UNFCCC, UNEP, and the World Climate Research Programme (WCRP) of the WMO on national policies is significant. The UNOOSA, with its ISWI initiative, also plays a potentially important role in this context.

1.2 Challenges and Opportunities

As we examine key authors who significantly influenced the discourse on global climate governance, they have provided diagnoses, published proposals, and highlighted the challenges they analysed. The Table 1 below summarizes their contributions.

Table 1. State-of-Art in terms of Climate Governance by major authors: a summary

	DIAGNOSIS	CHALLENGES
Ostrom E. (2009)	Its complex and polycentric nature emphasize multiple, overlapping layers at different scales rather than relying solely on top-down approaches	Coordinating actions across different levels of governance, ensuring the participation of diverse stakeholders, and overcoming the barriers to collective action
Victor D.G. (2011)	International climate agreements, e.g. The Kyoto Protocol, have been largely ineffective due to their reliance on legally binding commitments and lack of enforcement mechanisms	Ensuring ambition, participation in voluntary agreements, overcoming free-rider issues, scaling successful initiatives globally, and limits of top-down approaches
Bulkeley H. et al. (2014)	growing importance of cities and urban areas in climate governance: they argue that urban areas are key sites for both the causes and solutions to climate change	Securing adequate funding for urban climate initiatives, coordinating actions across different levels of government, and ensuring equitable outcomes for all urban residents
Biermann F. (2014)	Framework too fragmented and insufficient to address the scale and urgency of the climate crisis: need for transformative change in global governance structures	Gaining international support for new global institutions, ensuring their legitimacy and effectiveness, and transitioning from fragmented systems to integrated governance
Wurzel R.K.W. et al. (2020)	Roles of pioneers, leaders, and followers in multilevel, polycentric climate governance highlight the importance of diverse actors in climate change mitigation and adaptation	The empirical impact of leaders and pioneers as dependent on their actions and dynamics with followers and laggards in governance structures

(*continued*)

Table 1. *(continued)*

	DIAGNOSIS	CHALLENGES
Cashore B. et al. (2021)	The role of non-state actors and transnational governance in climate policy and point out that traditional state-centric approaches are insufficient to tackle global environmental problems	Ensuring the credibility and effectiveness of NSMD systems, maintaining accountability and transparency, and integrating these approaches with state-led policies
Pattberg P. et al. (2022)	The intricate interactions between public and private sectors in governing climate change, moving beyond state-centric approaches	Need to address governance complexity, fragmentation, and the integration of diverse actors and levels of governance

1.3 Climate Governance: What is at Stake?

The complex landscape of climate governance involves numerous challenges and a diverse array of stakeholders, including various levels of government, international organizations, NGOs, private sector entities, and local communities. Effective climate governance hinges on several crucial aspects, as highlighted by major authors in Table 1. These include coordination and integration by aligning policies across different levels and transitioning from fragmented to integrated structures; stakeholder participation and engagement by including diverse groups in decision-making and overcoming collaboration barriers; ambition and implementation by striving for ambitious climate commitments, encouraging broad voluntary participation, addressing free-riding issues, and scaling successful local programs globally; funding and resource allocation by ensuring adequate and equitable resources for city-level climate actions; institutional support and legitimacy by establishing credible and effective new global climate bodies and facilitating the transition to cohesive structures; accountability and transparency by ensuring reliable non-state market-driven systems and high standards of openness in governance processes; and leadership and dynamics by leveraging proactive leaders to drive governance and manage interactions among leaders, followers, and laggards.

2 The Solar Geoengineering Case

2.1 What is SRM: Solar Radiation Management

SRM technologies aim to reflect a portion of solar energy back into space to reduce global temperatures. Key SRM technologies include Stratospheric Aerosol Injection (SAI) dispersing aerosols, such as sulfur dioxide, into the stratosphere to reflect sunlight; Marine Cloud Brightening (MCB) that is spraying sea salt into clouds to increase their reflectivity; space-based reflectors for example placing huge mirrors or reflective particles in space to block or reflect sunlight; surface albedo modification to increasing the reflectivity of surfaces on Earth, such as painting roofs white or covering deserts with reflective materials; Cirrus Cloud Thinning (CCT) by reducing the presence of high-altitude cirrus clouds that trap heat, allowing more long-wave radiation to escape. Each technology has unique potential and risks, and their implementation requires careful consideration of ecological and geopolitical implications.

2.2 The Urgent Need to Regulate SRM

SRM involves techniques to reflect sunlight to cool the Earth, potentially using space-based or atmospheric interventions. UNOOSA oversees international space-related activities, ensuring peaceful use and cooperation in outer space. The ISWI, a UNOOSA initiative, focuses on space weather research and international collaboration. The Environmental Modification Convention (ENMOD, 1976) prohibits military or hostile use of

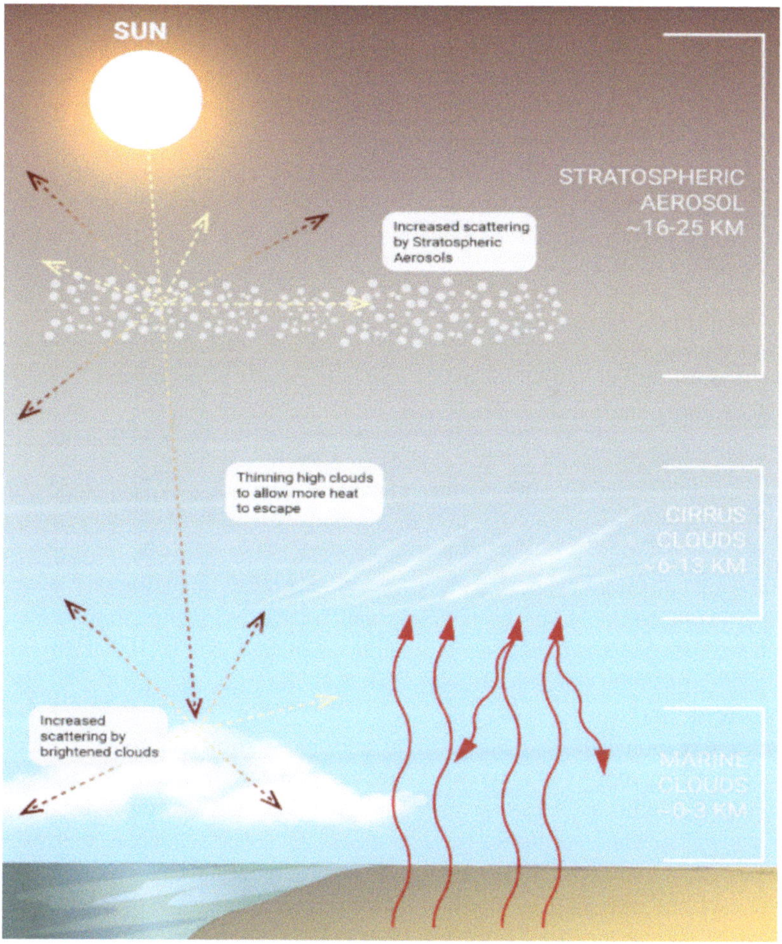

Fig. 2. Basic mechanisms involved in the three aerosol-based SRM approaches that have been studied using climate models: SAI, MCB and CCT. Source: UNEP, 2023 One Atmosphere: an independent expert review on Solar Radiation Modification research and deployment ()

environmental modification techniques with widespread, long-lasting, or severe effects. SRM technologies must be regulated to ensure alignment with UNOOSA's principles of peaceful space use and comply with ENMOD's restrictions on environmental modification for non-hostile purposes, ensuring ethical and safe implementation (Fig. 2).

Space weather refers to environmental conditions influenced by the Sun, impacting the magnetosphere, atmosphere (stratosphere and troposphere), ionosphere, and ground (Gopalswamy, 2022). The sun interacts with the earth's environment through two main channels: 1) electromagnetic emissions and 2) particles:

Solar Electromagnetic Emissions: The spectrum of solar electromagnetic emissions ranges from Gamma rays, X rays, Ultraviolet (UV), visible light, infrared, microwaves short waves etc. ... Only infrared and visible radiations reach the ground and play an important role in climate. In the stratosphere, between ~10 and 40 km, lies the ozone layer. The ozone layer absorbs most EUV (Extreme UV) and UV rays and acts as a shield against the EUV and UV radiations that causes skin cancer.

Particles: The Solar Wind Is the constant stream of solar coronal material flowing from the sun, mostly electrons, protons and alpha particles with energies usually between 1.5 and 10 keV. The solar wind directly affects the polar and auroral regions. **CME** (Coronal Mass Ejections) and coronal holes emitting fast winds disturbed the regular solar wind and cause geomagnetic storms. Auroral electrons (medium en-ergy electrons (MEE, 30–1000 keV), and relativistic electrons) penetrate into the lower thermosphere, mesosphere, and upper stratosphere.

SPE, Solar Protons Event Are highly energetic charged particles (protons, electrons, and heavy ions) coming from the atmosphere and solar wind. They are more sporadic than electron precipitation, but consist of highly energetic protons that can precipitate down to the upper stratosphere. Their energies range from a few tens of keV to many GeV.

Galactic Cosmic Rays Coming From Outside of our solar system, modulated by solar activity, have highest energies and thus mainly affect the lower stratosphere and tropo-sphere.

Solar geoengineering experiments in the stratosphere may interact with space weather phenomena, including cosmic rays and solar activity, affecting ozone chemistry and weather patterns. Understanding all these interactions is crucial for assessing and managing potential impacts of Solar Radiation Management (SRM), as discussed by Tilmes et al. (2022).

While international efforts focus on rapid emissions mitigation and adaptation, SRM is considered as an additional approach to prevent global temperatures from exceeding Paris Agreement limits by reflecting more sunlight into space. However, sustained SRM deployment for decades to centuries would be necessary for effectiveness, with estimated costs in the tens of billions of dollars annually per 1 °C of cooling. Operational SRM poses risks such as ozone layer damage, uneven regional climate impacts, rapid climate changes if abruptly halted, reduced mitigation efforts, and international power imbalances.

Given these risks and uncertainties, establishing an international scientific review process is crucial to evaluate SRM scenarios and consequences. Governance frameworks are essential for overseeing SRM research and deployment, including indoor studies, small-scale outdoor experiments, and potential large-scale implementations.

2.3 The SRM Aborted Resolution Discussed Informally at UNEA6

While SRM remains hypothetical and controversial, with no consensus on its implementation, recent informal discussions at UNEA6 (UNEP, 2024) focused on Switzer-land's resolution on SRM. The resolution aimed to enhance transparency and access to information, stressing that SRM should not substitute rapid GHG emissions reductions and precautionary principles. Many countries supported advancing scientific under-standing of SRM under bodies like the WCRP-WMO or the IPCC, while emphasizing caution about potential risks and prioritizing emissions reduction efforts. The resolution, withdrawn for further refinement, aimed to include diverse scientific expertise and ensure transparency, with potential for reconsideration at UNEA7 in late 2025.

3 Possible Contributions for UNOOSA and ISWI in SRM

The ISWI holds potential to contribute significantly to climate governance despite its primary focus on understanding and mitigating space weather impacts and supporting space-based technologies. By integrating ISWI's research on solar and magnetic storms into climate monitoring, it enhances climate model accuracy. Its efforts in disaster risk reduction improve space weather forecasting, aligning with UNFCCC goals on resilience. ISWI's interdisciplinary collaboration involves scientists across various fields, enriching UN scientific bodies' understanding of climate change. Education initiatives bolster capacity-building, particularly in developing countries for solar geoengineering technologies. ISWI's global network supports international cooperation on climate research, aligning with the UN's focus on technology transfer and shared knowledge to address climate change.

We conclude stating that the integration of ISWI's expertise into climate governance frameworks can enhance the understanding and management of both space weather and climate-related challenges, promoting resilience and informed decision-making on a global scale.

References

Biermann, F.: Earth System Governance: World Politics in the Anthropocene. The MIT Press (2014). https://doi.org/10.7551/mitpress/9780262028226.001.0001

Bulkeley, H., et al.: Transnational Climate Change Governance. Cambridge University Press (2014). https://doi.org/10.1017/CBO9781107706033

Cashore, B., et al.: Private Authority and Public Policy Interactions in Global Context (2021). https://doi.org/10.1111/rego.12395

Desert Research Institute (DRI): Can Scientific Ingenuity Turn the Clock Back on Climate Change? (2023). https://www.dri.edu/can-scientific-ingenuity-turn-the-clock-back-on-climate-change/ and https://www.eurekalert.org/multimedia/1007557

Gopalswamy, N.: The sun and space weather. Atmosphere **13**(11), 1781 (2022). https://doi.org/10.3390/atmos13111781

Green, J.F.: Climate change governance: past, present, and (hopefully) future. In: Barnett, M.N., Pevehouse, J.C.W., Raustiala, K. (eds.) Global Governance in a World of Change, pp. 109–129. Cambridge University Press (2021). https://doi.org/10.1017/9781108915199.004

ISWI International Space Weather Initiative (2024). https://iswi-secretariat.org/

Jordan, A., et al. (eds.): Governing Climate Change: Polycentricity in Action? Cambridge University Press (2018). https://doi.org/10.1017/9781108284646

Ostrom, E.: A Polycentric Approach for Coping with Climate Change (2009). https://documents1.worldbank.org/curated/fr/480171468315567893/pdf/WPS5095.pdf

Ostrom, E.: Polycentric systems for coping with collective action and global environmental change. https://doi.org/10.1016/j.gloenvcha.2010.07.004 (2010)

Pattberg, P., et al.: 20 years of global climate change governance research: taking stock and moving forward. Int. Environ. Agreements **22**(2), 295–315 (2022). https://doi.org/10.1007/s10784-022-09568-5

Tilmes, S., et al.: Impact of geoengineered aerosols on the troposphere and stratosphere (2009). https://doi.org/10.1029/2008JD011420

Tilmes, S., et al.: Stratospheric ozone response to sulfate aerosol and solar dimming climate interventions (2022). https://doi.org/10.5194/acp-22-4557-2022

UNEP: One Atmosphere: An Independent Expert Review on Solar Radiation Modification Research (2023). https://wedocs.unep.org/bitstream/handle/20.500.11822/41903/one_atmosphere.pdf?sequence=3&isAllowed=y

UNEP: UNEP/OECPR.6/L.14 – Draft Resolution on Solar Radiation Modification (2024). https://resolutions.unep.org/resolutions/?q=node/394

United Nations: Convention on the Prohibition of Military or Any Other Hostile Use of Environmental Modification Techniques (1976). https://treaties.un.org/doc/Treaties/1978/10/19781005%2007-19%20AM/Ch_XXVI_01p.pdf

Victor, D.G.: Global Warming Gridlock: Creating More Effective Strategies for Protecting the Planet. Cambridge University Press (2011). https://doi.org/10.1017/CBO9780511975714

Wurzel, R.K.W., et al. (eds.) Climate Governance Across the Globe (2020). https://doi.org/10.4324/9781003014249

Open Access This chapter is licensed under the terms of the Creative Commons Attribution 4.0 International License (http://creativecommons.org/licenses/by/4.0/), which permits use, sharing, adaptation, distribution and reproduction in any medium or format, as long as you give appropriate credit to the original author(s) and the source, provide a link to the Creative Commons license and indicate if changes were made.

The images or other third party material in this chapter are included in the chapter's Creative Commons license, unless indicated otherwise in a credit line to the material. If material is not included in the chapter's Creative Commons license and your intended use is not permitted by statutory regulation or exceeds the permitted use, you will need to obtain permission directly from the copyright holder.

Space Weather Prediction Using Various Techniques Including Machine Learning

Tackling Space Weather Forecasting Challenges with Machine Learning

Dario Del Moro(✉)

University of Rome Tor Vergata, Roma, Italy
delmoro@roma2.infn.it

Abstract. The development in the last decades in the sciences of complex systems and statistics has allowed us to define the limits of predictability better and -often- to extend those limits. The rapid progress of computing techniques and capabilities has even further bolstered those two branches of science to the point that it is now feasible both to compute the trajectory of plasma and magnetic field structures under the MHD equations in domains as large as the Heliosphere and to try to predict -just by analysing full disk images or the magnetograms- whether a given solar Active Region will release part of its stored energy as high energy photons and particles, or shoot out a coronal mass ejection.

Yet, a robust forecast of flare eruption still escapes us. Even the apparently more straightforward problem of the propagation of a coronal mass ejection in the interplanetary medium has not been solved to the limit we would like. At the same time, we fight with the uncertainties associated with the boundary conditions. Consequently, some of us turned to the dark side and applied this hybrid approach of numerical methods, complex system science and statistics, which usually goes under Machine Learning (ML).

Considering how large the ML field is, I can only share my personal experience of how ML contributes to the forecast of space weather and why it is probably here to stay. I will focus on some aspects, reporting on recent approaches that show how ML methods can go further than the common "black-box" approach.

In particular, I will zoom in on two recent ML branches: employing attention-based tools to help the interpretability of Deep Learning models and the possibility of enforcing some of the system physics in training, thus limiting the results to be adherent to physical laws.

Keywords: Space Weather · Machine Learning · Data Analysis

1 Introduction

Humanity's ability to forecast the future has been the key to its success in the evolutionary "survival of the fittest" game. This predictive capability first emerged through the long process of biological evolution, where anticipating threats and opportunities was essential for survival. Over millennia, this innate skill was further refined and encoded into cultural practices, rituals, and collective wisdom, allowing societies to plan for seasons, migrations, and other vital activities.

© The Author(s) 2026
N. Gopalswamy et al. (Eds.): ISWI 2024, SPPHY 431, pp. 205–210, 2026.
https://doi.org/10.1007/978-981-95-1121-1_23

The advent of the scientific method marked a significant leap in our forecasting abilities. By applying systematic observation, experimentation, and analysis, humans began to tackle everyday life problems with greater precision and reliability. Science enabled us to develop tools and techniques that improved our understanding of natural phenomena and enhanced our capacity to predict various outcomes.

Yet, it is only in recent times that we have acquired the conceptual tools to understand why some events remain difficult to forecast. The development of complex systems theory, chaos theory, and advanced computational models has provided new insights into the inherent unpredictability of certain phenomena. These tools have revealed that some events are influenced by a number of variables interacting in nonlinear ways, making precise prediction challenging.

The recent advances in the sciences of complex systems and statistics have enabled us to define better and often extend the limits of predictability. The rapid progress in computing techniques and capabilities has further strengthened these fields, making it possible to calculate the trajectory of plasma and magnetic field structures using Magneto-Hydro-Dynamic (MHD) equations in regions as vast as the Heliosphere [1, 2]. Additionally, it is now feasible to analyse solar full disk images or magnetograms to predict whether a specific solar Active Region will release part of its stored energy as high-energy photons and particles or trigger a coronal mass ejection.

Is everything solved, then? Not at all. In some scenarios, we can make almost exact predictions; in others, we can offer robust predictions with an estimation of the associated errors; sometimes, we can only speak in terms of probabilities. Our ability to predict is hindered either by intrinsic limitations or by the need for complete knowledge about the present state. Unfortunately, space weather predictions often suffer from both of these issues: we frequently deal with complex systems and only have a fragmentary understanding of what is occurring on the Sun and its surroundings. To mitigate these limitations, we have developed increasingly sophisticated numerical techniques to solve the differential equations that typically describe these problems. Additionally, we use ensemble methods to handle measurement errors and unknown variables [3, 4]. By extracting all relevant information from remote measurements and applying our understanding of the underlying physics, we feed this information into our forecasting algorithms [5]. However, with ongoing research and the continuous refinement of our methods, we are steadily getting better at forecasting space weather events.

Despite these advancements, accurately forecasting flare eruptions still escapes us. Even the seemingly simpler task of predicting the propagation of a coronal mass ejection (CME) through the interplanetary medium has not been perfected, primarily due to uncertainties associated with boundary conditions. As a result, to tackle these challenges, some researchers [e.g., 6, 7] have turned to the "dark side", i.e., the more unconventional approach of using a hybrid of numerical methods, complex system science, and statistics that usually goes under the name of "Machine Learning" (ML).

2 Machine Learning

ML approaches have revolutionised problem-solving by providing a level of speed and efficiency that traditional methods, such as solving MHD equations or standard data analysis, often struggle to achieve. Solving complex differential equations can be highly computationally intensive and time-consuming, mainly when dealing with intricate variables or vast datasets. In contrast, ML excels at identifying patterns without the need for explicit programming, enabling the rapid analysis of large datasets and the reproduction of complex simulation outcomes. The inherent capacity of ML models to learn from data rather than relying on predefined equations greatly enhances their ability to extract meaningful insights quickly [8, 9]. This adaptability makes ML especially effective in handling nonlinear relationships and evolving patterns. Moreover, ML's ability to handle nonlinear relationships and adapt to evolving patterns makes it particularly advantageous. In contrast, brute-force approaches to data selection involve exhaustive trial and error, consuming substantial time and resources. ML accelerates the process with its ability to efficiently process and analyse complex data sets, adapt to changing patterns, and handle nonlinear relationships.

Moreover, the rapid evolution of ML literature is driven by the swift pace of research and technological advancements across multiple disciplines that utilise ML. The competitive nature among researchers and the interdisciplinary integration of insights from computer science, mathematics, and various other fields contribute significantly to the growth of literature.

However, ML models are highly sensitive to the training dataset and loss functions used due to their fundamental reliance on pattern recognition. The model's learning process depends heavily on the representativeness of the training data, making it crucial for achieving accurate generalisation. The choice of loss functions is equally important as it guides the optimisation process, influencing how the model adjusts its parameters. Imbalanced or biased datasets can lead to skewed learning, while inappropriate loss function selection can result in suboptimal model performance. This sensitivity underscores the importance of careful data curation and precise loss function selection to ensure that ML models produce accurate and meaningful results [10].

Recently, Deep Learning (DL) models, which are a subset of ML models that use artificial neural networks with multiple layers to learn hierarchical data representations through numerous non-linear transformations, have shown impressive successes in forecasting space weather activity [11, 12]. Utilising the capabilities of Neural Networks (NNs), particularly recurrent and Convolutional architectures (CNNs), DL methods efficiently process extensive datasets of solar and geomagnetic observations. These models excel at identifying complex patterns in time-series data, enabling precise predictions of solar flares, geomagnetic storms, and other space weather phenomena [13]. However, these models face a common challenge in ML forecasting: a potential disconnect from the physical understanding of system dynamics. While Deep Learning excels in making predictions, it often falls short in interpretability, making it difficult to understand how underlying physical quantities influence the outcomes. The "black-box" nature of certain Deep Learning models can obscure the cause-and-effect relationships essential for a comprehensive understanding. Striking a balance between prediction accuracy and maintaining a physical perspective is crucial for space weather physicists. This necessitates the development and use of models that not only forecast effectively but also offer transparent insights into the complex interplay of relevant quantities within the system.

3 Open up the Black Box

Recently, in physics and space weather research, a strong focus has been on closing the gap between ML predictions and the physical understanding of how systems work. Scientists are creating hybrid models that blend the predictive power of ML with traditional physics-based models. They are also working on making ML techniques more understandable so practitioners can trust the decisions made by DL models. This combined approach uses the best of both ML and physics to make more reliable and insightful predictions in complex areas like space weather.

3.1 Interpretability in Deep Learning

Looking at attention frames is one way to understand and explain DL model decisions, especially with NNs using attention mechanisms. These frames show which parts of the input data the model "focuses on" as most important for making predictions. Analysing attention frames helps us see which parts of the data the model looks at during different stages of its decision-making. This helps make the model's processes more transparent and more understandable, addressing the "black box" problem. For example, in flare forecasting, knowing which features of the input images the model prioritises can help us understand its decisions better, making DL models more transparent and focused on the key factors in the flaring process. See [14, 15] for examples of this approach.

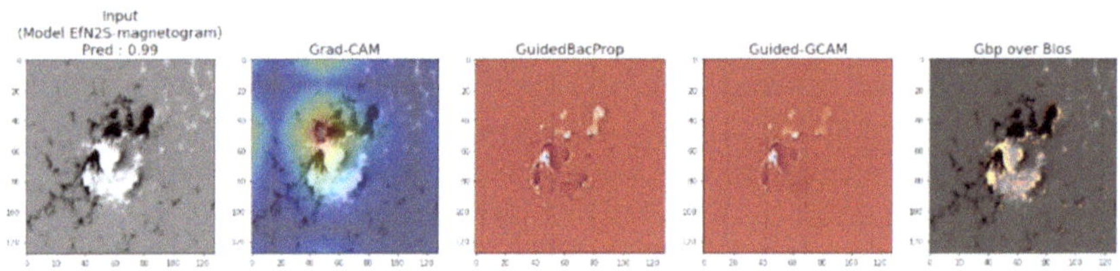

Fig. 1. Visual explanation of the Magnetogram-based predictions of an upcoming X1.5 flare (adapted from [14]). From left to right: input magnetogram of an Active Region, Grad-CAM image highlights the parts of the image most relevant for the prediction, Guided Back-propagation to show the most discriminant pixels of the image, Guided Grad-CAM combining the two methods, the Guided Grad-CAM values overlaid on the input magnetogram.

In Fig. 1, we show the output of several visual explainability methods: Grad-Class Activation Maps (Grad-CAM [16]) to localise the region of the input image that contributed the most to the prediction, Guided Back-propagation [17] to visualise the fine-grained details of the images that the model is using to make the prediction. Last, Guided Grad-CAM combines the Guided Back-propagation with Grad-CAM to visualise fine-grained pixel contribution to the gradients with a strengthened weight given to the most discriminating region of the picture.

3.2 Physics-Informed ML

Physics-informed ML (PI-ML) combines traditional physics with modern ML techniques. This approach integrates known physical laws into ML algorithms to guide their learning. By doing so, PI-ML ensures that predictions align with fundamental principles, making the models more accurate and trustworthy. This method is especially useful when data is limited, noisy, or expensive to obtain. It helps the models stay interpretable and reliable. PI-ML is particularly valuable in scenarios where data might be limited, noisy, or expensive to obtain. By enforcing physics-based constraints, the models become more interpretable, and predictions become more trustworthy as they adhere to known laws. This approach is being increasingly explored in various scientific fields, including fluid dynamics [18] and flare forecast [19], where a balance between data-driven insights and physical principles is crucial. Just as examples, [9, 20] both used a physics-driven ML approach to the prediction of ICME (Interplanetary Coronal Mass Ejections) travel times, exploiting a Drag-Based model [21] to improve the training and concluding that the use of physical information in the ML architecture significantly improved both the accuracy and the robustness of the predictions.

4 Conclusions

ML is becoming increasingly effective, accessible, and widely adopted in many scientific fields. In the space weather field, efforts have been made to address and rectify its major drawbacks, enhancing its performance and reliability. In particular, interpretability and physics-informed approaches for ML can help us produce more transparent models, which can be more dependable and also open the way to new interpretations of the data. It is also worth noticing that since the success and accuracy of ML models strongly depend on the data used to train and validate them, the preparation of high-quality datasets explicitly devised for ML purposes is of critical importance as well.

References

1. Odstrcil, D.: Modeling 3-D solar wind structure. Adv. Space Res. **32**(4), 497–506 (2003)
2. Pomoell, J., Poedts, S.: EUHFORIA: European heliospheric forecasting information asset. J. Space Weather Space Clim. **8**, A35 (2018). https://doi.org/10.1051/swsc/2018020
3. Napoletano, G., Forte, R., Del Moro, D., Pietropaolo, E., Giovannelli, L., Berrilli, F.: A probabilistic approach to the drag-based model. J. Space Weather Space Clim. **8**, A11 (2018). https://doi.org/10.1051/swsc/2018003
4. Dumbović, M., et al.: The drag-based ensemble model (DBEM) for coronal mass ejection propagation. Astrophys. J. **854**(2), 180 (2018)
5. Leka, K.D., Barnes, G.: Photospheric magnetic field properties of flaring versus flare-quiet active regions. IV. A statistically significant sample. Astrophys. J. **656**(2), 1173–1186 (2007). https://doi.org/10.1086/510282
6. Bobra, M.G., Couvidat, S.: Solar flare prediction using SDO/HMI vector magnetic field data with a machine-learning algorithm. Astrophys J **798**(2), 135 (2015)
7. Cicogna, D., et al.: Flare-forecasting algorithms based on high-gradient polarity inversion lines in active regions. Astrophys. J. **915**(1), 38 (2021)
8. Benvenuto, F., Campi, C., Massone, A.M., Piana, M.: Machine learning as a flaring storm warning machine: was a warning machine for the 2017 September solar flaring storm possible? Astrophys. J. Lett. **904**(1), L7 (2020)
9. Chierichini, S., et al.: A Bayesian approach to the drag-based modelling of ICMEs. J. Space Weather Space Clim. **14**, 1 (2024). https://doi.org/10.1051/swsc/2023032

10. Guastavino, S., Piana, M., Benvenuto, F.: Bad and good errors: value-weighted skill scores in deep ensemble learning. IEEE transactions on neural networks and learning systems. (2022)
11. Deshmukh, V., Flyer, N., Van der Sande, K., Berger, T.: Decreasing false-alarm rates in CNN-based solar flare prediction using SDO/HMI data. Astrophys. J. Suppl. Ser. **260**(1), 9 (2022)
12. Deshmukh, V., Baskar, S., Berger, T.E., Bradley, E., Meiss, J.D.: Comparing feature sets and machine-learning models for prediction of solar flares-topology, physics, and model complexity. Astron. Astrophys. **674**, A159 (2023)
13. Chierichini, S., Liu, J., Korsós, M.B., Del Moro, D., Erdélyi, R.: CME Arrival Modeling with Machine Learning. Astrophys J **963**(2), 121 (2024)
14. Francisco, G., et al.: Limits of solar flare forecasting models and new deep learning approach. JGR: Machine Learning and Computation (2024)
15. Sun, Z., et al.: Predicting solar flares using CNN and LSTM on two solar cycles of active region data. Astrophys. J. **931**(2), 163 (2022)
16. Selvaraju, R.R., Das, A., Vedantam, R., Cogswell, M., Parikh, D., Batra, D.: Grad-CAM: Why did you say that? (2016)
17. Springenberg, J. T., Dosovitskiy, A., Brox, T., Riedmiller, M.: Striving for simplicity: The all convolutional net (2014)
18. Kashinath, K., et al.: Physics-informed machine learning: case studies for weather and climate modelling. Philosophic. Trans. Royal Soc. A: Math. Phys. Eng. Sci. **379**, 2194 (2021)
19. Jiao, Z., Sun, H., Wang, X., Manchester, W., Gombosi, T., Hero, A., Chen, Y.: Solar flare intensity prediction with machine learning models. Space weather. (2020)
20. Guastavino, S., et al.: Physics-driven machine learning for the prediction of coronal mass ejections' travel times. Astrophys. J. **954**(2), 151 (2023)
21. Napoletano, G., Forte, R., Del Moro, D., Pietropaolo, E., Giovannelli, L., Berrilli, F.: A probabilistic approach to the drag-based model. J. Space Weather Space Clim. **8**, A11 (2018). https://doi.org/10.1051/swsc/2018003

Open Access This chapter is licensed under the terms of the Creative Commons Attribution 4.0 International License (http://creativecommons.org/licenses/by/4.0/), which permits use, sharing, adaptation, distribution and reproduction in any medium or format, as long as you give appropriate credit to the original author(s) and the source, provide a link to the Creative Commons license and indicate if changes were made.

The images or other third party material in this chapter are included in the chapter's Creative Commons license, unless indicated otherwise in a credit line to the material. If material is not included in the chapter's Creative Commons license and your intended use is not permitted by statutory regulation or exceeds the permitted use, you will need to obtain permission directly from the copyright holder.

AI with Large Model for Solar Activity Forecasting

Long Xu[1(✉)], Yao Zhang[1], Xinze Zhang[1], and Yihua Yan[2]

[1] Ningbo University, Ningbo 315211, China
lxu@nao.cas.cn
[2] National Space Science Center, Beijing 100109, China

Abstract. Solar activity forecasting plays a crucial role in various domains, including space weather prediction and satellite communications. With the emergence of large artificial intelligence (AI) models, there is a growing interest in exploring their potential to improve the accuracy of solar activity forecasting. This paper presents a study on the application of AI with large models in solar activity forecasting. We investigate the effectiveness of leveraging deep learning techniques, such as convolutional neural networks (CNNs), residual neural networks (ResNets), and Long Short-Term Memory (LSTM) networks, to capture complex patterns and dependencies in solar data. Additionally, we explore the benefits of using large models, such as Mask AutoEncoder, CLIP, Florence, and ALIGN, to integrate intelligent processing tasks and enhance predictive capabilities.

Keywords: Deep learning · large model · solar flare · solar activity forecast

1 Introduction

In recent years, artificial intelligence (AI) has made significant strides in solar astronomy [1–5], enhancing the analysis and understanding of solar phenomena. Key applications include solar image analysis for detecting features such as sunspots, flares, and coronal mass ejections; solar activity forecasting to predict events like flares, coronal mass ejections, and geomagnetic storms; data processing to extract insights and identify patterns; and automated solar observations and control for efficient data collection. The advent of large AI models, such as GPT-3 [6], DALL-E [7], and CLIP [8], further boosts these capabilities by providing unprecedented data processing and pattern recognition abilities.

The integration of large AI models in solar physics can significantly enhance solar activity forecasting and space weather prediction [1, 9, 10]. By utilizing their pattern recognition and time series forecasting capabilities, we can gain deeper insights into solar dynamics, leading to more accurate predictions of flares, coronal mass ejections, and solar wind variations.

These models also support the development of sophisticated prediction systems that consider various solar parameters and multimodal observations, improving our ability to anticipate impacts on high-tech infrastructure on Earth [11].

© The Author(s) 2026
N. Gopalswamy et al. (Eds.): ISWI 2024, SPPHY 431, pp. 211–221, 2026.
https://doi.org/10.1007/978-981-95-1121-1_24

Furthermore, large AI models facilitate the integration of diverse datasets—such as solar images, spectroscopic observations, and historical records—allowing for a comprehensive understanding of solar dynamics. This holistic approach aids in creating AI-driven tools for real-time monitoring and analysis. Ultimately, advancements in AI technology promise to enhance our understanding of the Sun and improve the accuracy and reliability of space weather forecasts.

2 Deep Learning Techniques for Solar Activity Forecasting

Deep learning is a pivotal implementation of AI, particularly shining in the realm of solar activity forecasting [4]. It offers sophisticated tools for dissecting intricate solar data and producing precise predictions. Among the key deep learning techniques employed for this purpose are:

(1) Convolutional Neural Networks (CNNs) [12–23]: These networks excel at analyzing solar images, identifying features such as sunspots, solar flares, and coronal mass ejections. By learning hierarchical representations of solar imagery, CNNs facilitate accurate detection and classification of solar phenomena, thereby enriching solar activity forecasting.

(2) Recurrent Neural Networks (RNNs) [24], particularly Long Short-Term Memory (LSTM) networks [25]: RNNs are utilized for time series analysis of solar data, adept at capturing temporal dependencies and patterns. This capability enables them to predict variations in solar activity over time, such as fluctuations in solar radio flux or sunspot numbers.

(3) Generative Adversarial Networks (GANs) [26]: GANs generate synthetic solar images and data, addressing data scarcity and augmenting existing datasets. They also enhance the quality and resolution of solar images used for forecasting through data denoising and super-resolution techniques [27, 28].

(4)(1) Transformer Models: Transformer-based architectures, like BERT [29], excel in processing sequential solar data and capturing long-range dependencies. These models can analyze and forecast solar activity patterns based on a wide array of input parameters.

(5) Hybrid Architectures [30]: Combining CNNs, RNNs, and attention mechanisms, hybrid architectures harness the strengths of each for comprehensive solar activity forecasting. These models proficiently handle diverse solar data types, including images, time series, and spectral data.

While each of these techniques holds immense promise in the field of solar activity forecasting, the most promising approach may vary depending on specific forecasting goals and available data. CNNs continue to be indispensable for image-based solar feature detection and classification. RNNs and LSTM networks, with their ability to capture temporal patterns, are crucial for time series forecasting of solar activity. GANs offer innovative solutions to data scarcity and image enhancement challenges. Transformer models and hybrid architectures, with their capacity to process sequential data and integrate multiple data types, represent a forward-thinking direction in solar activity forecasting. Ultimately, the most effective strategy may involve leveraging the strengths of multiple techniques within a cohesive forecasting framework, tailored to the unique demands of solar activity prediction.

3 Large Model for Enhanced Predictive Capabilities

Large models [31–33] refer to AI models with a significant number of parameters, often in the billions or trillions. They can process vast amounts of data, enabling complex tasks like language understanding, image recognition, and more. The introduction of the Transformer in 2017 revolutionized NLP. The release of GPT-2 [34] and GPT-3 [33] showed the capabilities of LLMs. GPT-4o [35] is astonishing in multimodality, seamlessly integrating multiple modalities; closer to human-like interaction; human-like emotions and sentiments. At the beginning, large models usually referred to language large models. Now, large models have entered the era of multimodal large models [35], which can handle various modalities of data and various types of outputs. This has greatly enhanced the application scope of large models. In the literatures, image large models have been well developed, such as CLIP [8], SAM [36], etc.

Large model usually concerns multiple tasks, integrating multiple tasks together to achieve the performance of the integrated model beyond each individual model. To train a large model, multimodal information usually is collected and cross-correlation between modalities can be mined through model training over multimodal information.

Large models like Mask Auto-Encoder [37], CLIP [8], Florence [38], and ALIGN [39] are trained on multimodal data, integrating text, video, and multispectral images. Solar astronomy offers a rich source of naturally aligned multimodal information, including magnetograms, images, and spectroscopic data. This enables the training of large models that can leverage the complementary information from different modalities, enhancing predictive performance.

We aim to integrate various common tasks in solar astronomy into a single, general large model, also referred to as a fundamental model [38]. These common tasks include object detection, radiation flux forecasting, image generation, image classification, and general forecasting. By combining these tasks, we seek to create a powerful and versatile model that serves as a fundamental feature extractor for specific downstream tasks using labeled data.

The framework of the proposed large model is illustrated in Fig. 1. The model accepts three types of input data: images, one-dimensional data, and structured data. These modalities are aligned within a multimodal heterogeneous domain, allowing the model to explore cross-spatio-temporal and cross-band correlations in a unified feature space.

To integrate the various modalities, we employ cross-attention mechanisms, enabling the model to learn from and leverage complementary information across different sources, thereby enhancing its predictive performance. This integration allows the model to perform a wide range of intelligent processing tasks in solar astronomy, including flare and coronal mass ejection (CME) forecasting, solar radio flux predictions, space weather index forecasting, as well as imaging and image processing. By consolidating these tasks within a single model, we aim to establish a comprehensive and effective framework for handling diverse applications.

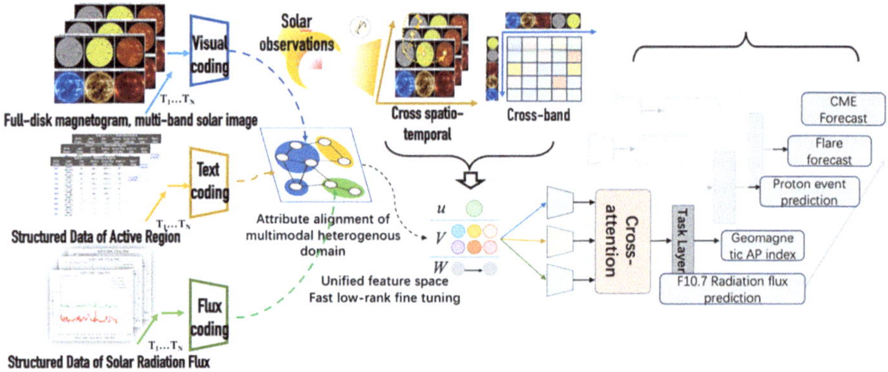

Fig. 1. The framework of the proposed large model of solar astronomy

In summary, we have developed a large model for solar astronomy that integrates multiple common tasks and utilizes cross-attention mechanisms to explore and leverage complementary information from different modalities. This approach enhances the model's versatility and predictive performance. The framework of the proposed model is illustrated in Fig. 1, providing a clear understanding of the methodology employed in this study.

4 Data Collection and Preprocessing

To train the large model, we have established a large dataset by collecting photospheric magnetograms, UV/EUV solar images, solar radio flux, and soft X-ray flux data from advanced ground and space telescopes [40–42, 46]. Figure 2 demonstrates the data collection and preprocessing for magnetogram and UV/EUV solar images.

The data undergoes rigorous preprocessing, standardization, cleaning, and alignment to ensure consistency and quality:

(1) Preprocessing: Preprocessing of SDO data typically begins at the Level 0 stage, where raw telemetry data is received and reconstructed into images. This stage involves minimal processing, such as removing duplicate data points and performing basic sorting and matching of data packets. However, the true preprocessing starts at Level 1, where images are further refined by removing bad pixels, applying flat-field corrections, and conducting quality assessment checks. For instruments like HMI and AIA, this level of processing ensures that the images are as close to their true form as possible, ready for further analysis.
(2) Standardization: Standardization of SDO data involves converting the preprocessed data into a common format that can be easily accessed and manipulated. For instance, JSOC image data is stored in compressed FITS files, a standard format for astronomical data. Metadata associated with the images is also stored in a standardized format, such as DRMS records, which can be exported in various standard protocols. Standardization makes it easier for researchers to work with the data, as they do not need to worry about different data formats or incompatible software.

(3) Cleaning: Cleaning of SDO data is an essential step in ensuring data quality. It involves removing any artifacts or anomalies that could affect the accuracy of scientific analysis. For example, Level 1.5 processing for both HMI and AIA involves additional cleaning steps, such as despiking (removing sudden, large pixel value changes that are not due to solar activity) and updating pointing keywords to remove the roll angle of the satellite. These steps help to produce images that are as free from errors as possible, allowing for more accurate scientific inferences.
(4) Alignment: Alignment of SDO data is crucial for comparing images from different channels or time points. For AIA, this typically involves promoting data from Level 1 to Level 1.5, which involves scaling the image to a standard resolution, translating the image to center the Sun, and updating the pointing keywords. This ensures that images from different channels or time points can be compared directly, without any issues related to differing resolutions, orientations, or centering.

This comprehensive dataset provides a solid foundation for training and evaluating the large model. The dataset includes 141888 flares of C-class, and 5949 M class and above flares. The distribution of samples of the dataset in chronological order is shown in Fig. 3

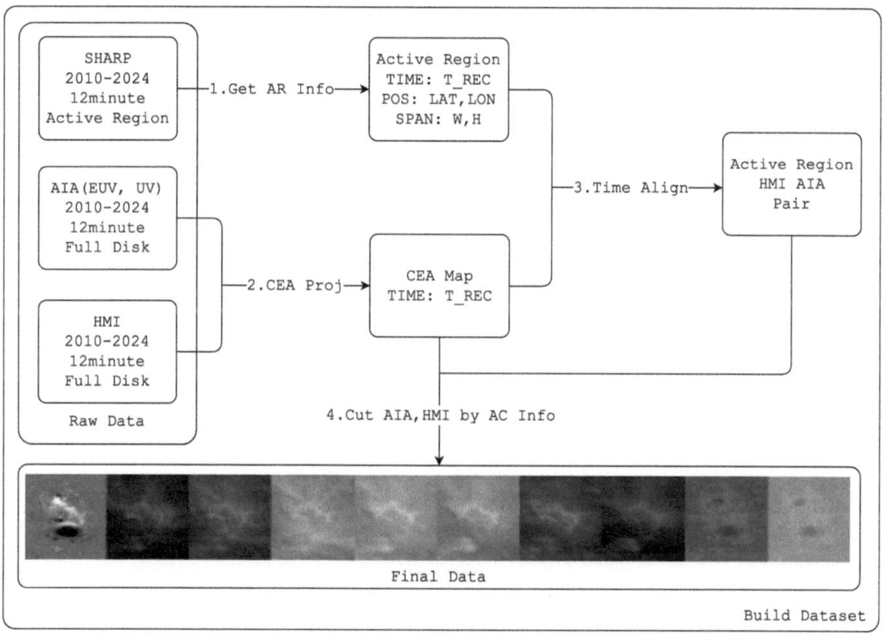

Fig. 2. Data collection for training large model

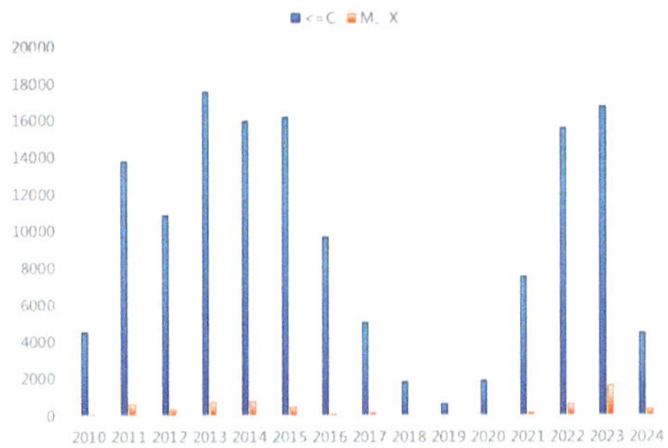

Fig. 3. Data distribution of our established flare forecast dataset

5 Experimental Results and Insights

At this moment, we don't have a mature large model of multimodality. Only an image large model taking the classic Mask-AutoEncoder [37] as baseline is trained over the observations of SDO/AIA and HMI [40–42]. Four Mask-AutoEncoder networks with different number of parameters are studied in this work, namely "Tiny", "Small", "Base" and "Large". The "Large" one has the most parameters, with the best performance at the same time. The configurations of these four pre-trained Mask-AutoEncoder models are listed in Table 1, where number of parameters and flops are provided for comparison.

Table 1. The configuration of the four Mask-AutoEncoders

Model volume	Parameters (M)	FLOPS (G)
Tiny	1.27	0.25
Small	18.85	3.67
Base	72.36	14.18
Large	250.11	49.01

Figure 4 shows an example of magnetogram generation, where UV/EUV images are the input, the model outputs magnetogram. In addition, flare forecasting is investigated by utilizing the proposed large model for feature extraction. In this experiment, we have a dataset as listed in Table 2. The forecasting accuracy measured by AUC, ACC, Precision and Recall is listed in Table 3. We note that the accuracy (ACC) of 0.87 surpasses the current state-of-the-art as reported in [43–45].

For evaluating a classifier, many evaluation indexes/metrics have provided, e.g., accuracy, error rate, precision and recall. They are defined on the basis of the following four basic terms.

(i) True positive (TP): if a positive instance is successfully predicted to be a position class;
(ii) False negative (FN): if a positive instance is wrong predicted to be a negative class;

(iii) False positive (FP): if a negative instance is predicted to be a positive class; (iv) True negative (TN): if a negative instance is successfully predicted to be a negative class.

Over these four terms, the percentage/ratio of TP among all positive instances (TP + FN) is named true positive rate (TPR), i.e.,

$$\text{TPR} = \text{TP}/(\text{TP} + \text{FN}), \tag{1}$$

representing how much percent of positive instances which are successfully retrieved (correctly classified into positive class) within all positive instances. Similarly, false positive rate (FPR) is defined as

$$\text{FPR} = \text{FP}/(\text{FP} + \text{TN}), \tag{2}$$

indicating how much percent of negative instances which are wrong decided as positive instances within negative instance category. Let P and N represent the number of positive instance and negative instance, respectively. Accuracy index is defined as

$$\text{Accuracy} = (\text{TP} + \text{TN})/(\text{P} + \text{N}), \tag{3}$$

precision or precision ratio is defined as

$$\text{Precision} = \text{TP}/(\text{TP} + \text{FP}), \tag{4}$$

recall or recall ration is defined as

$$\text{Recall} = \text{TP}/(\text{TP} + \text{FN}) = \text{TP}/\text{P} = \text{sensitive}. \tag{5}$$

From definitions above, accuracy give the percentage of instances successfully classified in all instances (P + N is the total number of instances). The numerator is composed of all instances classified correctly, no matter positive instance or negative instance, while the denominator consists of all instances.

AUC (Area Under the Curve) is a metric mainly used in binary classification. It represents the area under the Receiver Operating Characteristic (ROC) curve. The ROC curve plots the TPR against the FPR at different classification thresholds. AUC ranges from 0.5 (random guessing) to 1 (perfect classification). It measures the overall performance of a classification model across all possible thresholds.

Experimental results demonstrate the enhanced performance of AI with large models compared to traditional forecasting methods documented in [43–45]. The large model shows greater accuracy in predicting solar flares and other eruptive events, providing valuable insights into the mechanisms and triggers of solar bursts. These findings indicate an advancement in AI-driven solar activity forecasting, highlighting the predictive potential of utilizing large models.

Table 2. The number of samples for model training

Flare level		
	<C	M, X
Train	97623	3181
Test	44265	2768

Table 3. Solar flare forecasting performance

Index	Score
Auc	0.8943
Acc	0.8730
Precision	0.2710
Recall	0.6846

Figure 4 demonstrates the magnetogram generation, it can be observed that HMI-like magnetogram can be generated by the proposed large model from the inputs of UV/EUV solar images of SDO/AIA. The black colored area indicates no UV/EUV image input at this channel. It can be observed that more channels can benefit magnetogram generation with better image quality than others.

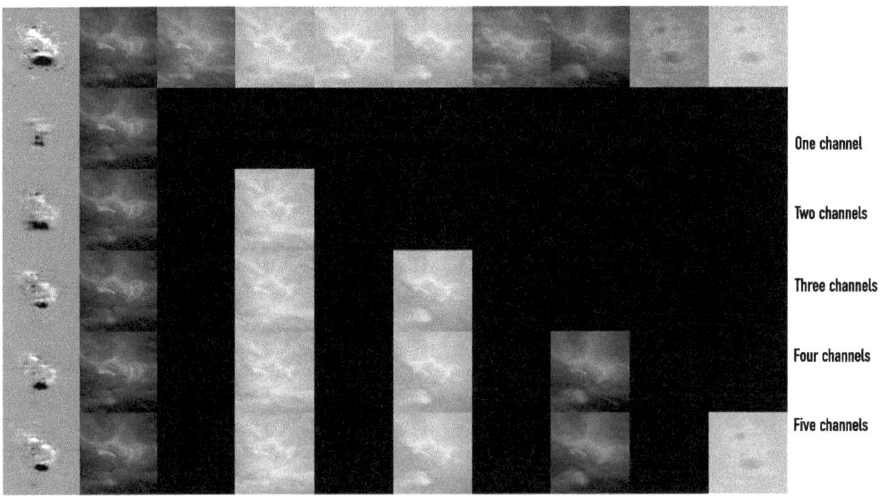

Fig. 4. Magnetogram generation from UV/EUV solar images of SDO/AIA

6 Conclusion

This article has presented a comprehensive overview of the application of large AI models for solar activity forecasting. By leveraging deep learning techniques and integrating multimodal information, we have demonstrated a promising example of the potential of large models to enhance predictive capabilities and provide valuable insights into solar activity. With ongoing research and advancements, the future of solar activity forecasting will possibly be revolutionized by the power of large AI models.

While the application of large models to solar activity forecasting shows promising results, several challenges remain. The scarcity of specifically designed large models for solar astronomy necessitates further research and development. Additionally, the computational demands of training and deploying such models pose practical limitations. Future work should focus on optimizing model architectures, improving data efficiency, and developing techniques to mitigate overfitting.

By leveraging these deep learning techniques, researchers and scientists can improve the accuracy, robustness, and timeliness of solar activity forecasting, ultimately contributing to a better understanding of solar dynamics and more effective space weather prediction.

References

1. Georgoulis, M., Yardley, S., Guerra, J., et al.: Prediction of solar energetic events impacting space weather conditions. Adv. Space Res. (2024)
2. Liu, H., Liu, C., Wang, J., Wang, H.: Predicting coronal mass ejections using SDO/HMI vector magnetic data products and recurrent neural networks. Astrophys J **890**(1), 1–9 (2020)
3. Huang, X., Zhao, Z., Zhong, Y., et al.: Short-term solar eruptive activity prediction models based on machine learning approaches: a review. Sci. China Earth Sci. **67**, 1–38 (2024)
4. Xu, L., Yan, Y., Huang, X.: Deep Learning in Solar Astronomy. Springer Nature Singapore, Singapore (2022)
5. Ramos, A., Cheung, M., Chifu, I., Gafeira, R.: Machine Learning in Solar Physics. Living Rev. Sol. Phys. **20**(4), 1–89 (2023)
6. Brown, T., Mann, B., Ryder, N., et al.: Language models are few-shot learners. In: Advances in Neural Information Processing Systems, vol. 34, pp. 1877–1901. Virtual (2020)
7. Ramesh, A., Pavlov, M., Goh, G., et al.: Zero-shot text-to-image generation. In: Proceedings of the 38th International Conference on Machine Learning, vol. 139, pp. 8821–8831. Virtual (2021)
8. Radford, A., Kim, J., Hallacy C., et al: Learning transferable visual models from natural language supervision. In: International Conference on Machine Learning, pp. 8748–8763 (2021)
9. Zhong, Y., Zhao, D., Huang, X., Xu, L.: CME arrival time prediction via fusion of physical parameters and image features. Astrophys. J. Suppl. Ser. **271**(1), 1–11 (2024)
10. Lin, R., Luo, Z., He, J., Xie, L., Hou, C.: Prediction of solar wind speed through machine learning from extrapolated solar coronal magnetic field. Space Weather **22**(6), 1–12 (2024)
11. Hosseinzadeh, P., Boubrahimi, S., Hamdi, S.: Toward enhanced prediction of high-impact solar energetic particle events using multimodal time series data fusion models. Space Weather **22**(6), 1–18 (2024)
12. Bengio, Y.: Learning Deep Architectures for AI. Foundations and Trends in Machine Learning (2009)

13. LeCun, Y., Boser, B., Denker, J., et al.: Backpropagation applied to handwritten zip code recognition. Neural Comput. **1**(4), 541–551 (1989)
14. Simonyan, K., Zisserman, A.: Very deep convolutional networks for large-scale image recognition. In: International Conference on Learning Representations. San Diego, CA, USA (2015)
15. Vincent, P., Larochelle, H., Bengio, Y., Manzagol, P.: Extracting and composing robust features with denoising autoencoders. In: Proceedings of the 25th International Conference on Machine Learning, pp. 1096–1103. Association for Computing Machinery, New York, NY, United States (2008)
16. Sohn, K., Jung, D., Lee, H., Hero, A.: Efficient learning of sparse, distributed, convolutional feature representations for object recognition. In: 2011 International Conference on Computer Vision, pp. 2643–2650. IEEE, Barcelona, Spain (2011)
17. Mohamed, A., Dahl, G., Hinton, G.: Acoustic modeling using deep belief networks. IEEE Trans. Audio Speech Lang. Process. **20**(1), 14–22 (2012)
18. Collobert, R., Weston, J., Bottou, L., et al.: Natural language processing (almost) from scratch. J. Machine Learn. Res. **12**, 2493–2537 (2011)
19. Krizhevsky, A., Sutskever, I., Hinton, G.: Imagenet classification with deep convolutional neural networks. Commun. ACM **60**(6), 84–90 (2017)
20. Szegedy, C., Liu, W., Jia, Y.: Going Deeper with Convolutions. In: 2015 IEEE Conference on Computer Vision and Pattern Recognition, pp. 1–9. IEEE, Boston, MA, USA (2015)
21. He, K., Zhang, X. Ren, S.; Sun, J.: Deep residual learning for image recognition. In: Proceedings of the IEEE Conference on Computer Vision and Pattern Recognition, pp. 770–778. IEEE, Las Vegas, NV, USA (2016)
22. Huang, G., Liu Z., Maaten, L., Weinberger, K.: Densely connected convolutional networks. In: 2017 IEEE Conference on Computer Vision and Pattern Recognition, pp. 2261–2269. IEEE, Honolulu, HI, USA (2017)
23. Russakovsky, O., Deng, J., Su, H., et al.: Imagenet large scale visual recognition challenge. Int. J. Comput. Vision **115**, 211–252 (2015)
24. Sherstinsky, A.: Fundamentals of recurrent neural network (RNN) and long short-term memory (LSTM) network. Physica D **404**, 1–28 (2020)
25. Hochreiter, S., Schmidhuber, J.: Long short-term memory. Neural Comput. **9**(8), 1735–1780 (1997)
26. Goodfellow, I., Pouget-Abadie, J., Mirza, M., et al.: Generative adversarial networks. Commun. ACM **63**(11), 139–144 (2020)
27. Dou, F., Xu, L., Ren, Z., et al.: Super-resolution of solar magnetograms using deep learning. Res. Astron. Astrophys. **22**(8), 1–12 (2022)
28. Dou, F., Xu, L., Zhao, D., Ren, Z.: A Multibranch deep neural network for the super-resolution of solar magnetograms. Astrophys. J. Suppl. Ser. **271**(1), 1–13 (2024)
29. Devlin, J., Chang, M., Lee, K., Toutanova K.: BERT: pre-training of deep bidirectional transformers for language understanding. In: Proceedings of the 2019 Conference of the North American Chapter of the Association for Computational Linguistics: Human Language Technologies, pp. 4171–4186. Association for Computational Linguistics, Minneapolis, Minnesota (2019)
30. Sun, Z., Bobra, M., Wang, X., et al.: Predicting solar flares using CNN and LSTM on two solar cycles of active region data. Astrophys J **931**(2), 1–23 (2022)
31. Minaee, S., Mikolov, T., Nikzad, N., et al.: Large language models: A survey. arXiv (2024)
32. Zhang, X., Chowdhury, R., Gupta, R., Shang, J.: Large language models for time series: a survey. In: Proceedings of the Thirty-Third International Joint Conference on Artificial Intelligence, pp. 8335–8343. International Joint Conferences on Artificial Intelligence, Jeju, Korea (2024)

33. Kalyan, K.: A survey of GPT-3 family large language models including ChatGPT and GPT-4. Nat. Lang. Process. J. **6**, 1–48 (2024)
34. Radford, A., Wu, J., Child, R., et al.: Language models are unsupervised multitask learners. OpenAI Blog **1**(8) (2019)
35. Islam, R., Moushi, O.: GPT-4o: The Cutting-Edge Advancement in Multimodal LLM. TechRxiv (2024)
36. Kirillov, A., Mintun, E., Ravi, N., et al.: Segment Anything. In: 2023 IEEE/CVF International Conference on Computer Vision. IEEE, Paris, France (2023)
37. He, K., Chen, X., Xie S., et al.: Masked autoencoders are scalable vision learners. In: Proceedings of the IEEE/CVF Conference on Computer Vision and Pattern Recognition, pp. 16000–16009. IEEE, New Orleans, LA, USA (2022)
38. Yuan, L., Chen, D., Chen, Y., et al.: Florence: A New Foundation Model for Computer Vision. arXiv (2021)
39. Jia, C., Yang, Y., Xia, Y., et al.: Scaling up visual and vision-language representation learning with noisy text supervision. In: Proceedings of Machine Learning Research, pp. 4904–4916. JMLR, Virtual, Cambridge MA (2021)
40. SDO Obseratory, https://sdo.gsfc.nasa.gov/. Last accessed 31 Oct 2020
41. SOHO Obseratory: https://sohowww.nascom.nasa.gov/. Last accessed 31 Oct 2020
42. Stereo Obseratory: https://stereo.gsfc.nasa.gov/. Last accessed 31 Oct 2020
43. Abduallah, Y., Wang, J.: A Deep Learning Approach to Operational Flare Forecasting. arXiv (2024)
44. Liu, S., Xu, L., Zhao, Z. et al.: Deep learning based solar flare forecasting model. II. Influence of image resolution. Astrophys. J. **941**(20), 1–11 (2022)
45. Zhang, X., Xu, L., Li, Z., Huang, X.: Causal attention deep-learning model for solar flare forecasting. Astrophys. J. Suppl. Ser. **274**(38), 1–11 (2024)
46. SPWC: https://www.swpc.noaa.gov/products/goes-x-ray-flux. Last accessed 31 Oct 2020

Open Access This chapter is licensed under the terms of the Creative Commons Attribution 4.0 International License (http://creativecommons.org/licenses/by/4.0/), which permits use, sharing, adaptation, distribution and reproduction in any medium or format, as long as you give appropriate credit to the original author(s) and the source, provide a link to the Creative Commons license and indicate if changes were made.

The images or other third party material in this chapter are included in the chapter's Creative Commons license, unless indicated otherwise in a credit line to the material. If material is not included in the chapter's Creative Commons license and your intended use is not permitted by statutory regulation or exceeds the permitted use, you will need to obtain permission directly from the copyright holder.

Deep Learning Prediction of Inter-storm Parameters Using Transformer Convolution Network

Samuel Ogunjo[1(✉)], Babatunde Rabiu[2], Ibiyinka Fuwape[1,3], and Oluwatoyin Atikekeresola[1]

[1] Federal University of Technology Akure, Akure 340101, Nigeria
stogunjo@futa.edu.ng

[2] National Space Research and Development Agency, Abuja, Nigeria

[3] Michael and Cecilia Ibru University, Ughelli, Delta State, Nigeria

Abstract. The global impact of space weather has created interest in understanding and prediction of geomagnetic storms. Most approaches have considered the prediction of geomagnetic storms based on the time series. In this study, two characteristics of geomagnetic storms will be considered – the time interval and difference in value between two consecutive storms. The Kp index was used as a proxy for geomagnetic activities with values greater than or equal to 4 considered as a storm. The Transformer Convolution Network (TCN) deep learning algorithm was used for the prediction. Optimal model parameters for the two indices were obtained using hyperparameter tuning. Results obtained showed that TCN has good performance in predicting the difference in value but limited performance in time intervals. The limitations of TCN in predicting the inter-storm interval can be attributed to its chaotic nature.

Keywords: inter-storm interval · geomagnetic storm · transformer convolution network

1 Introduction

The impact of space weather on the technological [1], climatic [2], and biological [3] aspects of human existence on earth has been established. The stratified layers, complexity of the layers, and dynamical interactions between layers make space weather a complex system. The complexity is further exacerbated by the influence of human and natural activities such as earthquakes [4] and gravity waves [5] influence geospace. The need to better understand the space above us and mitigate its impact on our existence has led to increased study of geospace.

The behaviour of space parameters is usually characterized by measures of central tendencies and dispersion. However, this approach is not suitable for modelling and prediction. Stochastic methods such as autoregressive integrated moving average [6] have been used to model and predict the ionosphere, mesosphere, magnetosphere. In recent times, machine learning algorithms such as Long short-term memory, decision

© The Author(s) 2026
N. Gopalswamy et al. (Eds.): ISWI 2024, SPPHY 431, pp. 222–226, 2026.
https://doi.org/10.1007/978-981-95-1121-1_25

trees, random forest, support vector machines have been applied to space weather studies [6, 7]. In recent years, it has been shown that several parameters across different layers of geospace have chaotic properties [8, 9]. This implies that their long-term prediction will be difficult. Hence, there is the need for better approach, measures, and algorithms for prediction of space parameters.

Previous studies have applied machine learning algorithms directly on space weather parameters. However, these space weather parameters have complex structures which evolve in time and space. In this study, we applied a deep learning algorithm to two features of a space weather parameter – the difference between two storm values and the difference between the time intervals of two storm events. The effective predictions of these parameters will give information about the timing and value of future storms.

2 Materials and Methods

The measure of geomagnetic activity considered in this study was the Kp index. Kp index values between January 1, 1932 to December 31, 2020 were obtained from the archives of the GFZ German Research Centre for Geosciences. A storm day is considered as the day with Kp index greater than 4. Days with Kp $>$ 4 were isolated. The difference between the values of two consecutive storm days is regarded as the value interval while the time difference between them is referred to as the time interval. The time series for these variables are shown in Fig. 1.

The Transformer Convolution Network (TCN) is a neural network architecture that combines the strengths of convolutional neural networks (CNNs) and transformer models. It employs a 1D fully convolutional network, where each hidden layer consists of two convolution operations: a causal convolution and a dilated convolution. The causal convolution ensures that the output at any time step depends only on the inputs from previous time steps, making it suitable for sequence modelling tasks. The dilated convolution enables an exponentially large receptive field, allowing the network to capture long-range dependencies efficiently. TCNs employ residual connections and normalization techniques for improved training and performance. Furthermore, they leverage positional encodings and attention mechanisms from transformer models, enabling them to capture long-range dependencies and learn complex patterns in sequential data. The transformer convolution network has been used in various forms to predict space weather and parameters [10, 11].

TCN has a large parameter space. The kernel size determines the temporal span of each convolution, affecting the network's ability to capture short-term and long-term dependencies. Large kernel size enables the network to directly model longer-range dependencies within each layer. The number of filters dictates the quantity of feature maps produced, thus impacting the model's capacity to learn diverse representations. Low number of filters reduce the model's complexity, which can be beneficial for smaller datasets or when computational resources are constrained. The dilation base governs the expansion rate of the receptive field across layers, balancing between local and global pattern recognition. Dropout, a regularization technique, mitigates overfitting by stochastically nullifying a fraction of input units during training, thereby enhancing the model's generalization capabilities. To obtain the optimal parameters, an hyperparameter search was used. The optimized parameters are kernel size (2 – 8), number of filters (1 – 5), dilation base (2 – 4), and dropout (0.0 – 0.4). The TCN algorithm was implemented using the Darts python package [12].

3 Results and Discussion

The time series of both time and value differences are shown in Fig. 1. The temporal variation in the value time series mimics a stochastic system with symmetry around 0. However, the time difference series did not show any known pattern.

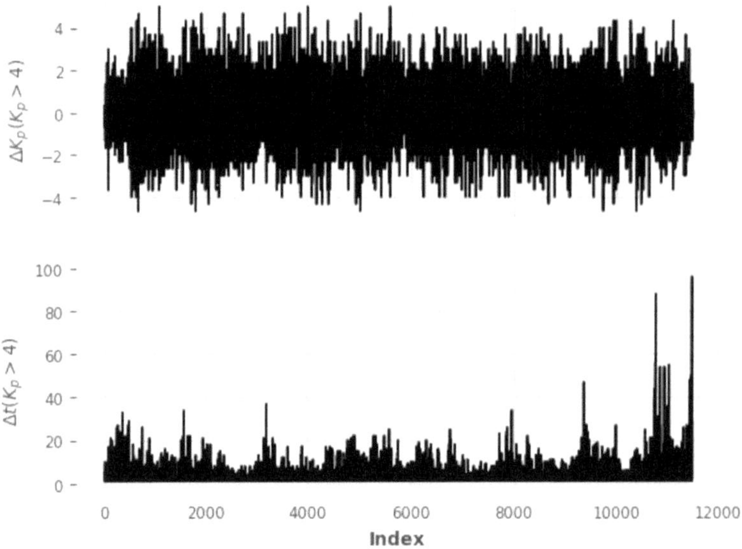

Fig. 1. Time series of the difference in value (top figure) and time interval (bottom) between two consecutive storms

For the value time series, the optimal parameters as reveal by the hyperparameter search were input length (42), length of kernel (4), number of filters (1), dilation base (3), and dropout (0.107). The optimal parameters for the time difference series were found to be input length (18), length of kernel (3), number of filters (3), dilation base (3), and dropout (0.1). Using these parameters, the computation took about 2 h on Google Colaboratory cloud computing platform. Based on the optimal parameters, predictions were made for the time series (Fig. 2). The predictions for the value time series were found to be in the range of the original time series with variations at some points. The root mean square error and mean absolute error for the value time series are 1.37 and 1.09 respectively. However, for the time difference series, the root mean square error and mean absolute errors are 16.29 and 9.39 respectively. The poor performance of TCN in the time difference were attributed to the chaotic nature [9]. In a chaotic system, the long-term prediction of future states is very difficult due to the system's sensitive dependence on initial conditions. This concept can be deployed to provide real-time

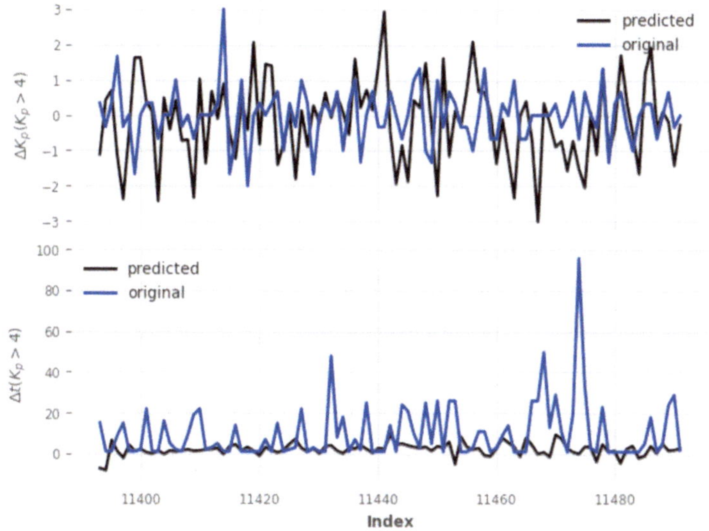

Fig. 2. Comparison of predicted values of the difference in value (top figure) and time interval (bottom) against original time series using TCN

forecasting system for strength and time to the next geomagnetic storms. One challenge with the real-time prediction is the accuracy of the forecast data. This can be overcome by comparing the output from different machine learning algorithm and finetuning the parameters over time.

4 Conclusion

Geomagnetic storms are important features of not only the space environment but the earth also. There have been several approaches to studying geomagnetic storms in literature. In this study, we explore a novel approach which uses the values and time differences between two consecutive storms. Predicting both the time and value between storms gives an indication of the timing and strength of the next storm. TCN with optimal parameters were used for the prediction of inter-storm values and times. The algorithm had a better performance in the value difference compared to the time difference. The poor performance in the time difference was attributed to the chaotic nature of the time series.

References

1. Brautigam, D.H.: CRRES in review: space weather and its effects on technology. J. Atmos. Solar Terr. Phys. **64**(16), 1709–1721 (2002)
2. Ogunjo, S.T., Rabiu, A.B.: Spatio-temporal influence of solar activity on global air temperature. Sun Geosphere **15**(2), 75–80 (2022)
3. Breus, T.K., Ozheredov, V.A., Syutkina, E.V., Rogoza, A.N.: Some aspects of the biological effects of space weather. J. Atmos. Solar Terr. Phys. **70**(2–4), 436–441 (2008)

4. Maekawa, S., et al.: A statistical study on the effect of earthquakes on the ionosphere, based on the subionospheric LF propagation data in Japan. Ann. Geophys. **24**(8), 2219–2225 (2006)
5. Laštovička, J.: Forcing of the ionosphere by waves from below. J. Atmos. Solar Terr. Phys. **68**(3–5), 479–497 (2006)
6. Saqib, M., Şentürk, E., Sahu, S.A., Adil, M.A.: Comparisons of autoregressive integrated moving average (ARIMA) and long short term memory (LSTM) network models for ionospheric anomalies detection: a study on Haiti (Mw = 7.0) earthquake. Acta Geod. Geophys. **57**(1), 195–213 (2022). https://doi.org/10.1007/s40328-021-00371-3
7. Iban, M.C., Şentürk, E.: Machine learning regression models for prediction of multiple ionospheric parameters. Adv. Space Res. **69**(3), 1319–1334 (2022)
8. Akerele, A., et al.: Complexity and nonlinear dependence of ionospheric electron content and doppler frequency shifts in propagating HF radio signals within equatorial regions. Atmosphere **15**(6), 654 (2024)
9. Ogunjo, S., Rabiu, B., Fuwape, I., Atikekeresola, O.: Multifractal and chaotic characteristics of geomagnetic inter-storm interval over eight solar cycles. Adv. Space Res. **73**(10), 5406–5413 (2024)
10. Abduallah, Y., Wang, J.T., Xu, C., Wang, H.: A transformer-based framework for geomagnetic activity prediction. In: International Symposium on Methodologies for Intelligent Systems, pp. 325–335. Springer International Publishing, Cham (2022). https://doi.org/10.1007/978-3-031-16564-1_31
11. Abduallah, Y., Wang, J.T., Wang, H., Jing, J.: A transformer-based framework for predicting geomagnetic indices with uncertainty quantification. J. Intell. Inf. Syst. **62**, 887–903 (2023)
12. Herzen, J., et al.: Darts.: user-friendly modern machine learning for time series. J. Mach. Learn. Rese. **23**(124), 1–6 (2022)

Open Access This chapter is licensed under the terms of the Creative Commons Attribution 4.0 International License (http://creativecommons.org/licenses/by/4.0/), which permits use, sharing, adaptation, distribution and reproduction in any medium or format, as long as you give appropriate credit to the original author(s) and the source, provide a link to the Creative Commons license and indicate if changes were made.

The images or other third party material in this chapter are included in the chapter's Creative Commons license, unless indicated otherwise in a credit line to the material. If material is not included in the chapter's Creative Commons license and your intended use is not permitted by statutory regulation or exceeds the permitted use, you will need to obtain permission directly from the copyright holder.

Space Weather Instrumentation

Solar Radio Imaging-Spectroscopic Observations of Fine Structures in a Flare Event

Yihua Yan[1,2,3](✉), Zhichao Zhou[1], Xin Yao[1], Chengming Tan[1], Wei Wang[1,2], and Linjie Chen[1,2]

[1] State Key Laboratory of Solar Activity and Space Weather, National Space Science Center, Chinese Academy of Sciences, Beijing, China
yyh@nao.cas.cn
[2] National Astronomical Observatories, Chinese Academy of Sciences, Beijing, China
[3] University of Chinese Academy of Sciences, Beijing, China

Abstract. Solar radio spikes and quasi-periodic pulsations (QPPs) can occur as fine structures in dynamic radio spectra. While their physical interpretation is still under debate, they are seen to contain detailed information about the complex energy release processes in solar flares. In this paper we investigate the relation of both phenomena in a case study of a M-class flare event on 21 June 2015. The imaging-spectroscopic analysis is based on observations by the Mingantu Spectral Radioheliograph (MUSER) and compared with EUV observation by the Atmospheric Imaging Assembly (AIA). The location of microwave (0.4–2 GHz) sources with respect to the flaring active region is discussed for selected times. Each group of spikes occurred simultaneously with quasi-periodic pulsations (QPPs). The spatially resolved data shows that during the flare process the radio sources were mainly located in the eastern bipolar region. For three groups of peculiar fine structures, we find that (1) the FS radio sources are mainly in the fan area converged to western spot instead of the eastern bipolar region; (2) the FS radio sources are in both the eastern bipolar region and the fan area; (3) the FS radio sources are mainly in the eastern bipolar region. This may be due to the interactions between the shearing flare loops in the eastern bipolar region and the fan area converged to the western sunspot.

Keywords: Solar Flares · Solar Radio Bursts · Radioheliograph

1 Introduction

Solar radio spikes associated with solar flares appear in very short, narrow band and bright structures in the dynamic spectrum [1–6], whereas quasi-periodic pulsations (QPPs) are short and very fast pulsations with broadband features [7–10]. Both fine structures are normally considered as manifestations of the energy release processes in the solar corona. The short duration and the narrow bandwidth of spikes suggest a small source size and therefore a high radio brightness temperature. A loss-cone instability of trapped electrons was proposed to produce electron cyclotron maser emission at the footpoints of flare loops [1], or the spike sources could be in the acceleration regions of flares and

© The Author(s) 2026
N. Gopalswamy et al. (Eds.): ISWI 2024, SPPHY 431, pp. 229–240, 2026.
https://doi.org/10.1007/978-981-95-1121-1_26

to result from waves produced by the acceleration process [3]. It was also suggested that the stochastic spike bursts were due to the termination shock front reacts dynamically to the arrival of the fast plasma downflows [6]. On the other side, during some solar flares, multiple layers of the solar atmosphere are involved in the process, making it possible to emit over wide wavelengths. QPPs allow to determine whether the observed features are the result of modulation of the acceleration process itself or by modulating the emission of already accelerated particles [9]. The spikes and QPPs in the decimeter wave range are therefore providing a solid opportunity for us to understand the plasma kinetic processes in solar flares with the radio imaging-spectroscopic observations, which are helpful for advancing our knowledge of these dynamic processes, magnetic fields and the atmospheric structures above the solar surface (see, e.g., recent review [11]).

The Mingantu Spectral Radioheliograph (MUSER, [12]), located in Inner Mongolia of China, is a solar-dedicated facility designed with characteristics to perform high-quality imaging spectroscopy of the Sun. The MUSER images were processed with a new calibration method [13] for processing MUSER observations. From both simulations and MUSER realistic data in solar observations, one obtains an accuracy of a few percents of the synthesized beam [13]. The method can be applied to general radio interferometry and aperture synthesis problems [13].

On 2015 June 21, many solar radio fine structures were observed by MUSER when two successive M-class flares occurred, among these fine structures, groups of spikes in 0.6–2 GHz range showed special features, i.e., each group of spikes were co-occurred on QPPs [14]. Here we present the first imaging-spectroscopic analysis for this event to understand the physic processes of these fine structures.

2 Methods

In this paper we use data from ground- and space-based instruments for a detailed analysis of fine structures in radio observations in the context of a flare event. A focus is set on MUSER radio observations, which require a specific pre-processing before analyzing them.

The MUSER images are processed with a new calibration method [13]. For the sake of clarity, we briefly summarize the method by Zhou et al. [13] here. A point-source is an ideal calibrator in radio interferometry and radio aperture synthesis. However, its position cannot be determined by the aperture synthesis theory exclusively but must be prescribed in advance by other means. The position of the point-source cannot be deduced from closure phase measurements alone [15].

In general, the radio image obtained by an aperture synthesis array can be expressed as ([15, 16]),

$$I^D(l,m) = \iint V(u,v)S(u,v)e^{2\pi j(ul+vm)}dudv, \text{ or } I^D(l,m) = I(l,m)*B(l,m), \quad (1)$$

where

$$I(l,m) = \iint V(u,v)e^{2\pi j(ul+vm)}dudv, \; B(l,m) = \iint S(u,v)e^{2\pi j(ul+vm)}dudv. \quad (2)$$

That is an inverse Fourier Transform approach, in which I^D is the observed radio map in the image plane (l,m), or dirty map, V and S are the calibrated visibility function of the target observed by the aperture synthesis array and the sampling function in the spatial frequency plane (u,v), respectively. I is the sky map and B is the dirty beam. Normally we can obtain the approximation of I with the observed I^D and known B from Eq. (1) through standard CLEAN algorithm [15].

If, however, there exists an offset term $(\Delta\, l,\ \Delta\, m)$, either known or unknown, of the point-like calibrator source from the origin, which would affect the calibrated visibility function V by introducing an additional phase term: $2\pi j(u_{cal}\Delta l + v_{cal}\Delta m)$. Thus, the final dirty image I^D obtained with $(\Delta\, l,\ \Delta\, m)$ as parameters will be blurred and modulated as follows,

$$I^D(l, m; \Delta l, \Delta m) = \iint V(u, v)e^{2\pi j(u_{cal}\Delta l + v_{cal}\Delta m)} S(u, v)e^{2\pi j(ul+vm)} dudv. \qquad (3)$$

With the decomposition below [13],

$$\frac{u_{cal}}{u} = \xi_0 + \frac{\xi(u, v)}{u}, \ \frac{v_{cal}}{v} = \eta_0 + \frac{\eta(u, v)}{v}, \qquad (4)$$

where (ξ_0, η_0) are constants, and substituting it into Eq. (3), one obtains the following result,

$$I^D(l, m; \Delta l, \Delta m) = I(i, m)*B(l + \xi_0\Delta l, m + \eta_0\Delta m)*H(l, m; \Delta l, \Delta m), \qquad (5)$$

in which

$$H(l, m; \Delta l, \Delta m) = \iint e^{2\pi j[\xi(u,v)\Delta l + \eta(u,v)\Delta m]} e^{2\pi j(ul+vm)} dudv, \qquad (6)$$

is a blurring modulation function [13]. For a given baseline in the (u,v) plane, it in general traces an ellipse varying with hour angle [15]. Therefore (ξ_0, η_0) are also changing with time although they are constants for a certain instant. As $H(l,m;\ \Delta l,\ \Delta m)$ has the form of $e^{j\phi}$ in the integrand in Eq. (6), its modulus identifies unity. The merit is that $H(l,m;\ \Delta l,\ \Delta m)$ may modulate the target dirty image distribution neither exceeding the maximum intensity of the original dirty image nor changing the total energy of the original signal [13]. Therefore, the influence of the point-source calibrator offset to the final dirty map ID is that the target solar image should now be convolved with a deviated dirty beam and the blurring modulation function $H(l,m;\ \Delta l,\ \Delta m)$ as expressed in Eq. (5). Furthermore, the inverse Fourier Transform of $H(l,m;\ \Delta l,\ \Delta m)$ becomes a δ-function under the condition of zero offset ($\Delta l = 0,\ \Delta m = 0$). Hence one can achieve the desired correct result based on Eq. (5), which provides a solid mathematical foundation. So, the position of any point-like calibrator can now be determined within the framework of the aperture synthesis theory. It was simulated for prescribed known radio sources with the offset of 5 to 10 times synthesized beam size and an accuracy of a few percents of the synthesized beam can be achieved based on the novel formula Eq. (5) [13].

3 Multi-Wavelength Flare Observations

3.1 GOES Observations

The Geostationary Operational Environmental Satellite (GOES) observed a M2.0 flare between 01:02UT and 02:00 UT (peak at 01:42UT). Successively, a M2.7 flare was observed from 02:04 UT to 03:11 UT (peak at 02:34UT).

3.2 SDO Observations

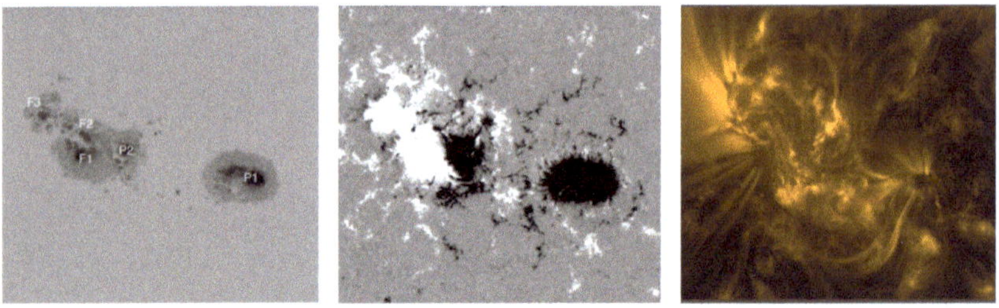

(a) Sunspots (b) Magnetogram (c) 171 Å EUV image

Fig. 1. The active region NOAA 12371 as observed by HMI/SDO and AIA/SDO on 21 June 2015 at 01:00:40 UT before the flare. The subfigures show sunspots (a), a line of sight magnetogram (b), and an 171 Å EUV image (c).

Both flares are originating from the same active region NOAA 12371 as shown in Fig. 1. The data shows the sunspots, the line-of-sight magnetogram and the extreme ultraviolet image at 171 Å observed by the Atmospheric Imaging Assembly (AIA) and the Magnetic Imager (HMI) onboard the Solar Dynamics Observatory (SDO). at 01:00:40 UT before the flare. The western sunspot group P1 is mainly characterized by negative magnetic polarity, whereas the eastern sunspot group is consisting of mixed magnetic polarities. The leading part P2 is of negative magnetic polarity and the rest sunspots F1, F2, and F3 are of positive magnetic polarity. The magnetic fields near the major polarity inversion line in the eastern sunspot group between P2 and F1, F2 must be in very strong shearing situation as outlined by EUV loops that are almost along with polarity inversion line there. The western P1 and eastern F1 sunspot groups seem to be magnetically connected through large scale loops. There seems also existing a supra-arcade fan area (e.g. [17]) or quasi-separatrix layers (e.g.[18]) from the shearing EUV loops converged into P1 spot. Figure 2 shows the evolution of the active region in the different flare phases as outlined by the AIA images at 171 Å at several instants from 01:00 UT to 03:10 UT. The M2 flare starts at shearing EUV loops along the polarity inversion line between P2 and the sunspots F1, F2. The loops reach their maximum around 1:42 UT with the whole shearing EUV loops brightening there. Then the shearing loops relax and expand for a few minutes. They continue to interact with the surrounding magnetic structures and reach the second M2.7 flare peak at 02:34 UT. The shearing EUV loops become more expanded and relaxed as witnessed by the increased orientation angles with the polarity inversion line of these loops and more F1 area were covered by the overlying bright EUV loops. The discrete bright points can be observed in the fan area during M2.0 flare process. The enhanced brightening features outlining the fan area can be observed during the whole process of the M2.7 flare.

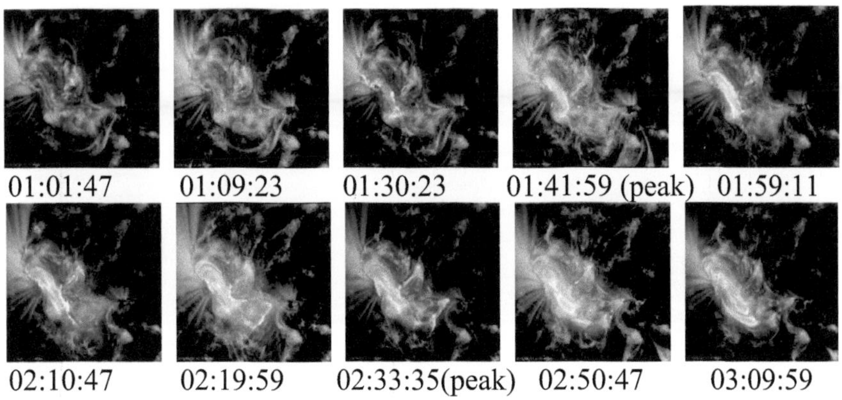

Fig. 2. The AIA/SDO EUV 171 Å images show the evolution of the M2.0 flare (started at 01:02, peaked at 01:42, and ended at 02:00 UT) and the M2.7 flare (started at 02:04, peaked at 02:34, and ended at 03:11 UT).

3.3 Observation of Solar Radio Burst Fine Structures

In the dynamic radio spectra recorded by MUSER on 2015 June 21, there are various radio fine structures including QPPs, lace, spikes, fibers and slow drifting bursts [19]. Among these fine structures, 64 groups of spikes are found with two different spectral features [14]:

- There are 21 groups of spikes in right-circular-polarization (RCP) spectra. Each group of spikes is divided into many quasi-periodic clusters and each cluster contains many individual spike bursts. These spikes are concurrent with the broadband QPPs.
- There are other groups of spikes randomly scattered in a broader frequency band. These include 34 groups in the left-circular-polarization (LCP) spectrum, 3 groups in the RCP spectrum and 6 groups in both LCP and RCP spectrums.

The latter category has the common spectral feature as studied in many previous literatures (e.g., [1–6]). For the first category, the spectral analysis of these 21 groups of quasi-periodic clusters of spikes during the M2.0 flare was studied in [14]. Here we present the spectroscopic imaging of some clusters of spikes.

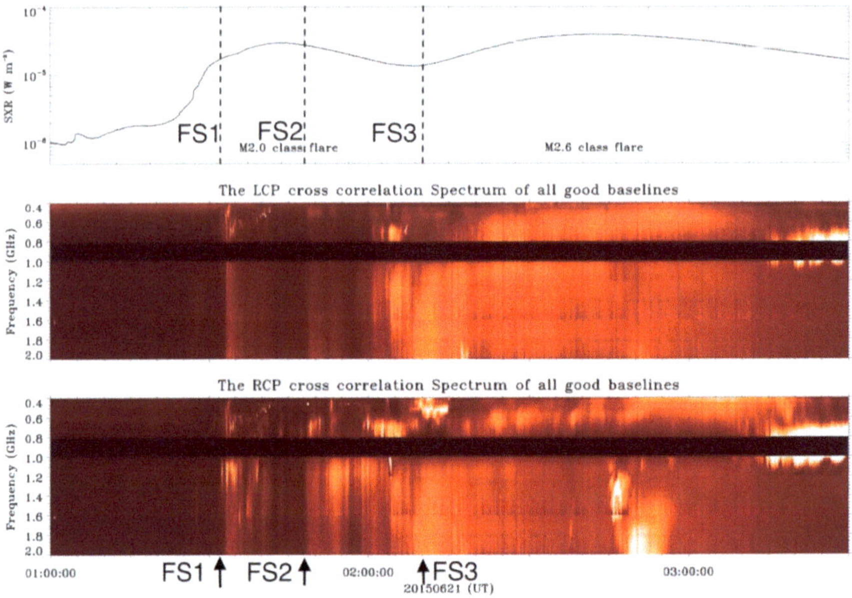

Fig. 3. The M-class flares in GOES SXR flux at 1–8 Å (top) and the cross-correlation dynamic spectra in left-circular polarizations (LCP, middle) and right-circular polarizations (RCP, bottom) by MUSER on 2015 June 21during 01:00–03:30 UT. FS1, FS2 and FS3 mark the instants around 01:32UT, 01:48 UT and 02:10UT where some peculiar fine structures occurred and are shown in Fig. 4.

There are quasi-periodic clusters of spikes co-occurred on QPPs (Fig. 3), and some peculiar examples of FSs (Fig. 4). These clusters of spikes are mainly RCPs. The radio fine structures FS1 occurred before the maximum of M2.0 flare, whereas FS2 occurred after the M2.0 flare peak. FS3 occurred after the end of M2.0 flare and just at the rising phase of M2.7 flare. In the following we will present the spectroscopic imaging of these three groups of clusters of spikes.

3.4 Radio Imaging-Spectroscopy

The MUSER imaging-spectroscopic observations on 21 June 2015 are thus processed with the above calibration method (see Sect. 2). The co-alignments of MUSER and AIA data are in agreement with each other (Fig. 5). Then we demonstrate the MUSER imaging-spectroscopic observations of the radio fine structures during the flare processes.

We selected some instants in the time interval from 01:32:01UT to 01:32:09 UT closing to the M2.0 flare maximum at 01:34 UT when the quasi-periodic clusters of spikes (marked as FS1) occurred and obtained the corresponding MUSER multiple frequency images as shown in Fig. 6(a) by different contours. For some instants covering the same period of FS1 but without the occurrences of these fine structures as well as some instants before the flare onset when there were no fine structures, the corresponding MUSER multiple frequency contours are shown in Fig. 6(b). By comparing Fig. 6(a) and (b) we found that the radio sources were mainly located in the eastern bipolar region when there were no occurrences of FS1. But when the peculiar fine structures FS1 occurred the radio sources also appeared in the fan area approaching to the western spot P1 in addition to the eastern main flare region. Though the FS1 sources are distinct from the main flare source, a situation once mentioned in [3], they still agree with the scenario that the spike emission is produced near the footpoints of the fan loops connecting the shearing flare loops. In this case, an electron-cyclotron maser instability may arise from the loss-cone distribution of the coronal plasma. The interaction between the shearing flare loops and the fan loops may account for the QPPs. We conducted the same analyses for the radio fine structures FS2 and FS3. For FS2 from 01:47:37UT to 01:48:02 UT after the M2.0 flare peak, the corresponding MUSER multiple frequency images with and without the occurrences of FS2 are shown in Fig. 6(c) and (d) with different contours. We find that the radio sources are in both the eastern bipolar region and the fan area approaching to the western spot P1, no matter with or without the occurrences of FS2. This may be caused by the interaction between the shearing flare loops and the fan loops just after the M2.0 flare maximum. For FS3 from 02:10:42UT to 02:10:53 UT between the two M-class flares, the corresponding MUSER multiple frequency images with and without the occurrences of FS3 are shown in Fig. 6(e) and (f) by different contours. We find that the radio sources are mainly in the eastern bipolar region with or without the occurrences of FS3. This may be due to the occurrence of FS3 after the ending of the first M2.0 flare and just at the rising phase of the second M2.7 flare. The relaxed shearing flare loops likely weakened the interaction between the shearing flare loops and the fan loops.

These MUSER multiple frequency images may help to understand some peculiar examples of the quasi-periodic clusters of spikes: FS1, FS2 and FS3.

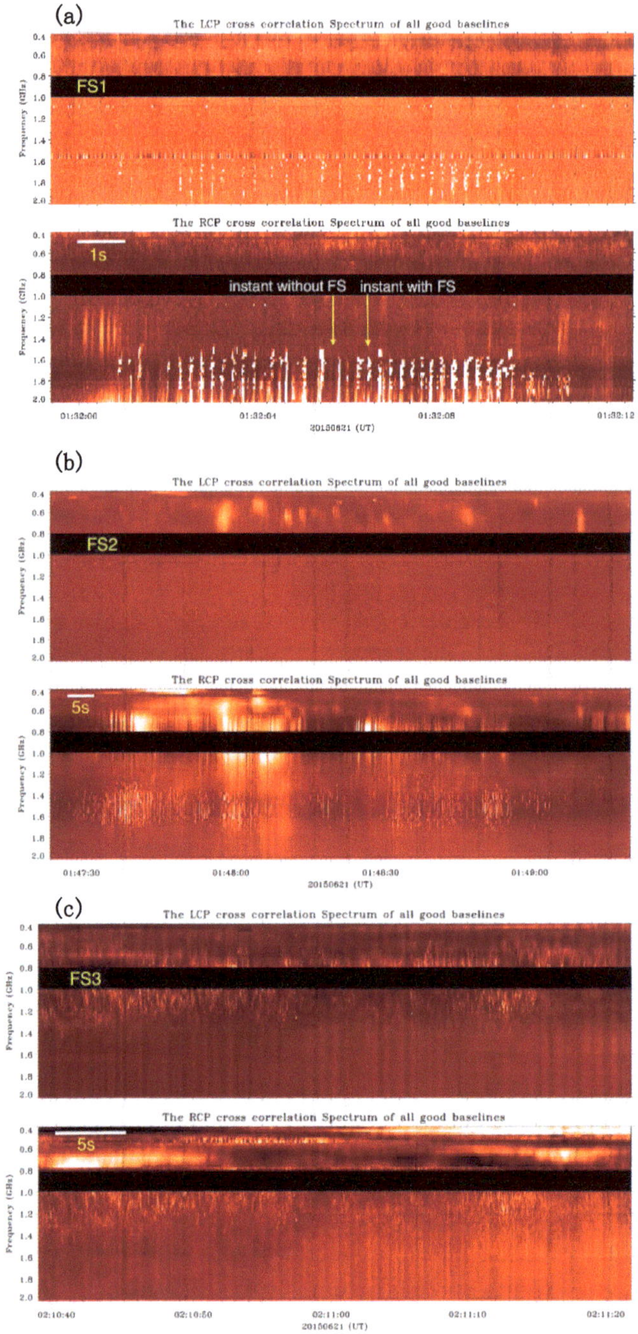

Fig. 4. The cross-correlation dynamic spectra of some peculiar quasi-periodic clusters of spikes co-occurred on QPPs around (a) 01:32 UT (FS1), (b) 01:48 UT (FS2), and (c) 02:10 UT (FS3) with the white bar indicating different time scales.

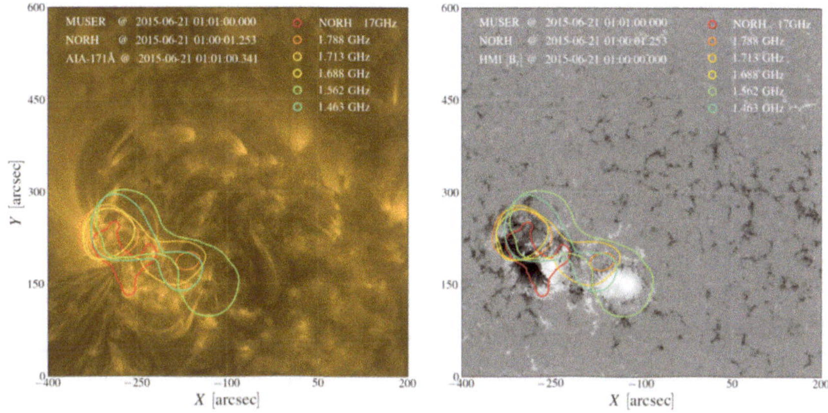

Fig. 5. MUSER multiple frequency contours (from green to orange) and the NoRH 17 GHz contours (in red) for the quiet Sun around 01:00 UT overlaid with AIA/SDO EUV image (Left panel) and HMI/ SDO magnetogram (Right panel).

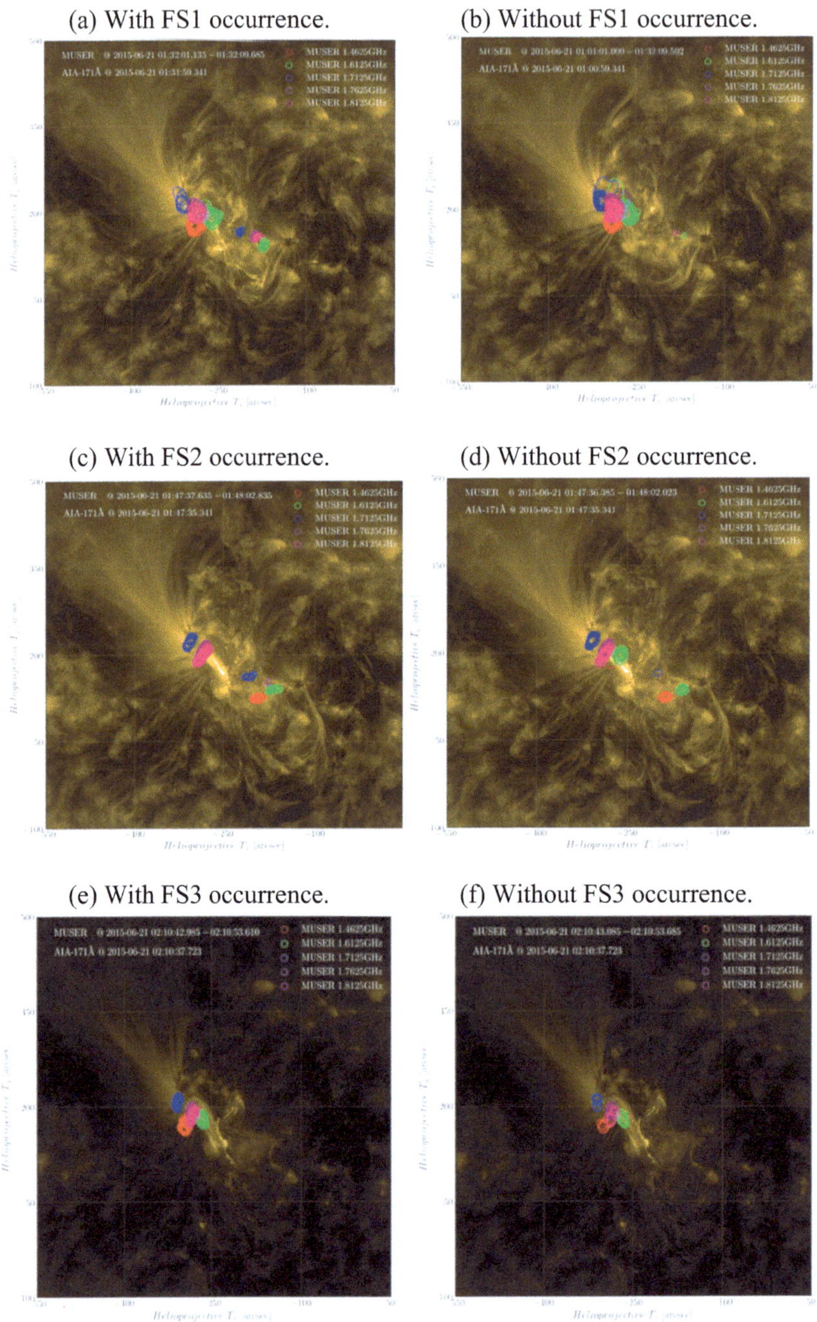

Fig. 6. MUSER multiple frequency contours for the peculiar examples of the fine structures: FS1, FS2, and FS3 with (left panels) and without (right panels) the presence of the spike enhancement.

4 Conclusion

The presented MUSER multi-frequency observations of the M-class flare event on 21 June 2015 contained peculiar fine structures. It shows that during the flare process the radio sources were mainly located in the eastern bipolar region. When the radio fine structures FS1 occurred closing to the M2.0 flare peak, the corresponding radio sources were located elsewhere in the western fan area as demonstrated above. For the radio fine structures FS2 occurred after the M2.0 flare peak, the corresponding radio sources were in both eastern bipolar region and western fan area, whereas for the radio fine structures FS3 between the two M-class flares, the radio sources were mainly located in the eastern bipolar region. These RCP burst fine structures are probably excited in extraordinary mode. The spike emission is produced near the footpoints and interpreted as a result of arising electron-cyclotron maser instabilities in the loss-cone distribution of the ambient plasma. The interaction between the shearing flare loops and the fan loops may account for the QPPs. The calibration method by Zhou et al. [11] was successfully applied. The MUSER data analysis shows that solar radio imaging-spectroscopy is a powerful method to observe small-scale processes in solar flares, in order to extended our knowledge about space weather.

Acknowledgement. The work of YY was supported by the National Key R&D Program of China (2021YFA1600500 and 2021YFA1600503) and partially supported by Tianchi Talent Program of Xinjiang Uygur Autonomous Region of China. This research is based on observations made with MUSER and AIA/HMI on NASA's SDO satellite.

References

1. Melrose, D.B., Dulk, G.: Electron-cyclotron masers as the source of certain solar and stellar radio bursts. Astrophys. J. **259**, 844–858 (1982)
2. Benz, A.O.: Radio spikes and the fragmentation of flare energy release. Sol. Phys. **96**, 357–370 (1985)
3. Guedel, M., Benz, A.O.: Time profiles of solar radio spikes. Astron. Astrophys. **231**, 202–212 (1990)
4. Karlický, M., Sobotka, M., Jiřička, K.: Narrowband dm-Spikes in the 2 GHz frequency range and MHD cascading waves in reconnection outflows. Sol. Phys. **168**(2), 375–383 (1996)
5. Benz, A.O., Saint-Hilaire, P., Vilmer, N.: Location of narrowband spikes in solar flares. Astron. Astrophys. **383**, 678–684 (2002)
6. Chen, B., Bastian, T.S., Shen, C., Gary, D.E., Krucker, S., Glesener, L.: Particle acceleration by a solar flare termination shock. Science **350**, 1238–1242 (2015)
7. Tan, B., Yan, Y., Tan, C., Liu, Y.: The microwave pulsations and the tearing modes in the current-carrying flare loops. Astrophys. J. **671**(1), 964–972 (2007)
8. Nakariakov, V.M., Melnikov, V.F.: Quasi-periodic pulsations in solar flares. Space Sci. Rev. **149**, 119–151 (2009)
9. Kupriyanova, E.G., Kashapova, L.K., Van Doorsselaere, T., Chowdhury, P., Srivastava, A.K., Moon, Y.J.: Quasi-periodic pulsations in a solar flare with an unusual phase shift. MNRAS **483**, 5499–5507 (2019)
10. Chen, X., et al.: Quasi-periodic pulsations before and during a solar flare in AR 12242. Astrophys. J. **878**, 78 (2019)

11. Dale, D.E.: New insights from imaging spectroscopy of solar radio emission. Annu. Rev. Astron. Astrophys. **61**, 427–472 (2023)
12. Yan, Y., et al.: Mingantu spectral radioheliograph for solar and space weather studies. Front. Astron. Space Sci. **8**, 20 (2021)
13. Zhou, Z., Yan, Y., Chen, L., Wang, W., Ma, S.: A new position calibration method for MUSER images. Res. Astron. Astrophys. **22**(10), 105019 (2022)
14. Huang, J., et al.: Clusters of solar radio spikes modulated by quasi-periodic pulsations in a confined flare. Universe **8**, 348 (2022)
15. Thompson, A.R., Moran, J.M., Swenson, G.W.J.: Interferometry and Synthesis in Radio Astronomy, 3rd edn. Springer, Berlin (2017)
16. Fomalont, E.B., Perley, R.A.: Calibration and editing. In: Taylor, G.B., Carilli, C.L., Perley, R.A. (eds.) ASP Conf. Ser. 180, Synthesis Imaging in Radio Astronomy II, pp.79–110. ASP, San Francisco (1999)
17. Innes, D.E., Guo, L.-J., Bhattacharjee, A., Huang, Y.-M., Schmit, D.: Observations of supra-arcade fans: instabilities at the head of reconnection jets. Astrophys. J. **796**(1), 27 (2014)
18. Demoulin, P., Bagala, L.G., Mandrini, C.H., Henoux, J.C., Rovira, M.: G. Quasi-separatrix layers in solar flares II. Observed magnetic configurations. Astron. Astrophys. **325**, 305–317 (1997)
19. Minghui, Z., Yin, Z., Yihua, Y., et al.: Observational results of MUSER during 2014–2019. Res. Astron. Astrophys. **21**(11), 284 (2021)

Open Access This chapter is licensed under the terms of the Creative Commons Attribution 4.0 International License (http://creativecommons.org/licenses/by/4.0/), which permits use, sharing, adaptation, distribution and reproduction in any medium or format, as long as you give appropriate credit to the original author(s) and the source, provide a link to the Creative Commons license and indicate if changes were made.

The images or other third party material in this chapter are included in the chapter's Creative Commons license, unless indicated otherwise in a credit line to the material. If material is not included in the chapter's Creative Commons license and your intended use is not permitted by statutory regulation or exceeds the permitted use, you will need to obtain permission directly from the copyright holder.

Developments in Ground-Based Space Weather Monitoring: GIFDS and CALLISTO for Event Analysis

Daniela Banyś[1(✉)], David Wenzel[1], Lutz Heinrich[1], Frank Tandler[1], Christian Monstein[2], and Malte Bröse[1]

[1] Institute for Solar-Terrestrial Physics, German Aerospace Center (DLR e.V.), Kalkhorstweg 53, 17235 Neustrelitz, Germany
daniela.banys@dlr.de

[2] Ricerche Solari Locarno IRSOL, Via Patocchi 57, 6605 Locarno Monti, Switzerland

Abstract. The International Space Weather Initiative (ISWI) features several instruments, two of which this paper is focusing on: GIFDS and CALLISTO. The original receivers have been further developed aiming a small network of 4 combined stations. Moreover, a comparison of the event reports from CALLISTO and RSTN (Radio Solar Telescope Network) shows that the ISWI instrument can compete with larger facilities. Finally, a brief overview on different approaches for automated event detection is given.

Keywords: space weather instruments · solar flares · radio bursts

1 Introduction

Ground-based radio instrumentation offers good possibilities for investigating space weather events and their impact on technological systems, as shown in [1, 2]. Although there is a high demand and interest, the maintenance of a reliable and solid network is challenging but important in order to derive continuous space weather information.

Amongst others, DLR operates two ISWI instruments: the Global Ionospheric Flare Detection System (GIFDS) and the Compound Astronomical Low frequency Low cost Instrument for Spectroscopy and Transportable Observatory (CALLISTO). In Neustrelitz, there are several GIFDS receivers, as well as CALLISTO receivers (HF/VHF: 10–80 MHz, VHF/UHF: 100–800 MHz, L-Band: 1–1.6 GHz) for monitoring solar flares or solar radio bursts (SRBs), respectively. While CALLISTO allows characterizing the propagation of flare-accelerated electrons and coronal mass ejections close to their origin at the Sun, GIFDS shows the impact of solar X-ray flares on the D-region ionosphere. The original receivers have since been further developed in hardware and software to minimize noise, ease maintenance work, and provide improved comparability. The results are compact receivers for GIFDS, as well as for the three bands of CALLISTO with a built-in touchscreen and an industrial mini PC forming the basis for a wide range of space weather event analyses. A small ground-based network of these 4 combined stations, referred to as Solar Ionosphere Global Network (SIGN), shall enable a continuous monitoring of solar flares and radio bursts over a broad spectrum (from HF to L-band).

© The Author(s) 2026
N. Gopalswamy et al. (Eds.): ISWI 2024, SPPHY 431, pp. 241–247, 2026.
https://doi.org/10.1007/978-981-95-1121-1_27

2 Ground-Based Space Weather Monitoring

2.1 GIFDS

Main objective of GIFDS is the immediate and continuous detection of solar flares, see [3] for an introduction. Since X-rays cause an enhanced electron density of the D-region ionosphere, VLF measurements propagating within the Earth-ionosphere waveguide inherit useful information on flares and sudden ionospheric disturbances (SIDs). As these events can only be detected during daytime, 4 original GIFDS receivers were installed in Neustrelitz, Boston, Stanford, and Taiwan, and are currently being updated by the compact GIFDS. The compact GIFDS consists of a Perseus SDR (Software Defined Radio) receiver, a frequency standard, an industrial mini PC, and a touchscreen. The system provides 10 Hz VLF amplitude and phase measurements at multiple frequency channels ranging from 10 kHz to 470 kHz via a connected MiniWhip antenna.

2.2 CALLISTO

Main objective of the international e-CALLISTO network is the continuous monitoring of solar radio bursts. We recommend [4] for general details. The compact CALLISTO consists of two heterodyne receivers, the same industrial mini PC, as well as touch-screen, a preamplifier and depending on the frequency bands heterodyne up- or down-converters. The type of antenna depends on the chosen frequency range: a DecaMeter Antenna (DMA) for the HF/VHF band and Log-Periodic Dipole Antennas (LPDA) for the VHF/UHF and L-band.

Since 2006, over 190 CALLISTO instruments [5] have been deployed worldwide in the framework of the International Heliophysical Year in 2007 (IHY2007) and ISWI. Figure 1 shows the growth of installed stations with published data [cf. 5]. Figure 1 (top) illustrates the overall increase of up to about 80 available CALLISTO stations, mainly VHF/UHF receivers although HF/VHF receivers are also gaining interest for the last two decades. On the other hand, Fig. 1 (bottom) shows that the number of locations is stagnating at 50 for the last 5 years. The global distribution of the CALLISTO receivers is illustrated in Fig. 2 (left), along with the different compact receivers for CALLISTO, as well as GIFDS (right).

2.3 RSTN

The Radio Solar Telescope Network (RSTN) gathers wideband spectral radio observations using the Solar Radio Spectrograph (SRS) monitoring SRBs from 25–85 MHz. Event reports are provided manually via NOAA/SWPC in near-real time [6]. There are several observatories that operate RSTN telescopes with currently 4 reporting telescopes in Australia, Hawaii, Massachusetts and Italy providing 24/7 coverage.

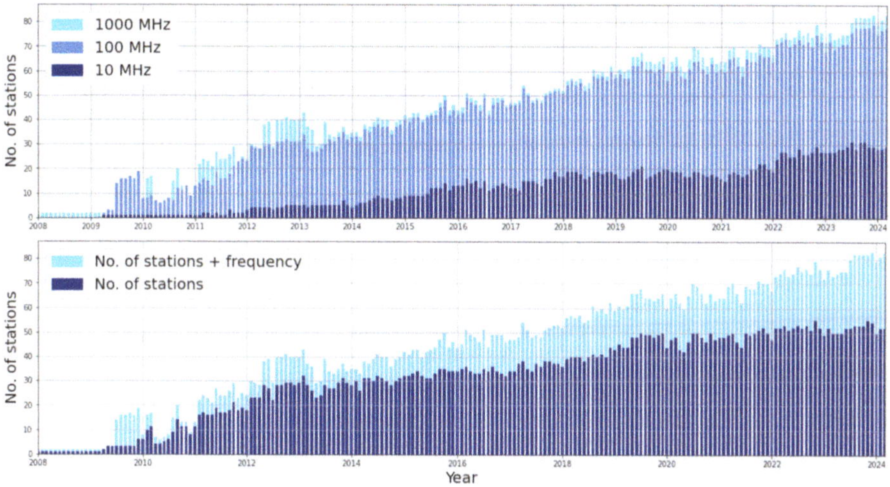

Fig. 1. No. of available CALLISTO stations colored by frequency (top) and location (bottom).

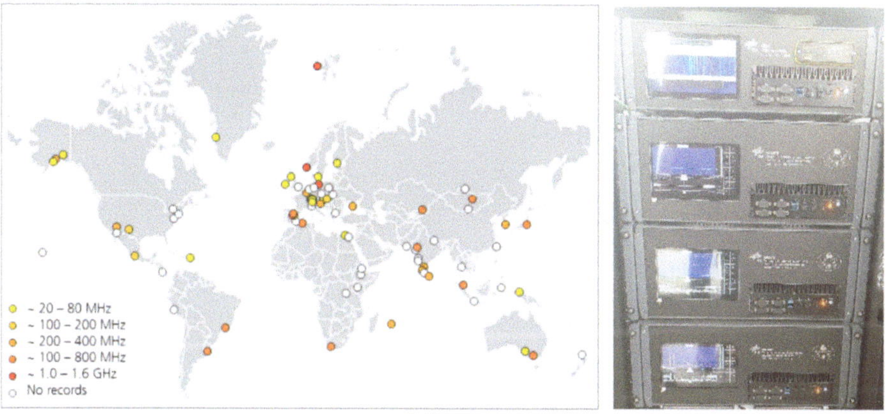

Fig. 2. Global distribution of CALLISTO receivers colored by frequency (left), and compact receivers of GIFDS, L-Band-, VHF/UHF-, and HF/VHF-CALLISTO (from top to bottom, right).

3 Manual Event Reports and Automated SRB Detection

A comparison of the SRB event reports based on CALLISTO by Christian Monstein [5] and RSTN via NOAA/SWPC (i.e. RSP reports, sweep-frequency radio burst) [6] over the last 4 years is shown in Fig. 3. Considering all SRB events, both reports match well and reflect the activity of solar cycle 25. Looking at individual types of bursts, we can see NOAA reporting more broadband long-lived decametric continuum (CTMs). It may appear as a lower sensitivity of CALLISTO measurements. However, the reason is simple: Our understanding is that reporting CTMs is tedious as they take hours or even days, hence only CTMs with spectral caustics or strong circular polarization are reported via CALLISTO, where background levels may be judged differently.

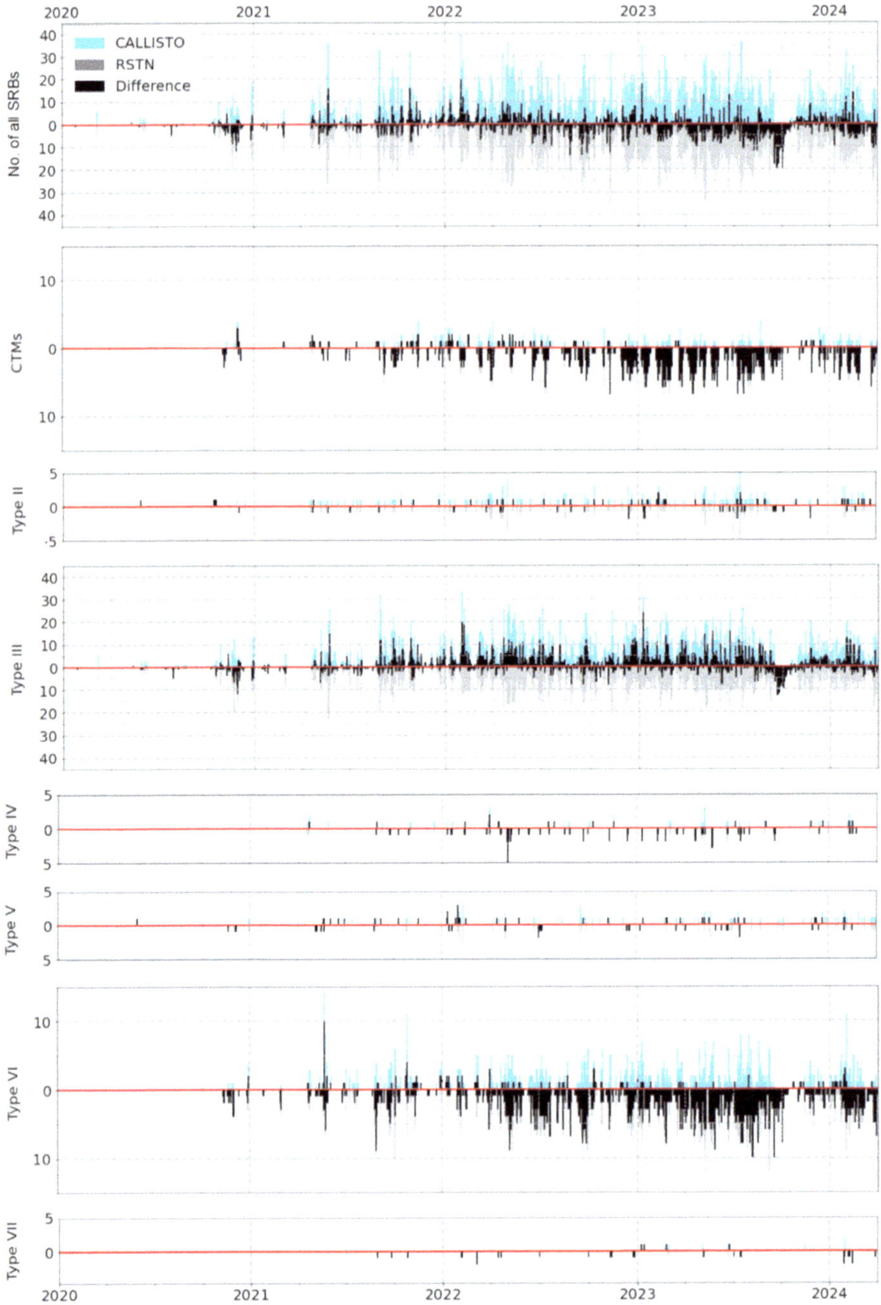

Fig. 3. Number of reported SRBs over time based on CALLISTO (light blue) and RSTN (grey), as well as its difference (black) for all SRBs, CTMs, and Type II – VII (from top to bottom)

Table 1. Overview on selected automatic radio burst detection methods.

Publication	Algorithm name	Approach	Data set	Specifications
Lobzin et al. 2009, 2010 [7, 8]	ARBIS (Automated Radio Burst Identication System)	Statistical		70% acc. For Type III
Zhang et al. 2018 [9]	image binarization	Statistical	Nançay Decameter Array, 2012 – 2017	
Salmane et al. 2018 [10]	CFAR-like approach (Constant False Alarm Rate)	Statistical	Nançay Decameter Array, June 2012 + June 2014	81% acc. For Type III Possible events: Type II, III, IV + regions of interest
Xu et al. 2019 [11]	LSTM neural network (long short-term memory)	AI	Huairou Solar Observing Station (HSOS)	True Positive Ratio (TPR) 85.4% False positive Ratio 6.7%
Singh et al. 2019 [12]	Via area slope index (ASI)	Statistical	e-Callisto, Gauribudanur, 2013 – 2014	50 – 61.7% prec. For Type III
Afandi et al. 2020 [13]	Burst-Finder	Statistical	e-Callisto, 11 February 2014	89% acc
Liu et al. 2021 [14]	DBSCAN clustering algorithm	AI		Type III
Guo 2022 [15]		AI		
Bussons et al. 2023 [16]	deARCE (deep Automatic Radioburst Compilation Engine)	AI	e-Callisto, ASSA, Glasgow, Humain, Landschlacht, 2021	88% acc., 8% false alarms
Höfig et al. 2023 [17]	ROBUST	Statistical		
Scully 2021, 2023 [18, 19]	You Only Look Once (YOLO)	AI		
Zhang et al. 2024 [20]	YOLOv7	AI	RSTN, Learmonth 2001–2022	73.5% prec. For Type II + III

In contrast to this, Fig. 3 shows that CALLISTO provides more Type III but less Type VI reports. NOAA tends to form groups while the CALLISTO reports rather log individual Type III bursts proving again the necessity of clear thresholds for SRB detection and classification. The overall good agreement of reports and moreover fairly similar numbers on Type II show that the ISWI instrument offers important, sensitive measurements for space weather monitoring. Automated SRB detection and classification is very much needed. Different approaches, statistical or artificial intelligence (AI), have been proposed over the last years (see Tab. 1). While not all SRBs are harmful for our infrastructure, reliable warnings on Type II and III are crucial. Harmonized constraints for SRB detection and classification are required and are investigated at DLR.

4 Conclusions

Space weather influences the performance and reliability of our technological infrastructure, such as electronics, wireless systems of communications and navigation, power plants. The increasing use and dependence on such sensitive systems demands scientific investigations of current and potential space weather effects in a wide range of radio frequencies by continuous monitoring of the terrestrial electromagnetic spectrum. Continuous and broadband radio spectra via the CALLISTO network provide valuable information on the occurrence of SRBs, solar activity, and the origin of radio emissions from the Sun (e.g., frequency drifts, velocity of CMEs, proton events, etc.) [21]. In combination with other measurements, like GIFDS VLF measurements, we are able to derive the direct impact of SRBs on communication and navigation signals, and also the ionospheric impact due to solar flares.

Manual detection and classification of SRBs are challenging and strongly biased on the operator, as shown in this paper. We are therefore aiming an automatic event detection with standardized thresholds. Data and derived products of SIGN will become available in due course via the IMPC (Ionosphere Monitoring and Prediction Centre).

Acknowledgement. We thank all hosts of GIFDS and CALLISTO stations for their good collaboration, maintenance work, and data provision.

References

1. Berdermann, J., et al.: Ionospheric response to the X9.3 flare on 6 September 2017 and its implication for navigation services over Europe. Space Weather **16**(10), 1604–1615 (2018)
2. Sato, H., et al.: Solar radio burst events on 6 September 2017 and its impact on GNSS signal frequencies. Space Weather **17**, 816–826 (2019)
3. Wenzel, D., Jakowski, N., Berdermann, J., Mayer, C., Valladares, C., Heber, B.: Global ionospheric flare detection system (GIFDS). J. Atmos. Solar Terr. Phys. **138–139**, 233–242 (2016)
4. Benz, A.O., Monstein, C., Meyer, H.: Callisto – a new concept for solar radio spectrometers. Sol. Phys. **226**, 143–151 (2005)
5. e-Callisto Homepage: https://www.e-callisto.org/. Last accessed 30 May 2024
6. NOAA FTP Server: ftp://ftp.swpc.noaa.gov/pub/. Last accessed 30 May 2024
7. Lobzin, V.V., Cairns, I.H., Robinson, P.A., Steward, G., Patterson, G.: Automatic recognition of type III solar radio bursts: Automated Radio Burst Identification System method and first observations. Space Weather **7**(4) (2009)

8. Lobzin, V.V., Cairns, I.H., Robinson, P.A., Steward, G., Patterson, G.: Automatic recognition of coronal type ii radio bursts: the automated radio burst identification system method and first observations. Astrophys. J. Letters **710**, L58–L62 (2010)
9. Zhang, P.J., Wang, C.B., Ye, L.: A type III radio burst automatic analysis system and statistic results for a half solar cycle with Nançay Decameter Array data. A&A **618**(A165), 11 (2018)
10. Salmane, H., Weber, R., Abed-Meraim, K., et al.: A method for the automated detection of solar radio bursts in dynamic spectra. J. Space Weather Space Clim. **8**, A43 (2018)
11. Xu, L., Yan, Y.-H., Yu, X.-X., Zhang, W.-Q., Chen, J., Duan, L.-Y.: LSTM neural network for solar radio spectrum classification. Res. Astron. Astrophys. **19**(9), 135 (2019). https://doi.org/10.1088/1674-4527/19/9/135
12. Dayal Singh, K., Raja, S., Prasad Subramanian, R., Ramesh, C.M.: Automated detection of solar radio bursts using a statistical method. Solar Phys. **294**(8), 1–14 (2019). https://doi.org/10.1007/s11207-019-1500-0
13. Afandi, N.Z.M., Sabri, N.H., Umar, R., et al.: Burst-Finder: burst recognition for E-CALLISTO spectra. Indian J. Phys. **94**, 947–957 (2020)
14. Liu, S., Yuan, G., Tan, C., Zhou, H., Cheng, R.: Automatic Detection of Type III Solar Radio Burst, pp. 553–562. Advances in Swarm Intelligence, Springer International Publishing (2021)
15. Guo, J.C., Yan, F.B., Wan, G., Hu, X.J., Wang, S.: A deep learning method for the recognition of solar radio burst spectrum. PeerJ Comput. Sci. **1**, 36 (2022)
16. Bussons Gordo, J., Fernández Ruiz, M., Prieto Mateo, M., et al.: Automatic burst detection in solar radio spectrograms using deep learning: deARCE method. Solar Phys. **298**(82) (2023)
17. Höfig L., Temmer M., Koller F., Drescher L., Monstein C.: ROBUST - a radio burst identification algorithm using the e-CALLISTO station at University of Graz. Article in EGU conference proceedings (2023)
18. Scully, J., Flynn, R., Gallagher, P.T., Carley, E.P., Daly, M.: Improved Type III solar radio burst detection using congruent deep learning models. A&A **674**(A218), 12 (2023)
19. Scully, J., Flynn, R., Carley, E., Gallagher, P., Daly, M.: Simulating Solar Radio Bursts Using Generative Adversarial Networks. Solar Phys. **298** (2023)
20. Zhang, W., Wang, B., Wu, Z., Chen, Y., Yan, F.: Identification and extraction of type II and III radio bursts based on YOLOv7. Astronomy Astrophys. 683 (2024)
21. Klein, K.-L., Salas Matamoros, C., Zucca, P.: Solar radio bursts as a tool for space weather forecasting. C R Phys. **19**, 36–42 (2018)

Open Access This chapter is licensed under the terms of the Creative Commons Attribution 4.0 International License (http://creativecommons.org/licenses/by/4.0/), which permits use, sharing, adaptation, distribution and reproduction in any medium or format, as long as you give appropriate credit to the original author(s) and the source, provide a link to the Creative Commons license and indicate if changes were made.

The images or other third party material in this chapter are included in the chapter's Creative Commons license, unless indicated otherwise in a credit line to the material. If material is not included in the chapter's Creative Commons license and your intended use is not permitted by statutory regulation or exceeds the permitted use, you will need to obtain permission directly from the copyright holder.

Use of Low-Cost GNSS Receivers for Ionospheric Monitoring

Bruno Nava[1(✉)], Francisco Azpilicueta[2], Anton Kashcheyev[3], Dinesh Manandhar[4], and Sharafat Gadimova[5]

[1] The Abdus Salam International Centre for Theoretical Physics, Trieste, Italy
bnava@ictp.it

[2] Univesidad Nacional de La Plata – CONICET, La Plata, Argentina
azpi@fcaglp.unlp.edu.ar

[3] Department of Physics, University of New Brunswick, Fredericton, NB 3B5A3, Canada
anton.kashcheyev@unb.ca

[4] Center for Spatial Information Science, The University of Tokyo, 5-1-5 Kashiwanoha, Kashiwa 277-8568, Chiba, Japan
dinesh@csis.u-tokyo.ac.jp

[5] United Nations Office for Outer Space Affairs (UNOOSA), Vienna, Austria
sharafat.gadimova@un.org

Abstract. Electromagnetic waves propagating through the Earth's ionosphere are subjected to changes in group and phase velocities, refraction, dispersion, and diffraction. For systems relying on the usage of radio signals affected by the ionosphere, it is of crucial importance to account for these effects. As far as ionospheric monitoring is concerned, high-grade multi-frequency and multi-constellation Global Navigation Satellite System (GNSS) receivers are commonly used. The existing ground-based GNSS receiver networks can provide information on a global scale. Nevertheless, geographic regions still exist where ionospheric monitoring facilities are limited. To improve our knowledge about ionosphere in areas where the existing networks coverage is sparse, an increase in the receivers' density would be desirable. The cost of the devices has indeed constituted a major obstacle, however, with the significant advance of software-defined radio, the situation is changing through the availability of low-cost solutions. The present work will discuss the performance of the available off-the-shelf low-cost dual-frequency GNSS receivers, testing them in the controlled environment and comparing the derived data with scientific-grade instruments. The accuracy of the collected ionospheric data, such as uncalibrated and calibrated total electron content (TEC) will be analyzed.

Keywords: GNSS low-cost receivers · Ionosphere · Total Electron Content

© The Author(s) 2026
N. Gopalswamy et al. (Eds.): ISWI 2024, SPPHY 431, pp. 248–257, 2026.
https://doi.org/10.1007/978-981-95-1121-1_28

1 Introduction

1.1 Introduction

Electromagnetic waves propagating through the Earth's ionosphere are subjected to changes in group and phase velocities, refraction, dispersion, and diffraction. For systems relying on the usage of radio signals affected by the ionosphere, it is of crucial importance to account for these effects. As far as ionospheric monitoring is concerned, high-grade multi-frequency and multi-constellation Global Navigation Satellite System (GNSS) receivers are commonly used. The existing ground-based GNSS receiver networks can provide information on a global scale. Nevertheless, geographic regions still exist where ionospheric monitoring facilities are limited. To improve our knowledge about ionosphere in areas where the existing networks coverage is sparse, an increase in the receivers' density would be desirable. The cost of the devices has indeed constituted a major obstacle, however, with the significant advance of software-defined radio [1], the situation is changing through the availability of low-cost solutions. Indeed, Okoh et al. [2] have demonstrated a good correlation comparing TEC measurements obtained with a low-cost U-blox ZED-F9P and a high-cost receiver installed in a low-latitude region. Dan et al. [3] have compared a Javad Triumph LS with a U-blox ZED-F9P (respectively a costly geodetic and a low-cost) receiver to explore the potential of the low-cost devices for ionosphere probing, especially in terms of total electron content (TEC). Socola and Rodrigues [4] in addition to demonstrating the possibility to estimate TEC, indicated the feasibility to detect amplitude scintillations with a low-cost U-blox ZED-F9P.

In this paper we evaluate the capabilities in estimating TEC of two low-cost GNSS receivers by comparing their data with those of two scientific grade devices. Particular attention will be given to the methodology used to assess the receives performance.

This work is carried out in the framework of the Working Group on Information Dissemination and Capacity-building of the International Committee on GNSS (ICG) to explore low-cost GNSS receivers and software that could be used to compute total electron content-related parameters and to design a prototype low-cost GNSS receiver for space weather related applications.

2 Data Acquisition

2.1 Hardware Setup

To compare multi-frequency multi-constellation low-cost and scientific grade GNSS receivers, the hardware illustrated in Fig. 1 has been deployed. Figure 1 shows the GNSS data acquisition method to compare TEC parameters from high-end and low-cost GNSS receivers. The GNSS antenna is located on the roof of the university building in Kashiwa, Japan. The coordinates of the antenna location are 35.90304517 (Latitude), 139.93930567 (Longitude), and 93.810 m (Height). The antenna cable is connected to a splitter to split the signal into four different types of receivers shown in Table 1. NetR9 and PolaRx5 are high-end GNSS receivers. Mosaic-X5 and ZED-F9P are low-cost GNSS receivers. Mosaic-X5 costs about $700 and ZED-F9P costs about $300. Mosaic-X5 is a triple frequency (L1, L2, L5 or L1, L2, L6) receiver whereas ZED-F9P is dual frequency receiver (either L1/L2 or L1/L5). All four receivers log data from the same antenna located at the same place. This allows to compare the TEC values from different receivers using the same algorithm.

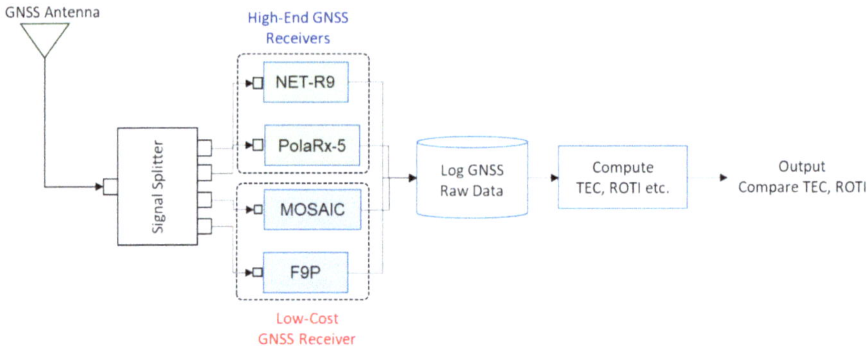

Fig. 1. GNSS data acquisition system.

Table 1. GNSS receiver types

Maker	Brand	Frequency	Format	Price Range	Category
Trimble	NetR9	L1, L2, L5 L6	T02	>7K$	High-End
Septentrio	PolaRx5	L1, L2, L5, L6	SBF	>7K$	High-End
Septentrio	Mosaic-X5	L1, L2, L5	SBF	<1K$	Low-Cost
U-Blox	ZED-F9P	L1, L2	UBX	<500$	Low-Cost

2.2 Data Used

To evaluate the receivers' performance, GNSS data for the days 16, 17, 18, and 19 June 2023 have been logged at 1-Hz sampling rate with an elevation mask angle of zero degree. All data were logged in the proprietary formats of the receiver (see Table 1). All these proprietary data have been converted into RINEX Version 3.04 using RTKLIB's RTKCONV program [5]. The RINEX output includes observation and navigation data files. The observation file includes data for different types of GNSS signals. In this paper, only GPS L1 and civilian L2 (L2C) signals are used.

3 Data Analysis

3.1 GNSS Ionospheric Observables

Following [6], the expressions for the geometry-free combination (ionospheric combination) of pseudorange (PI_r^s) and carrier-phase (LI_r^s) are the following:

$$PI_r^s \equiv P_{r,2}^s - P_{r,1}^s = KSTEC_r^s + DCB_r - DCB^s + \left(\varepsilon_{r,2}^s - \varepsilon_{r,1}^s\right) \tag{1}$$

$$LI_r^s \equiv L_{r,1}^s - L_{r,2}^s = K\ STEC_r^s + \left(\lambda_1 N_{r,1}^s - \lambda_f N_{r,2}^s\right) + \left(\xi_{r,1}^s - \xi_{r,2}^s\right) \tag{2}$$

where: r → receiver; s → satellite; f → frequency,

$P_{r,f}^s$ → pseudorange observation at receiver r to satellite s at frequency f,

$L_{r,f}^s$ → accumulated phase at receiver r to satellite s at frequency f,

$STEC_r^s = \int_s^r Neds$, is the slant TEC between satellite s and receiver r and constitutes the ionospheric information provided by GNSS measurements,

$\left(\lambda_1 N_{r,1}^s - \lambda_f N_{r,2}^s\right)$ is the geometry-free combination of the ambiguities,

$\varepsilon_{r,f}^s$ is the thermal noise of the pseudorange measurement at frequency f,

$\xi_{r,f}^s$ is the thermal noise of the carrier phase measurement at frequency f,

$K = 40.3\left(\frac{1}{f_2^2} - \frac{1}{f_1^2}\right)$, DCB_r and DCB^s are the Differential Code Bias (DCB) for the receiver and the satellite. Equation (1) is unambiguous and requires the estimation of the DCB constants (i.e. it requires to be calibrated) and its measurement noise is on the same level as the pseudorange noise. In contrast, Eq. (2) noise level is on the order of the carrier-phase noise level (approximately two orders of magnitude lower than the pseudorange noise level) but it is ambiguous, i.e. it requires the estimation of the arc-dependent term $\left(\lambda_1 N_{r,1}^s - \lambda_f N_{r,2}^s\right)$. One alternative way to exploit the advantages of each observable is to use the carrier-to-code leveling method [7] which consists of computing the uncalibrated leveled carrier phase slant TEC $\widetilde{LI}_r^s$ as:

$$\widetilde{LI}_r^s \equiv LI_r^s - \langle LI_r^s - PI_r^s \rangle = K\ STEC_r^s + DCB_r - DCB^s + \Xi_r^s \tag{3}$$

where $\Xi_r^s = \left(\xi_{r,1}^s - \xi_{r,2}^s\right)$ and $\langle LI_r^s - PI_r^s \rangle$ is the average of the difference between carrier-phase and pseudrange geometry-free observations for a complete arc. The leveling method implies two assumptions: 1) the DCBs are constant along one day (widely used in the community) and 2) the measurements from a continuous arc have no cycle-slips. When processing GNSS measurements in a configuration where two receivers are connected to the same GNSS antenna through a splitter, and computing the difference, $\widetilde{LI}_1^s - \widetilde{LI}_2^s$, between measurements from the two receivers to the same satellite we obtain:

$$\widetilde{LI}_1^s - \widetilde{LI}_2^s = K\ STEC_1^s + DCB_1 - DCB^s - K\ STEC_2^s - DCB_2 + DCB^s + \Xi_1^s - \Xi_2^s \tag{4}$$

As the ionospheric term and the satellite DCB are the same for both receivers, it follows:

$$\widetilde{LI}_1^s - \widetilde{LI}_2^s = DCB_1 - DCB_2 + \Xi_1^s - \Xi_2^s \tag{5}$$

Then, using Eq. (3) and (5) it is possible to define ΔTEC, with the following equation:

$$\Delta TEC \equiv STEC_1^s - STEC_2^s - \langle STEC_1^s - STEC_2^s \rangle \tag{6}$$

$$\Delta TEC = \frac{1}{K}\left(\widetilde{LI}_1^s - \widetilde{LI}_2^s\right) - \frac{1}{K}\langle \widetilde{LI}_1^s - \widetilde{LI}_2^s \rangle = \Xi_1^s - \Xi_2^s \tag{7}$$

Figure 2a shows the behavior of $\widetilde{LI}_r^s - \langle \widetilde{LI}_r^s \rangle$ in TECU as a function of UT corresponding to PRN 31 for the two receivers Septentrio PolaRx5 (KSX5, red in the figure) and U-blox ZED-FP9 (KSF9, blue in the figure). This graphic clearly illustrates the variation of the $STEC_r^s$ of the two receivers and the matching between them. Figure 2b shows the behavior of ΔTEC as defined in Eq. (7), at full scale, for PRN 31 and the pair of receivers PolaRx5 and the ZED-FP9, as a function of UT, while Fig. 2c shows the same as the previous figure but in a zoomed-in scale. This last plot clearly illustrates the expected scatter of values around zero according to Eq. (7). At each end of the arc, the dispersion increases due to the increase in the noise level at low elevation angles. The standard deviation of the series is of the order of magnitude of the carrier-phase measurements. Returning to Fig. 2b after this analysis, the full scale is needed to represent some unexpected large values of ΔTEC found at the end of the arc which cannot be explained by Eq. (7) The explanation for this behavior is a discontinuity on the ambiguity of the carrier-phase measurements of one of the receivers. This event turns Eq. (7) invalid and degrades the accuracy of the carrier-to-code leveling strategy. Figure 3 shows ΔTEC series for all the visible PRNs over the site for four days, as a function of DOY, for the pair of receivers PolaRx5 and Septentrio Mosaic-X5 (KSMS in the figure). The plot in the left panel is at full scale, while the one on the right is at a reduced scale. The plot on the left shows events where the carrier-phase ambiguity is discontinued and produces large jumps on the uncalibrated TEC (see Eq. 2). The analysis of the plot at reduced scale reveals other effects. Equation (7) indicates that series corresponding to each PRN, should consist of a scatter of values centered at zero. Figure 3b shows that this is the case for most of the PRN but there are few series with centers displaced by small values. These deviations from zero could be explained by the occurrence of cycle slips in the carrier-phase measurements of one of the receivers.

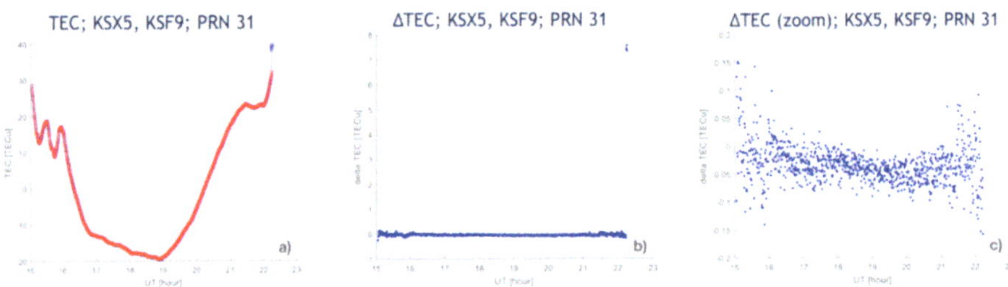

Fig. 2. a) uncalibrated TEC (minus its average) as a function of UT corresponding to PRN 31 for receivers KSX5 (red) and KSF9 (blue); b) ΔTEC as a function of UT corresponding PRN 31, between receivers KSX5 and KSF9, at full scale, c) the same as b) but at a reduced scale.

Figure 4 shows ΔTEC series for all the visible PRNs over the site for four days, as a function of DOY, for the pair of receivers PolaRx5 and ZED-FP9. The plot on panel a) is at full scale, while the one on the panel b) is at a reduced scale.

The analysis for this figure is similar to the one for the previous figure, but the principal differences between them is due to the frequency of occurrence of the ambiguity discontinuities (panel a) and the frequency of occurrence of cycle slips (panel b). Considering its high performance, the PolaRx5 has been considered as the reference for

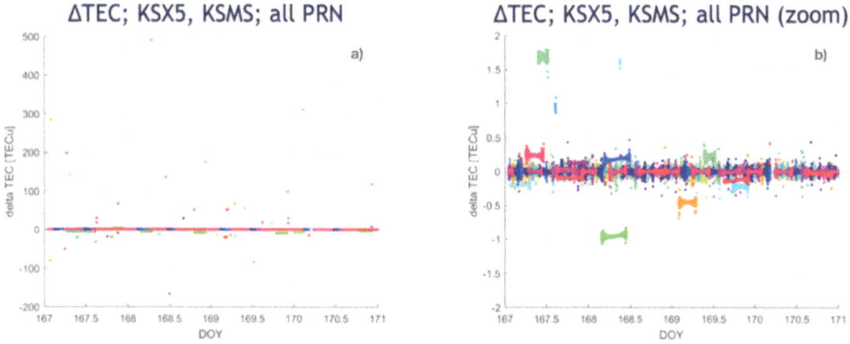

Fig. 3. a) ΔTEC for the pair of receivers KSX5 and KSMS, at full scale, as a function of DOY, where each satellite is identified by a different color; b) the same as a) but at a reduced scale.

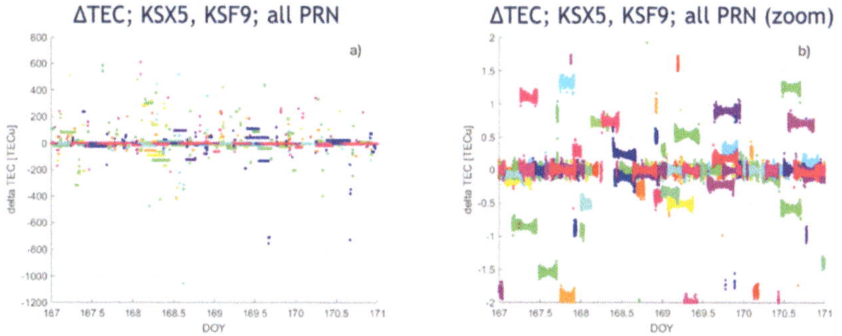

Fig. 4. a) ΔTEC for the pair of receivers KSX5 and KSF9, at full scale, as a function of DOY, where each satellite is identified by a different color; b) the same as a) but a reduced scale.

the assessment of receivers ZED-FP9 and Mosaic-X5. The comparison of the Figs. 3 and 4 therefore indicates that: i) ZED-FP9 is affected by more recurrent carrier phase ambiguity discontinuities than Mosaic-X5 (panels a); ii) ZED-FP9 is affected by more recurrent and larger cycle slips than Mosaic-X5 (panels b). These conclusions reflect the fact that Mosaic-X5 is more expensive than ZED-FP9. Nevertheless, it has to be noted that, when ZED-FP9 is not affected by cycle slips, the accuracy of TEC estimates stays within the intrinsic accuracy of TEC estimates of the other receivers. In addition, the discontinuities and cycle slip events are not so frequent to prevent the use of this type of receiver for ionospheric TEC monitoring, as will be illustrated in the following sections.

3.2 Uncalibrated Slant TEC Statistics

To assess the performance of the low-cost GNSS receivers in terms of the ionospheric data retrieval characteristics quality, the following tests have been performed. The geometry-free combinations of L1 and L2 GPS signal phase (Eq. 2) have been calculated for all four receivers under test using dual frequency observables to obtain uncalibrated slant TEC values. As two out of the four receivers under test did not track L2P(Y) signal, only those satellites that transmit L2C signals have been considered. Elevation mask of 15 degrees has been used to remove the detrimental effect of multipath on TEC estimates. The results of the TEC comparison between four receivers under test are presented in Figs. 5 and 6. Figure 5, panel a, shows an example of uncalibrated TEC calculated for PRN 17. The TEC curves are intentionally plot with a bias of 5 TECU to avoid overlapping. As can be seen from the plot, all four TEC curves are qualitatively matching each other very well. To quantitatively estimate the agreement between each receiver and the Septentrio PolaRx5 that is considered as the reference, the TEC difference (Eq. 7), for PRN 17, between ZED-FP9 and PolaRx5, NetR9 and PolaRx5, Mosaic-X5 and PolaRx5 are presented in Fig. 5 on panel b in blue, red and yellow respectively. To assess the performance during a longer period of time, the statistics of the TEC difference between all four receivers during June 17th, 2023, as well as the corresponding TEC values leveled to zero average (panel a) are given in Fig. 6. Panels c, d and e show the relative frequency distribution of the TEC difference between each receiver under test and the PolaRx5 considered as the reference. The standard deviation of the TEC difference between all four receivers is below 0.03 TECU that is well below the intrinsic accuracy of an individual receiver (0.1 TECU).

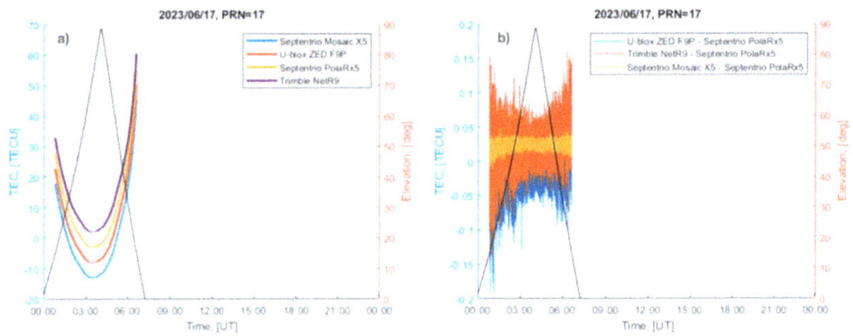

Fig. 5. Uncalibrated STEC for GPS PRN17 obtained from Septentrio Mosaic-X5, U-blox ZED-F9P, Septentrio PolaRx5 and Trimble NetR9 receivers on June 17th, 2023 are shown on panel a. Panel b contains the TEC difference (Eq. 7).

Nevertheless, as can be seen on Fig. 6 panel a, some of the receivers have a different number of cycle-slips. As mentioned in 3.1, cycle-slips are a nuisance for the TEC calibration process and can lead to significant errors in estimating absolute TEC values if not accounted for correctly. Therefore, Fig. 6, panel b shows the statistics of the cycle slips encountered during the one-day period of June 17th, 2023. U-blox ZED-F9P receiver shows the highest number of cycle slips (13) among all the receivers under test followed by the Trimble NetR9 (11 cycle slips), while Septentrio Mosaic-X5 and PolarRx5 receivers did not encounter any cycle-slips during this period. In conclusion, the results of the performed tests demonstrate that low-cost GNSS receivers can be used to estimate ionospheric TEC with the same or better level of accuracy compared to their high-cost counterparts.

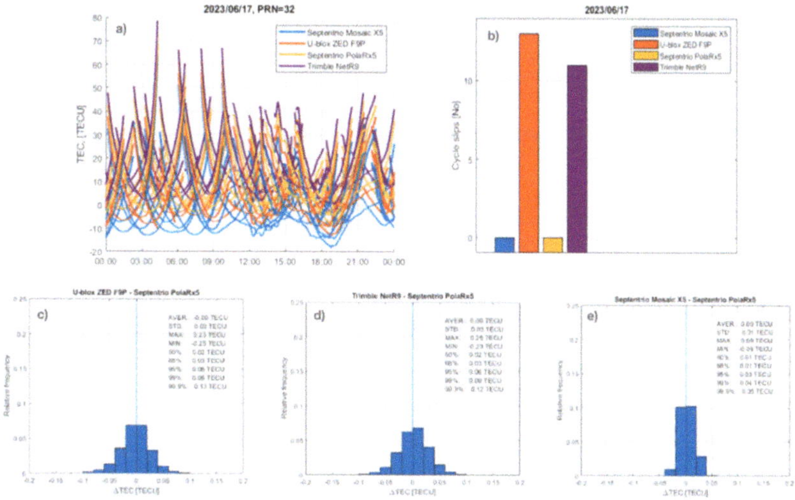

Fig. 6. Uncalibrated STEC for GPS PRN17 obtained from Septentrio Mosaic-X5, U-blox ZED-F9P, Septentrio PolaRx5 and Trimble NetR9 receivers on June 17th, 2023 are shown on panel a. Panel b contains number of cycle slips encountered, while panel c, d and e - statistics of the TEC difference.

3.3 Calibrated Slant TEC

To assess the capability of the low-cost receivers to monitor the ionospheric electron content, the same technique has been applied to all receivers under test to compute calibrated slant and vertical TEC for days 17, 18, 19 - June 2023. The calibration technique has been implemented by one of the authors in accordance with Azpilicueta et al. [8]. Indeed, specific pre-processing algorithms have been implemented to the dual frequency combination of signal phase to eliminate or mitigate as much as possible the effects of phase jumps and cycle-slips. To illustrate the results obtained, in Fig. 7 calibrated vertical TEC for the four receivers are plotted (red, green, blue and cyan curves corresponding to the U-blox ZED-F9P, Septentrio Mosaic-X5, Trimble NetR9, and Septentrio PolaRx5 receivers, which in the plot label are indicated as ZED-FP9, KSMS, KSR9, KSX5 respectively) for the period under investigation.

As can be concluded by visual inspection, the vertical TEC values obtained for the four receivers are very similar and the TEC differences among receivers is well below the accepted accuracy (few TEC units) of the TEC estimate. These results confirm that at the selected location and for the given heliogeophysical conditions, the low-cost GNSS receivers perform similarly to scientific grade receives in term of ionospheric TEC monitoring capabilities.

To identify possible scientific applications based on multi-frequency multi-constellation low-cost GNSS receivers, an additional test has been performed using NeQuick-2 [9], the ionospheric electron density model developed at The Abdus Salam International Centre for Theoretical Physics in Trieste, Italy in collaboration with the Institute for Geophysics, Astrophysics and Meteorology of the University of Graz, Austria. In particular, vertical TEC values have been computed with NeQuick for the same

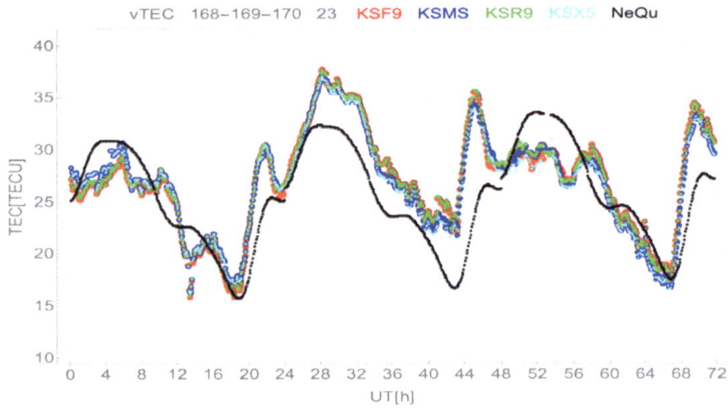

Fig. 7. Calibrated vertical TEC (red, green, blue and cyan curves corresponding to the U-blox ZED-F9P, Septentrio Mosaic-X5, Trimble NetR9, and Septentrio PolaRx5 receivers) for the low-cost and scientific grade receivers. The same calibration technique has been used in all cases. The NeQuick-derived TEC (black curve) is added for comparison purposes.

period during which experimental data from the GNSS receivers under test were available. These modeled data have been included in Fig. 7 (black curve) for comparison purposes. Considering that the differences between the modeled and the experimental TEC values are much greater than the difference among experimental TEC values, the result summarized in this figure also demonstrate that low-cost GNSS receivers can be used for models' validation. In addition, after examining the results (not shown here) obtained in terms of slant TEC estimates, it can be easily argued that low-cost GNSS receivers could also provide reliable data for assimilative models.

4 Conclusions

Four multi-frequency multi-constellation GNSS receivers have been deployed and tested, among which two receivers were low-cost, and the other two were high-cost, or scientific-grade. Three days of data have been analyzed in terms of uncalibrated phase slant TEC and calibrated vertical TEC. The relevant statistics have been therefore provided to assess performance and effectiveness of the low-cost receivers as compared to the scientific-grade devices. Even if few days have been used, the results obtained (due the specific kind of analysis and methodology employed) clearly confirmed that the tested low-cost receiver can be used for ionospheric TEC monitoring and related studies, including ionospheric modelling and analysis of space weather effects. In the next future, additional studies are foreseen to further assess the receivers' performance in diverse heliogeophysical conditions and to apprise the potential of having a network of low-cost GNSS receivers to carry out relevant scientific research.

References

1. Linty, N., Romero, R., Dovis, F., Alfonsi, L.: Benefits of GNSS software receivers for ionospheric monitoring at high latitudes, In: 1st URSI Atlantic Radio Science Conference (URSI AT-RASC), Gran Canaria, Spain, pp. 1–6 (2015)
2. Okoh, D., Obafaye, A., Rabiu, B., Seemala, G., Kashcheyev, A., Nava, B.: New results of ionospheric total electron content measurements from a low-cost global navigation satellite system receiver and comparisons with other data sources. Adv. Space Res. (2021)
3. Dan, S., Santra, A., Mahato, S., Koley, C., Banerjee, P., Bose, A.: On use of low cost, compact GNSS receiver modules for ionosphere monitoring. Radio Sci. (2021)
4. Gomez Socola, J., Rodrigues, F.S.: ScintPi 2.0 and 3.0: low-cost GNSS-based monitors of ionospheric scintillation and total electron content. Earth Planet. Space (2022)
5. RTKLIB: https://www.rtklib.com/. Last accessed 07 June 2024
6. Teunissen, P., Montenbruck, O.: Springer Handbook of Global Navigation Satellite Systems. Springer (2017)
7. Mannucci, A.J., Wilson, B.D., Yuan, D.N., Ho, C.H., Lindqwister, U.J., Runge, T.F.: A global mapping technique for GPS-derived ionospheric total electron content measurements. Radio Sci. **33**(3), 565–582 (1998)
8. Azpilicueta, F., Brunini, C., Radicella, S.M.: Global ionospheric maps from GPS observations using modip latitude". Adv. Space Res. **38**(11), 2324–2331 (2006)
9. Nava, B., Coïsson, P., Radicella, S.M.: A new version of the NeQuick ionosphere electron density model. J. Atmos. Solar Terr. Phys. **70**(15), 1856–1862 (2008)

Open Access This chapter is licensed under the terms of the Creative Commons Attribution 4.0 International License (http://creativecommons.org/licenses/by/4.0/), which permits use, sharing, adaptation, distribution and reproduction in any medium or format, as long as you give appropriate credit to the original author(s) and the source, provide a link to the Creative Commons license and indicate if changes were made.

The images or other third party material in this chapter are included in the chapter's Creative Commons license, unless indicated otherwise in a credit line to the material. If material is not included in the chapter's Creative Commons license and your intended use is not permitted by statutory regulation or exceeds the permitted use, you will need to obtain permission directly from the copyright holder.

Application of the Relativistic Electron Alert System (REleASE) to instruments on board of STEREO-A

H. Dröge[1(✉)], B. Heber[1], O. Malandraki[3], J. Martens[1,2], and A. Posner[4,5]

[1] Christian-Albrechts-Universität zu Kiel, Kiel, Germany
droege@physik.uni-kiel.de
[2] Institut für Solar-Terrestrische Physik, Neustrelitz, Germany
[3] National Observatory of Athens, IAASARS, Athens, Greece
[4] NASA Headquarters, Washington, United States
[5] NASA Johnson Space Center, Houston, United States

Abstract. SEP events can pose a significant radiation hazard for human and robotic space exploration activities. The Relativistic Electron Alert System [1] utilizes the fact that near relativistic electrons travel faster than energetic ions (30 MeV protons have 25% of the speed of light) and are always present in hazardous SEP events. Their early arrival can be used to warn against high proton fluxes near Earth. Originally REleASE uses real time data from the SOHO EPHIN and was later expanded to use measurements from the ACE EPAM called HESPERIA/REleASE [2]. We now applied the method to the HET and the SEPT on board of STEREO-A in order to develop a STEREO/REleASE system. This gives us the possibility to test and validate the method with different kind of instruments and energy ranges, which will help to adapt the system to upcoming missions (e.g. NASA's IMAP). The recent fly-by of STEREO-A past the Earth in 2023 allowed for a direct comparison between the predictions of STEREO/REleASE and the original REleASE.

Keywords: Solar Energetic Particles · Instrumentation · Radiation Hazard

1 Introduction

Solar Energetic Particle (SEP) events are a major threat for human exploration of the Moon and Mars. A. Posner [1] introduced the "Relativistic Electron Alert System for Exploration" (REleASE) forecasting scheme utilizing measurements of relativistic electrons from the "Electron Proton Helium INstrument" (EPHIN) aboard the "SOlar and Heliospheric Observatory" (SOHO) spacecraft [4] in order to forecast the flux of several tens of MeV protons. While the radiation hazard has a strong dependency on event characteristics such as the energy spectrum and abundances of the various particle types as well as the location of the observer with respect to the surrounding shielding, it can be generally assumed that the relativistic (~ 1MeV) electrons have a significantly lower impact than ions. However, since electrons and protons are often accelerated and injected

© The Author(s) 2026
N. Gopalswamy et al. (Eds.): ISWI 2024, SPPHY 431, pp. 258–262, 2026.
https://doi.org/10.1007/978-981-95-1121-1_29

into the interplanetary medium at the same time (see e.g. [3]), the shorter propagation time of the electrons is utilized to forecast the tens of MeV proton flux with a delay between 30 to 90 minutes utilizing an empirical relationship between the expected proton and the measured electron flux (for details see [1]). Due to limited down link provided by the "Deep Space Network" (DSN) after the launch of the "Solar Dynamics Observatory" (SDO), SOHO's EPHIN near real time measurements are only available for a couple of hours each day [2]. In order to overcome that limitation near real-time measurements of near relativistic electrons up to 315 keV from the "Electron Proton Alpha Monitor" (EPAM) aboard "Advanced Composition Explorer" (ACE) [5] at Lagrangian point 1 were utilized as input to REleASE. However, SOHO as well as ACE were launched in 1995 and 1998, respectively. Aging of the instrumentation as well as failures of the satellites would cause the end of the current REleASE forecast. Thus, efforts are underway to replace the electron flux measurements with the ones from HIT aboard the "Interstellar Mapping and Acceleration Probe" (IMAP), to be launched in February 2025. In order to facilitate the adaptation of the system towards modern instrumentation and further test and improve the method, we followed the general approach described in [1] using as input beacon data from the "High Energy Telescope" (HET) and the "Solar Electron and Proton Telescope" (SEPT) aboard the "Solar Terrestrial Relations Observatory-A" (STEREO-A) satellite.

For the HET the data that is available in real time is limited. There is only one electron channel (0.7–4.0 MeV) and three proton channels. From these three channels the 20.8–40.5 MeV energy range is the closest to the one utilized from EPHIN and will therefore be used here for comparison. In lower electron energy ranges (e.g. 125–255 keV) the SEPT can be used. It uses the magnet/foil technique to separate electrons and ions. An absorption foil in front of the electron detector is able to stop about 400 keV protons [6]. Higher energy ions can deposit energy in the electron detector. Careful handling of this contamination problem is necessary in real time to avoid a large number of false alarms from the instrument. The details are outside of the scope of this paper and will follow in a future publication.

2 REleASE Methodology and Application to STEREO-A HET

The basic methodology of REleASE was initially described in [1]. Therein it was shown that all hazardous solar ion events are accompanied by an increase of relativistic electrons. The onsets of these electrons are on average 1 hour earlier than the onset of the tens of MeV protons. That is most protons arrive within $\Delta t = 30$–90 mins of the electron onset. Furthermore, it was found that the increase of the electron and proton intensities as well as the maximum intensities are well correlated. Hence the forecasting of the proton flux I_p at the time $t + \Delta t$ based on relativistic electron measurements can be utilized:

A forecast matrix (FCM) **M** was developed on the basis of two parameters: the electron flux I_e and the rise parameter Φ_e, described in detail in [1]:

$$I_p(t + \Delta t) = \mathbf{M}(I_e(t), \Phi_e(t)) \tag{1}$$

To create a FCM that can be used with the HET we analyzed the STEREO-A beacon measurements from 2008 to May 2023. To determine the rise parameter Φ_e automatically we applied the algorithm introduced in [1] for every minute.

The I_e and Φ_e pairs for every minute of the data set are binned in a 14 × 19 matrix covering the range from 10^{-3} to 0.6 and from 10^{-2} to $4 \cdot 10^4$ (s sr MeV cm^2)$^{-1}$, respectively. In each bin the averaged values of I_e and Φ_e are used together with the corresponding averaged proton flux measured 30, 60 or 90 mins later to form the FCM **M**. These three different times are used to cover the range of the most likely proton delays depending on the magnetic connection and particle transport conditions of the observer to the Sun.

As an example, the 60-minute FCM for the HET proton flux in the energy range from 21 to 40 MeV is shown in Fig. 1. The x- and y-axes are the rise parameter Φ_e and the electron flux I_e, respectively. The color code gives the expected proton flux 60 minutes in advance. With increasing electron flux and rise parameter the statistic in each bin decreases because there were fewer or no events observed. However, the goal of the forecast is to give a warning if the proton flux will cross a threshold of 10^{-1} (MeV sr s cm^2)$^{-1}$ and not to make precise peak flux predictions. For that purpose, the relevant areas of the matrix are well covered. An application of the method is given in Fig. 2 as discussed in what follows.

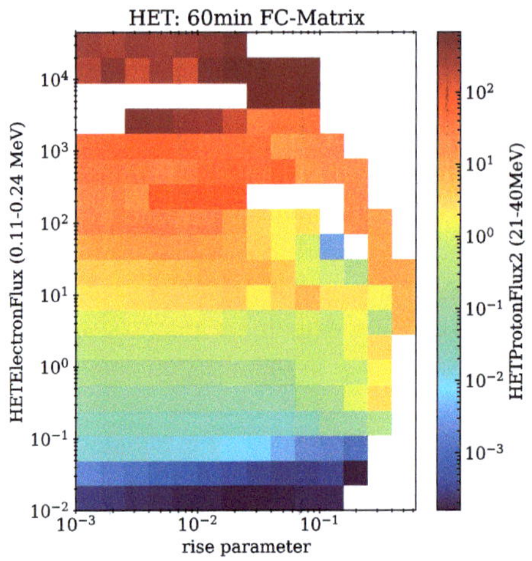

Fig. 1. The 60 minute HET forecast matrix resulting from the ~15 years of analyzed data.

3 Comparison during STEREO-A's flyby

On 2023-08-12 STEREO-A passed the Earth and is since then moving again ahead of Earth. During the period from June to November 2023 the magnetic footpoints to the Sun had a small longitudinal separation angle (<10°) between STEREO-A and SOHO. The occurrences of five SEP events exceeding the proton flux threshold of $>10^{-1}$ (cm^2 sr s MeV)$^{-1}$ were measured allowing a direct comparison of the SOHO/EPHIN REleASE with the newly developed STEREO/REleASE. As an example, the 30 min forecasts for the event on Sep 1, 2023, are shown in Fig. 2. In this case both systems triggered an alarm 76 min (blue, SOHO) and 81 min (red, STEREO) prior to the protons exceeding the threshold as indicated by the two dashed dotted lines in the figure.

Overall, the SOHO REleASE and the STEREO REleASE systems were able to warn before four and three of the five events and had one and two false alarms in this time period, respectively. In general, the forecasts from both systems are very similar. The warning times are between 22 and 132 min with a median of ~76 min.

The problem with the HET is that the only available electron channel has quite a high background level. Therefore, smaller events cannot always be properly forecasted. However, as mentioned earlier, it is also possible to use the SEPT in a lower electron energy range to assist in such cases. Studies to implement this method are ongoing.

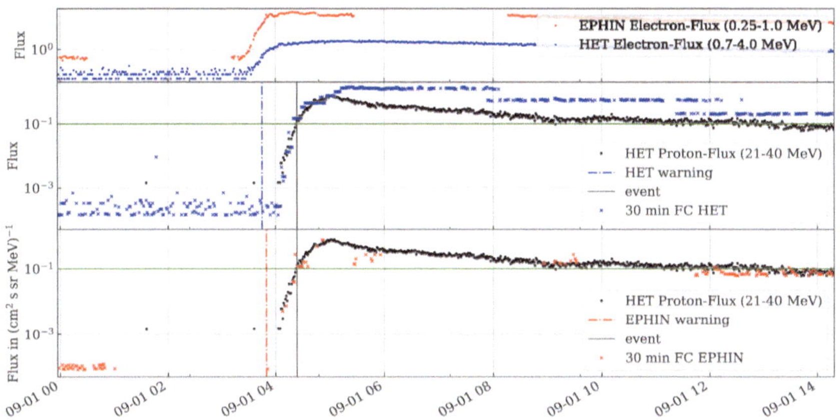

Fig. 2. Example 30 min HET and EPHIN forecasts for the September 1st event. The times of the alarms are marked by the two dashed-dotted vertical lines.

4 Summary and Outlook

We successfully implemented the REleASE forecasting method utilizing the HET on STEREO-A, thus establishing a STEREO/REleASE system. We show that its performance is similar to the original one (more details will be given in a future publication). Further studies are needed to improve the methodology with respect to the following topics:

- General improvements of the REleASE system

  optimized warning thresholds and conditions
  forecast matrix inter- and extrapolation

- Preparation for future missions

  transition REleASE to IMAP
  accuracy of forecasting near, but not on the magnetic field line where the forecast is issued (ARTEMIS)

- Development of a 'STEREO/REleASE +'

  utilizing remote sensing of solar radio bursts to reduce false alarms

References

1. Posner, A.: Up to 1-hour forecasting of radiation hazards from solar energetic ion events with relativistic electrons. Space Weather (2007)
2. Malandraki, O.E., Crosby, N.B.: Solar Particle Radiation Storms Forecasting and Analysis. Springer, Cham (2018)
3. Strauss, R.D., et al.: On the onset delays of solar energetic electrons and protons: evidence for a common accelerator. Astrophys. J. (2023)
4. Müller-Mellin, R., et al.: COSTEP - comprehensive suprathermal and energetic particle analyser. Solar Phys. (1995)
5. Gold, R.E., et al.: Electron, proton, and alpha monitor on the advanced composition explorer spacecraft. Space Sci. Rev. (1998)
6. Müller-Mellin, R., et al.: The solar electron and proton telescope for the STEREO mission. Space Sci. Rev. (2008)

Open Access This chapter is licensed under the terms of the Creative Commons Attribution 4.0 International License (http://creativecommons.org/licenses/by/4.0/), which permits use, sharing, adaptation, distribution and reproduction in any medium or format, as long as you give appropriate credit to the original author(s) and the source, provide a link to the Creative Commons license and indicate if changes were made.

The images or other third party material in this chapter are included in the chapter's Creative Commons license, unless indicated otherwise in a credit line to the material. If material is not included in the chapter's Creative Commons license and your intended use is not permitted by statutory regulation or exceeds the permitted use, you will need to obtain permission directly from the copyright holder.

Equatorial Ionospheric Irregularity Detection and Analysis Using 2-D ROTI Maps and VHF Radar Images During the Upcoming Solar Maximum

P. Supnithi[1(✉)], L. M. M. Myint[1], N. Tongkasem[1], T. Thanakulketsarat[1], and M. Nishioka[2]

[1] School of Engineering, King Mongkut's Institute of Technology Ladkrabang, Bangkok 10520, Thailand
pornchai.su@kmitl.ac.th

[2] National Institute of Information and Communications Technology, Koganei, Tokyo 184-8795, Japan

Abstract. In this work, we analyze the ionospheric irregularities at Chumphon station, Thailand, using observational data from GNSS receivers as well as VHF radar and ionosonde at Chumphon station, Thailand. The ionospheric irregularity event on 20 March 2020 and the super solar storms during 8–12 May 2024 are studied. Both instruments show traces the irregularities and interesting daytime fluctuation in total electron content over Thailand area. The statistics of ionospheric irregularities from 2020 to 2024 show that as we enter the solar maximum of the 25th solar cycle, more occurrences of ionospheric irregularities are clearly seen.

Keywords: ionospheric irregularity · equatorial plasma bubbles · total electron content · VHF radar · solar storms

1 Introduction

Equatorial plasma bubbles (EPB) refer to ionospheric irregularity which often occurs after sunset and may last until early morning [1, 2]. Both local and global conditions (geomagnetic storm) may cause EPB [3, 4]. Global geomagnetic storms tend to cause prolonged ionospheric irregularities. The scale size of the EPB scale size can vary from a few kilometers to hundreds of kilometers. The EPBs often travel from West to East and they can expand via the magnetic fluxtube to higher latitude region [5]. The effects of plasma bubbles on Global navigation Satellite System (GNSS) signals include scintillation and even loss-of-lock events, therefore, the technologies which rely on GNSS such as precise positioning and navigation systems will be affected as well [6]. Various sensors can be used to study the equatorial ionospheric irregularities or EPBs include GNSS receivers, VHF radar, ionosonde among others.

© The Author(s) 2026
N. Gopalswamy et al. (Eds.): ISWI 2024, SPPHY 431, pp. 263–271, 2026.
https://doi.org/10.1007/978-981-95-1121-1_30

The objectives of this work are to analyze multi-sensor instrument including the GNSS receivers, VHF radar images and ionograms from ionosonde station at Chumphon, Thailand. The occurrence statistics of ionospheric irregularities are to be analyzed. In addition, we will analyze the ionospheric irregularities during the recent May 2024 solar storms.

2 Methodology and Experimental Setup

2.1 VHF Radar Station

The Chumphon VHF radar station has been operational since January 2020. It operates by transmission of radar signal in the VHF band and receive the echo signals. The ionospheric irregularities such as EPBs can be clearly observed. This station, located on the Chumphon campus (Geographic: 10.72°N, 99.73°E and Geomagnetic: 1.33°N) of KMITL, is a complement to the two northern and southern conjugate stations with ground-based instrument such as GNSS receivers and VHF radar station: Chiangmai station (Northern crest) and Kototabang, Indonesia (Southern crest). The aim of the Chumphon radar station is to study 'fresh' bubbles at the magnetic equator. Main sponsors of this radar station include the National Institute of Information and Communications Technology (NICT) and King Mongkut's Institute of Technology Ladkrabang (KMITL).

Figure 1(a) shows the location of Chumphon radar station near the magnetic equator, whereas Fig. 1(b) shows the array of antenna at the station. The VHF radar station has three-component 18 yagi antennas positioned in the east to west direction at 5-m apart. The designed VHF frequency is 39.65 MHz with 25 kHz bandwidth, dictated by the state regulation. The essential parameters of the VHF radar system are described in Table 1.

(a) (b)

Fig. 1. Illustration of Chumphon VHF radar station at Chumphon, Thailand.

There are five beams transmitted upward to the ionosphere, i.e., Fd1 to Fd5, where the Fd3 is directed directly overhead (Fig. 2).

2.2 GNSS Station

The ionosphere is about 50 to 1,000 km from the ground. Since the region is ionized, we often describe using electron density information. Slant total electron density (TEC) is

Table 1. Technical description of the Chumphon VHF radar station.

Item	Description
Frequency	39.65 MHz (25 kHz BW)
$P_{transmitter}$	1.1 kW per module for 20 kW peak power
Antenna	18 Yagi-antennas, 130-m separation, 9 beams ± 60° (azimuth), off-zenith pointing angle = 8.3°
Range	140–860 km (Latitudinal coverage up to ~ 20°N)
Beam cycle	Scanning in E-W direction, the range of zenith angle 0° to 28°
Range resolution	4.8 km
No. of coherent integrations	2
Number of FFT pts	256

* **Fd3** as an overhead beam

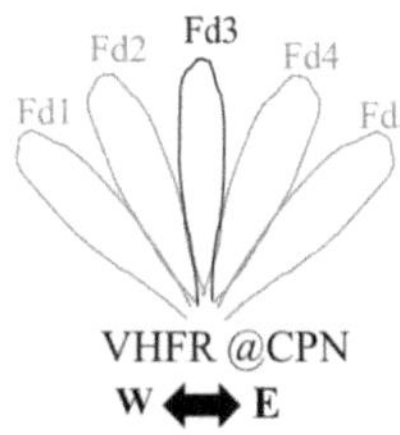

Fig. 2. The five beams of VHF radar.

defined as the amount of electron density in the cylindrical area of 1 m^2. GNSS receivers are often used to study the ionosphere. The slant total electron content (STEC) can be computed from the difference in code pseudorange values (or carrier phase information) of two frequencies. After phase leveling, satellite removal and receiver bias estimation, the resulting absolute STEC can be converted to vertical TEC (VTEC), with the unit of TECU (10^{16} electrons/m^2). Figure 3 shows examples of the GNSS receiver and antenna at Chumphon station. Here, GNSS receivers receive signals from multi-constellation satellites, multi-frequencies typically L1, L2 and L5. The antenna is a choke-ring antenna.

The ionospheric irregularity can be detected from the rate of TEC Change Index (ROTI) [7, 8]. As shown in Eq. (1), it can be computed from the standard deviation of rate of TEC (ROT).

$$ROTI = \sqrt{\frac{1}{N}\sum_{i=1}^{N}(ROT(i) - ROT)^2} \quad (1)$$

where

$$ROT(\mathrm{i}) = STEC(i+1) - STEC(i) \quad (2)$$

Figure 3 shows an example of ROTI plot. From 0 UTC (07 LT) to 1300 UTC (20 LT), the ionosphere is quiet with ROTI level of about 0.05, but from 1300 UTC (20 LT) to 1700 UTC (0 LT), the ROTI becomes very high and fluctuated indicating ionospheric irregularities or disturbances. The irregularities last until early morning.

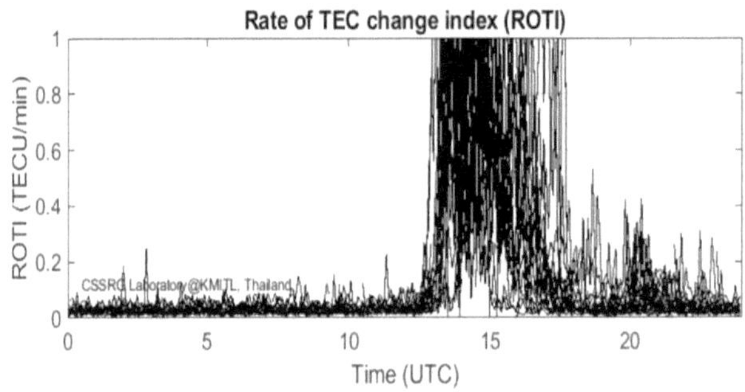

Fig. 3. An example of ROTI plot

2.3 Ionosonde Station

An ionosonde station has been installed on the Chumphon campus since 2005. It is mainly used to measure basic ionospheric parameters such critical frequency of F2 layer (foF2), bottom height of F2 layer (h′F) and spread F event, for example. The ionosonde system works by transmitting the HF signals between 2 and 30 MHz upward to the ionosphere. The echo signals of each frequency are measured and analyzed in the form of ionogram images. An important parameter known as spread F is used to study the ionospheric irregularity at the bottom of the ionosphere especially during the 'fresh' bubble occurrences. In some works, spread F and EPB are interchangeably used.

The Frequency Modulated Continuous Wave (FMCW) ionosonde system at Chumphon station has been conducted under the Southeast Asia Low-latitude Ionospheric Network (SEALION) project since 2005. The details of the ionosonde station is shown in Fig. 4 and Table 2.

Fig. 4. Illustration of an ionosonde station.

Table 2. Technical parameters of the Chumphon ionosonde station.

Item	Description
System	FM/CW with pseudo-random Tx/Rx switching
Frequency	2–30 MHz
$P_{transmitter}$	20 W (Max.), 10 W (Avg.)
Antenna	Delta antenna (tower height 27 m)
Sweep rate	100 kHz/sec
Sweep Period	5 min

3 Results and Discussions

3.1 Comparison of Multi-Sensor Data on March 20th, 2020

On March 20th, 2020, the ionosphere is in a quiet state with the Kp index of 3. During equinoctial months (March, April, September, October), local ionospheric irregularities such as EPBs are often seen. Figure 5 compared the ROTI maps, VHF radar images and ionograms during 13.00 h and 15.00 h. At the top panels, high ROTI are shown in red and yellow points at the ionospheric piercing points (IPPs). The IPPs with low ROTI are omitted on the ROTI maps. From the maps, we can see the high ROTI started to appear at 13.30 h, then peaked at 14.00 to 14.30 h. The direction was from West to East. The middle panels are the VHF radar imaged from Fd3 (overhead direction) beam. At 13.30 h, we see irregularities at the altitude of about 300 km, then at 14.30 h, the irregularity reached 600 km, in the next hour, the irregularities tended to decrease in size and altitudes. At the bottom panels, the ionograms are shown where the bottomside irregularities are seen from 300 to 400 km at the same time as those on the radar images. The ionosonde is often used to study the bottomside of the ionosphere since the space of propagating frequencies coincide with electron densities at each altitude up to the F2 region height. All three types of instruments can be used to study the various aspect of the ionospheric irregularities.

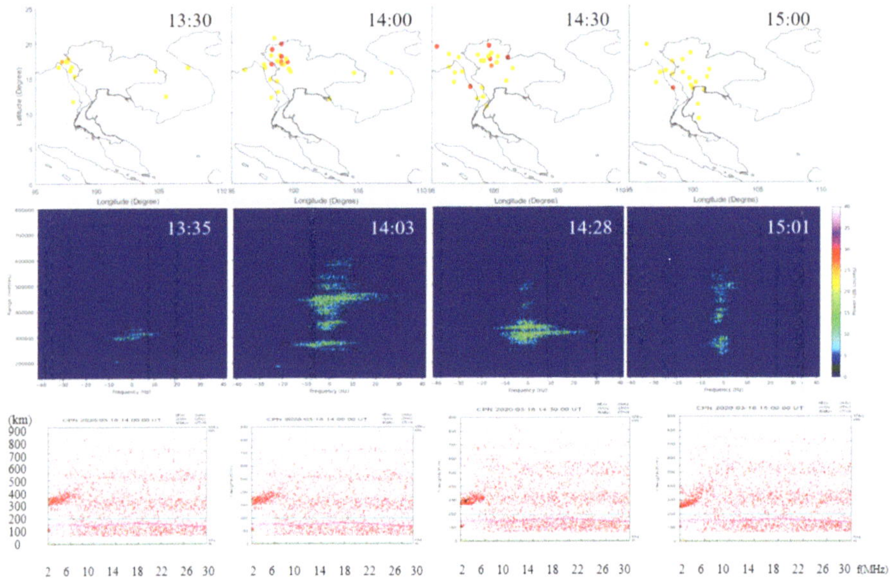

Fig. 5. Comparison of (top) ROTI map, (middle) VHF radar images and (bottom) ionogram (quick-look) at Chumphon station on 20 March 2020.

3.2 Comparison of Muti-Sensor Data on May 8 to 14, 2024 Storms

Figure 6 shows the space weather parameters during 8–12 May, 2024. The Dst index in Fig. 6(a) dipped below −100 nT on the night 10 May 10, reaching the minimimum near −400 nT then recovered during the next 2 days. The corresponding Kp index in Fig. 6(b) started to rise on 10 May reaching the peak of Kp = 9 and hovered at high Kp until 12 May. From Fig. 6(c), the critical frequency foF2 was very much below the 3-day mean foF2 on the night of 10 May reaching about 1–2 MHz, similar dipped occur on the May 11 as well. Similarly, on Fig. 8(d), the h′F highly fluctuated on 11 May. On the other hand, in Fig. 6(d), the spread F events were observed on 8–10 May, but no spread F was observed on 11 May.

Figure 7 shows vertical TEC plots from 10 to 12 May, 2024. From Fig. 7(a), the daytime TEC on 10 May looks normal, but the fluctuation occurred after sunset followed by a large drop of about 40 TECU from 1700 to 1900 UTC (2400-0200 LT). The ROTI levels in Fig. 7(b) are increased to 1 during 1200-1800 UTC (1900-0100 LT). On 11 May, we see interesting results on day-time TEC with high fluctuation of 30 TECU, the TEC appeared normal again on 12 May. From Fig. 7(c), the nighttime ROTI maps at 16:25 UTC (23:25LT) show clear high ROTI on 10 May, but no nighttime fluctuation is seen on 11–12 May.

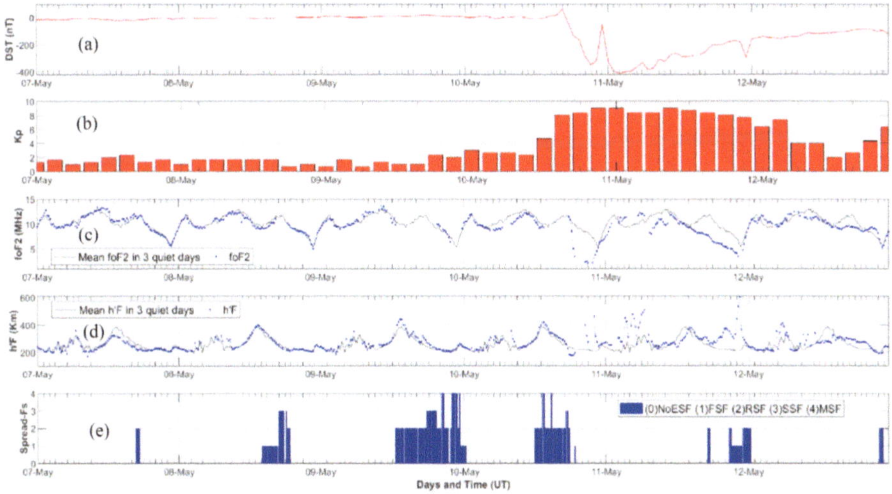

Fig. 6. Space weather parameters during 8–12 May, 2024.

3.3 Monthly Statistics of Ionospheric Irregularities (2021–2024)

We have analyzed the statistics of ionospheric irregularities during 2021–2024 as shown in Fig. 8. The graph shows the number of days when ionospheric irregularities are analyzed high ROTI (ROTI > 0.5) and VHF radar images. In 2021, there was a ack of VHF radar data so only high ROTI statistics are shown. In 2024, we show the results up to 24 May 2024. The equinoctial months see more irregularities, and during 2022–23, high occurrences of ionospheric irregularities are seen. The occurrences are higher as we reach 2024 where high occurrences are seen starting from February 2024. The discrepancy in statistics obtained from VHF radar images is due to the cases when short-lived irregularities appeared, then the ROTI was not much affected.

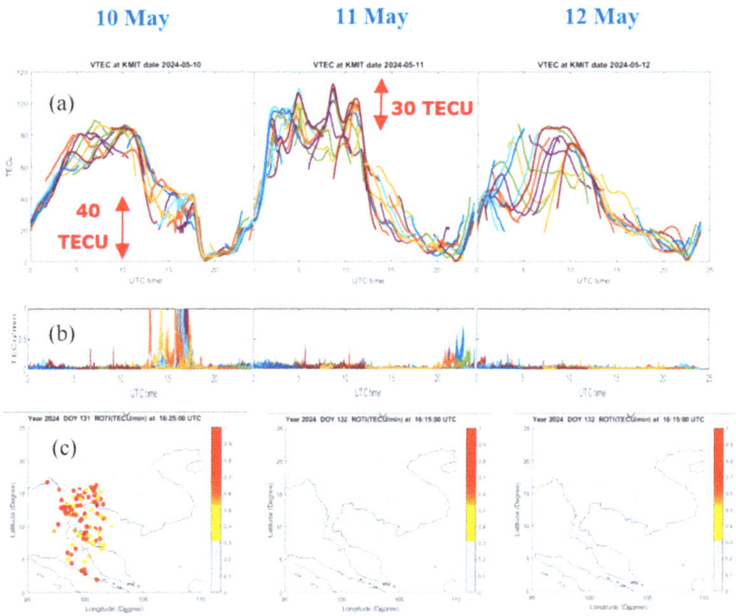

Fig. 7. The vertical TEC, ROTI and ROTI maps from 10 to 12 May 2024.

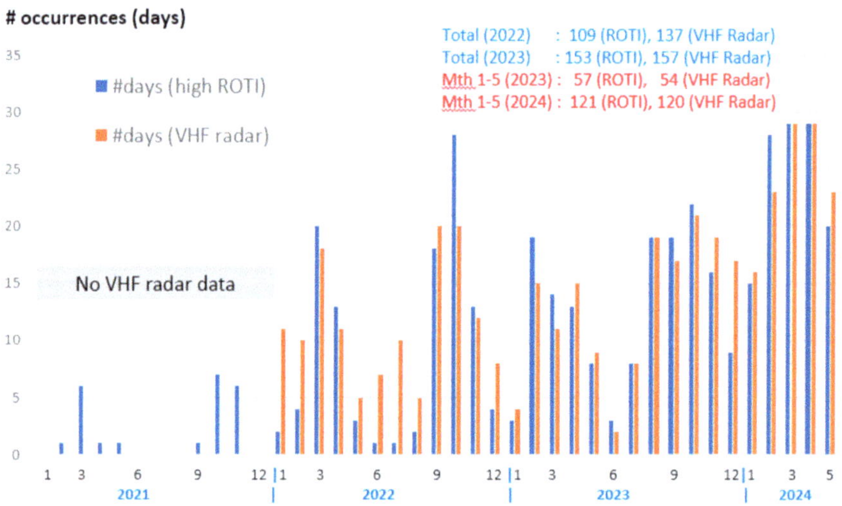

Fig. 8. Monthly statistics of ionospheric irregularities based on ROTI and VHF radar images.

4 Conclusions

We find increasing ionospheric irregularities from 2021 to 2024. The ROTI map, VHF radar images, ionograms agree on the ionospheric irregularity occurrences. During the recent May 2024 storms, TEC fluctuations and foF2 depressions are found.

Acknowledgements. This work was financially supported by the NSRF via the Program Management Unit for the Human Resources and Institutional Development, Research and Innovation (Grant no. B39G660029).

References

1. Kelly, M.C.: The Earth's Ionosphere: Plasma Physics and Electrodynamics, 2nd edn. Academic Press, San Diego (2009)
2. Abdu, M.A., de Souza, J.R., Kerani, E.A., Batista, I.S., MacDougall, J.W., Sobral, J.H.: Wave structure and polarization electric field development in the bottomside f layer leading to postsunset equatorial spread f. J. Geophys. Res. Space Phys. **120**(8), 6930–6940 (2015)
3. Abadi, P., Saito, S., Srigutomo, W.: Low-latitude scintillation occurrences around the equatorial anomaly crest over Indonesia. Ann. Geophys. **32**, 7–17 (2014)
4. Deepak, K.K., Richard, W.E., Robert, E.D., Carlos, R.M., William, E.M.: GOLD mission's observation about the geomagnetic storm effects on the nighttime equatorial ionization anomaly (EIA) and equatorial plasma bubbles (EPB) during a solar minimum equinox. Space Weather **21**(3), 5 (2023). https://doi.org/10.1029/2022SW003321
5. Huba, J.D., Joyce, G., Krall, J.: Three-dimensional equatorial spread F modeling. Geophys. Res. Lett. (2008). https://doi.org/10.1029/2008GL033509
6. de Alison, O., et al.: GPS availability and positioning issues when the signal paths are aligned with ionospheric plasma bubbles. GPS Solut. **22**(4), 1 (2018)
7. Pi, X., Mannucci, A.J., Lindqwister, U.J., Ho, C.M.: Monitoring of global ionospheric irregularities using the worldwide GPS network. Geophys. Res. Lett. **24**(18), 2283–2286 (1997)
8. Carmo, C.S., et al.: Evaluation of different methods for calculating the ROTI index over the Brazilian sector. Radio Sci. **56**, 1–12 (2021)

Open Access This chapter is licensed under the terms of the Creative Commons Attribution 4.0 International License (http://creativecommons.org/licenses/by/4.0/), which permits use, sharing, adaptation, distribution and reproduction in any medium or format, as long as you give appropriate credit to the original author(s) and the source, provide a link to the Creative Commons license and indicate if changes were made.

The images or other third party material in this chapter are included in the chapter's Creative Commons license, unless indicated otherwise in a credit line to the material. If material is not included in the chapter's Creative Commons license and your intended use is not permitted by statutory regulation or exceeds the permitted use, you will need to obtain permission directly from the copyright holder.

Operational Space Weather

Space Weather Within ESA's Space Safety Programme

Alexi Glover[1,2(✉)], Melanie Heil[1], Stefan Kraft[1], and Juha-Pekka Luntama[1]

[1] European Space Agency, Space Safety Programme Office, Robert Bosch Strasse 5, Darmstadt, Germany
Alexi.Glover@esa.int
[2] ESA Headquarters, 8-10 Rue Mario Nikis, Paris, France

Abstract. The objective of the ESA Space Safety Programme is the protection of our planet, humanity and assets in space and on Earth from hazards originating in Space. The programme incorporates activities in the areas of space weather, planetary defense, space debris and clean space technologies. This paper will focus on the space weather system developments ongoing within the Programme encompassing new measurement systems for space weather phenomena, service development and pre-operational provision.

This paper presents an overview of both space and service system developments currently ongoing and next steps.

Keywords: services · monitoring · research-to-operations

1 Introduction

Within the domain of space weather, the Space Safety Programme focuses on the development of an integrated space weather monitoring and service system targeting provision of actionable information to a diverse range of end users in the following domains:

- Spacecraft Designers
- Spacecraft Operators
- Human Spaceflight Mission Operators
- Launch Operators
- Communication and Navigation
- Space Traffic Coordination
- Power System Operators
- Pipeline Operators
- Aviation
- Resource Exploration: geomagnetic surveying, oil and gas prospecting and exploitation
- Aurora Observation and Forecast
- General Data Services geared towards third party downstream service providers, expert users and the scientific community

© The Author(s) 2026
N. Gopalswamy et al. (Eds.): ISWI 2024, SPPHY 431, pp. 275–284, 2026.
https://doi.org/10.1007/978-981-95-1121-1_31

Development activities also include space segment developments, aiming to ensure that necessary data is available to underpin overall service development and development of the data processing and provision system required to deliver timely and reliable information to service users along with expert support where required.

Based on a common set of underpinning customer requirements [1], new measurement systems are developed, and space missions implemented to provide the data needed to enable major enhancements in service provision capabilities.

User needs are gathered through regular interactions with the end user community and form the basis for a robust system development process whereby each individual element is targeted towards ultimately meeting the needs of the end users.

2 Space Weather Service Development and Testing

The ESA Space Weather Service (SWE) Network builds on the principle of federating and further developing European capabilities in the space weather domain. The network is based on a federated architecture where service provision is carried out by five Expert Service Centres (ESCs) in the Programme Member States together with a centralised Space Weather Coordination Centre (SSCC) located at the Space Pole in Brussels, Belgium.

Fig. 1. The ESA Space Weather Service Network

Each Expert Service Centre brings together European expertise and assets in their respective domain, through a network of participating Expert Groups (EGs) and their activities are coordinated by a Coordinating Expert Group. Five ESCs focusing on Solar Weather, Heliospheric Weather, Space Radiation Environment, Ionospheric Weather and Geomagnetic Conditions have been established and are being further developed. The

Coordinating Expert Groups active during Period 2 of the Space Safety Programme are identified in Table 1 below. At the present time, approximately 50 Expert Groups representing institutes and organisations providing data, modelling capabilities and Space Weather expertise collaborate within the ESC framework in order to provide more than 300 products [2] which underpin 29 user-driven services. Figure 1 illustrates the geographic distribution of the SWE Service Network.

Table 1. Expert Service Centres, Coordinating Expert Groups in the Space Safety Programme

Expert Service Centres	Coordinating Expert Groups
Solar Weather	Royal Observatory of Belgium
Space Radiation	Belgian Institute of Space Aeronomy
Ionospheric Weather	DLR Neustrelitz, Germany
Geomagnetic Conditions	DTU, Denmark
Heliospheric Weather	STFC RAL Space, UK

In addition to the Expert Service Centres, the Space Weather Coordination Centre (SSCC) maintains a centralised service helpdesk and provides first line support to end user enquiries via email, online help request or telephone during normal office hours. In case further expert support is required in order to respond to the user enquiry, the SSCC will forward the enquiry to the corresponding Expert Group within the service network as part of the network's overall second line support agreement. A centralised ticketing system is used by the helpdesk in order to maintain clear traceability of enquiries raised and responses received. The centre is also responsible for overall service availability monitoring and end user training and engagement activities (Fig. 2).

Fig. 2. The Space Weather Coordination Centre (SSCC) hosted at Space Pole, Brussels

3 New and Enhanced Measurement Systems

See Fig. 3.

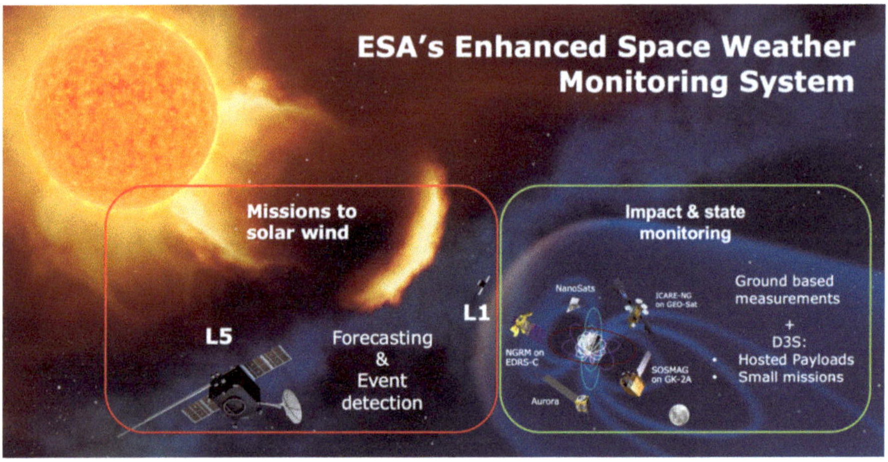

Fig. 3. ESA's Enhanced Space Weather Monitoring System

3.1 Vigil

Vigil will be Europe's first deep space operational space weather mission and will provide critical space weather data enabling early warning of extreme solar events. The Vigil Mission shall position a satellite at the 5th Lagrangian point (L5) with the objective to perform continuous observations of the Sun and the space between the Earth and the Sun to provide measurement data for space weather nowcasting and forecasting and for event-based warnings and alerts when solar events take place. The observations from L5 will complement space weather monitoring from Sun-Earth line and enable more accurate space weather impact predictions and early warnings of potentially hazardous solar weather conditions emerging. In space weather monitoring from deep space ESA is cooperating in space weather monitoring with the US (NOAA and NASA) who are planning to continue maintaining operational space weather monitoring on the Sun-Earth line by missions to the 1st Lagrangian point (L1) and by instruments onboard satellites orbiting the Earth.

The field of view from L5 allows monitoring of the onset of Coronal Mass Ejections (CMEs) with a coronagraph from a different angle than coronagraphy from the Sun-Earth line. Combination of the observations from two directions is foreseen to provide better estimates of the CME direction and speed, and to allow detection of faint "stealth" CMEs and CMEs that are ejected rapidly one after another. Vigil mission will also be able to monitor the entire space between Sun and Earth with a heliospheric imager allowing mid-course tracking of solar wind features including CMEs as they travel towards Earth.

The Vigil mission will enable observation of the solar disc several days before it becomes visible from the Earth. Longer monitoring of the active regions and their flaring is expected to make statistical flare forecasting more accurate. Vigil observations also prevent situations where strong flaring can be detected behind the solar horizon but without means to investigate the seriousness of the threat further.

In-situ measurements in L5 will allow monitoring of high-speed solar wind streams several days in advance before they rotate towards the Earth. Magnetograph observations from L5 will provide fresh solar magnetic field data for numerical solar wind models used in CME propagation estimation and enable more precise predictions of the CME arrival times on Earth. Magnetograph data is also expected to improve the solar flare and CME onset forecasting accuracy.

As the baseline payload the Vigil spacecraft will carry 5 instruments, with 3 instruments dedicated to remote sensing of the Sun and the interplanetary space between the Sun and the Earth, and 2 instruments in-situ measuring the interplanetary magnetic field (IMF) and the solar wind characteristics at the L5 position. ESA is implementing Vigil mission in close collaboration with NOAA and NASA. NOAA will provide the CCOR coronagraph, which is one of the 5 baseline instruments. NASA will provide an EUV imager that is an additional instrument on top of the baseline instruments.

The aim of the Vigil mission is to demonstrate the benefits from space weather observations away from the Sun-Earth line for operational applications. For that purpose, the mission is designed to carry out observations at all times including severe space weather events and to provide data about the current space weather conditions in a 24/7 operational mode. The data for operational space weather applications shall be delivered to the users with low latency.

In addition to the baseline measurements for operational space weather applications, Vigil mission will support "bonus science" by making all measurement data available for research purposes through the data archive in the ESA Space Weather Data System.

3.2 Distributed Space Weather Sensor System

The objective of ESA's Distributed Space Weather Sensor System (D3S) is to monitor the space environment near Earth and its impacts from space weather. D3S observations include monitoring of the magnetic field, neutral/charged particles, plasma environment, auroral oval, upper atmosphere, and other parameters required by space weather services. During the coming years, the implementation of D3S will be continued by utilizing hosted payloads, dedicated SmallSat, and nanosat missions, as well as international collaboration to exchange data [3].

The first two dedicated missions are currently being implemented, namely Aurora and the first nanosatellite mission, further described below.

Fig. 4. The Lunar Gateway with ERSA mounted on the power and propulsion element, which is the right-most one in the picture. (credit: NASA)

Several hosted payload missions are also in preparation, like a radiation monitor on Lunar Pathfinder and the European Radiation Sensor Array (ERSA) on the Lunar Gateway, as shown in Fig. 4, consisting of multiple radiation monitors, including NGRM and ICARE-NG, dosimeters and magnetometers. Radiation monitoring in orbits around the Moon is considered very important for space weather services for the next steps in human and robotic space exploration.

Current hosted payloads in operation include SOSMAG, the service oriented magnetometer instrument launched with the Korean satellite GEO-KOMPSAT-2A in 2018 and the NGRM radiation monitor, launched with EDRS-C in 2019 and Sentinel-6 in 2020. Additional space environment data is currently in preparation from the two ICARE-NG units on EUTELSAT's HOTBIRD 13 F&G satellites launched in 2022 with the data expected to become available in mid-2024, and the NGRM unit on MTG-I1 launched in December 2022. New observations to be addressed with hosted payloads proposed for the coming years include radiation measurements in LEO, GEO and other under-sampled regions.

3.3 Aurora

Monitoring of the Aurora is an essential and central target for space weather nowcasting and forecasting since it enables indirectly the observation of the solar wind variations and the impact of Coronal Mass Ejections (CMEs) on the Earth. Auroral emissions are a direct manifestation of physical processes occurring when the magnetosphere responds to the solar wind and CME plasma streams that strongly alter the interplanetary and geomagnetic field. The emissions and their time sequence contain information about the characteristics of the particle flow, and physical and chemical processes occurring locally and in the upper atmosphere, namely the ionosphere. Observations of the auroral oval may be used as input to the modelling of geomagnetic storm conditions and provision of timely information about space weather conditions to end users operating in high latitude regions. Additional instrumentation dedicated to in-situ measurements to monitor the radiation environment close to the inner radiation belt and the local magnetic field conditions are planned (Fig. 5).

Fig. 5. Artistic view of the Auora satellites monitoring the Auroral Oval. The insert on the top left shows the to scale orbital configuration of the four satellites

The Aurora mission targets provision of data leading to substantial improvement of services dependent on the modelling of the Earth's magnetic and electric field and currents; services that are used for example for aviation, transport in high latitude regions, telecommunication, search and rescue, and power grid operation.

The Aurora small satellites constellation will constantly monitor the entire Auroral Oval from a Medium Earth Orbit at about 6500 to 7000 km altitude enabling short refresh rates and allowing the identification and monitoring of locations of ongoing geomagnetic storms and sub-storms. The provided information will help improving services available to the locally affected users.

In the proposed mission scenario, a two-step approach is foreseen with the first step the implementation of a Demonstration mission (Aurora-D) focusing on the Auroral Oval imaging and demonstration of the mission concept. In a second step, a Constellation mission (Aurora-C) of 4 satellites is proposed. The full constellation will then enable continuous (24/7) monitoring of the Auroral oval.

The observations are based on two wide field-of-view optical instruments, the Aurora Optical Spectral Imager (AOSI) and the Aurora far UV Imager (AUI) covering in night-time condition the optical lines at 391 nm ($N2+$), 557.7 nm (O ^1S), 427.8 nm ($N2+$), 630 nm (O ^1D), 636 nm (O ^1D), and 639 nm (O ^1D) and in day and night time condition the far UV Nitrogen band emissions in the range of the short and long Lyman-Birge-Hopfield bands located between 140 and 175 nm. Alternatively, the Oxygen line emissions at 130 nm and 135.6 nm are considered. For the in-situ measurements a combination of a radiation monitor and a magnetometer is being developed. The instrument can be equipped with up to three solid state telescopes measuring the electron, proton and ion flux and will carry two flux gate magnetometers, one of them on a boom with approximately 1.4 m length measuring the magnetic field vector.

The objective of the demonstration mission is to explore the observational capabilities and to ensure that the observational system will be optimized for the constellation.

3.4 Space Weather Nanosatellites

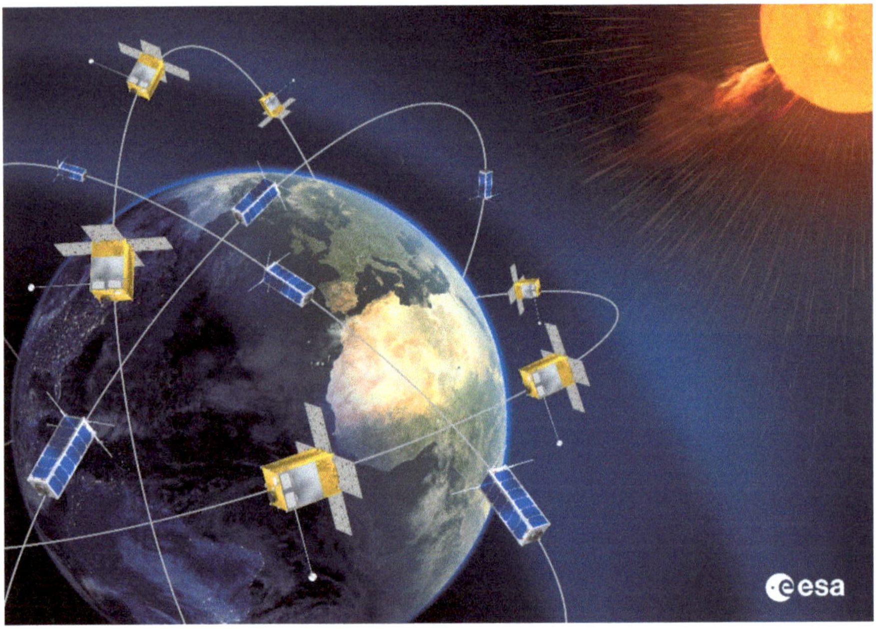

Fig. 6. Sketch of ESA's D3S Nanosatellite System.

The objectives of utilising nanosatellites for the implementation of D3S are cost optimisation for the collection of multi-point measurements and allowing for faster technology and instrumentation adoption. The first nanosatellite mission will demonstrate the use of nanosatellites for space weather monitoring for operational applications as well as demonstrate the mission/data-as-a-service concept in this context. The mission will provide data on the status of Earth's ionosphere, through measurements of the solar x-ray flux, the local electron density as well as radio occultation profiles, complemented with radiation monitor data. A second mission study and subsequent implementation is currently in preparation to complement the measurements collected from low Earth orbit (Fig. 6).

Further dedicated missions are planned to complement the targeted measurement data as requested by the community e.g. near real-time data on the radiation belts.

4 Conclusions and Outlook

The ESA Space Weather service network currently provides pre-operational services for spacecraft operations on the basis of a federated service provision model, providing end users with a very wide range of products and information about current and upcoming space weather conditions along with post event analysis support and access to expertise provided by leading domain experts from across the European region. This is provided along with the first data from ESA's Distributed Space Weather Sensor system via the ESA Space Weather Portal: https://swe.ssa.esa.int.

Over the course of the Space Safety Programme, the SWE service network will continue to develop, further expand and improve its capabilities to provide timely and reliable information to end users in a range of domains.

The SWE service network will continue to provide information on a pre-operational basis in the short term, with additional developments ongoing to establish both system design and a roadmap for transition of the current network to a framework able to provide timely and actionable information on an operational (24/7) basis in future, while maintaining a distributed service provision model.

In parallel, the Space Weather space segment developments taking place within the Space Safety Programme aim to strengthen and further guarantee the availability of critical data to underpin essential space weather services and also, through missions such as Vigil, to enable substantial steps forward in space weather service capabilities globally.

References

1. Customer Requirements for the ESA Space Safety Programme. https://swe.ssa.esa.int/documents/d/guest/esa_s2p_swe_crd
2. The ESA Space Weather Service Network Product Catalogue. https://swe.ssa.esa.int/documents/d/guest/product_catalogue_summary
3. Heil, M., et al.: ESA's space weather monitoring system. In: Proceedings of the 17th International Conference on Space Operations (2023). https://star.spaceops.org/user_manudownload.php?doc=652__9bv6w5vl.pdf

Open Access This chapter is licensed under the terms of the Creative Commons Attribution 4.0 International License (http://creativecommons.org/licenses/by/4.0/), which permits use, sharing, adaptation, distribution and reproduction in any medium or format, as long as you give appropriate credit to the original author(s) and the source, provide a link to the Creative Commons license and indicate if changes were made.

The images or other third party material in this chapter are included in the chapter's Creative Commons license, unless indicated otherwise in a credit line to the material. If material is not included in the chapter's Creative Commons license and your intended use is not permitted by statutory regulation or exceeds the permitted use, you will need to obtain permission directly from the copyright holder.

Operational Space Weather Services in Norway

Wojciech J. Miloch(✉)

Department of Physics, University of Oslo, Postboks 1048, 0316 Blindern, Norway
w.j.miloch@fys.uio.no

Abstract. Space weather can have important consequences for both safety and performance of operations in the polar regions. Examples of space weather effects are the degradation of quality of trans-ionospheric radio signals and reduced reliability of satellite-based navigation services during increased geomagnetic activity. Thus, there are worldwide efforts to study the space weather effects and develop robust nowcasting and forecasting services for end-users. The geographic location of Norway is favorable for such studies in the Arctic, but at the same time the air, sea, and terrestrial operations in Norway are subject to various space weather effects. This paper gives an overview of Norwegian space weather research infrastructure and activities, as well as insight into operational space weather services in Norway.

Keywords: Polar regions · Space weather · Ionosphere · Norway

1 Background

The understanding of space weather and associated phenomena is of increasing importance for the modern society. Effects due to space weather can impact the performance of technological systems both in space and on the ground [1]. One of examples is the impact of ionospheric irregularities on the propagation of trans-ionospheric radio waves, which can undergo diffraction and refraction on irregularities in ionospheric plasma density [2]. Consequently, signals from satellites received on the ground can be impacted, and the performance and reliability of systems that rely on such signals, such as the Global Navigation Satellite Systems (GNSS), can be deteriorated.

Increasing human activity in the polar regions, particularly in the Arctic, often relies on accuracy and reliability of positioning services. Economic activity in the polar regions is increasing worldwide. This includes fisheries and offshore industry, oil and gas exploration, shipping, aviation, and tourism. All these activities require reliable communication and positioning services, which can be impacted during increased geomagnetic activity. Thus, there are ongoing efforts to understand space weather and its effects, and to develop services for monitoring and forecasting space weather effects in the polar regions, which will eventually increase safety and reliability of operations [1].

Due to its geographic location, Norway is subject to all space weather effects that are characteristic for high geomagnetic latitudes: subauroral phenomena in the south of the country, aurora and related effects in the north of continental Norway, and the polar cap

© The Author(s) 2026
N. Gopalswamy et al. (Eds.): ISWI 2024, SPPHY 431, pp. 285–294, 2026.
https://doi.org/10.1007/978-981-95-1121-1_32

and auroral processes in the high Arctic [3]. Especially, activities in the polar regions, both on land and sea can be severely impacted by the space weather effects.

2 Space Weather Research in Norway

Modern space weather research in Norway started in late 19th century. One of the pioneers in space research was Kristian Birkeland (1867–1917), who was the first to propose that aurora should be associated with the field aligned current (FAC) systems, see Fig. 1(a) [4]. The Birkeland (FAC) currents and his theory were later confirmed based on in-situ measurements by satellites [5]. In years 1897–1903, Birkeland organized three auroral expeditions to northern Norway, Svalbard, Iceland, and Novaya Zemlya, where he studied geomagnetic disturbances, auroral dynamics, and tried to estimate the altitude of aurora. He also established plasma laboratory to understand the basic processes behind formation of aurora in the terrella experiment, see Fig. 1(b). The permanent auroral observatory was established in Haldde, near Kaafjord in northern Norway in 1912, and then in Tromsø in 1928. Nowadays, diverse research infrastructure in Norway allows for comprehensive studies of ionospheric phenomena at high geomagnetic latitudes.

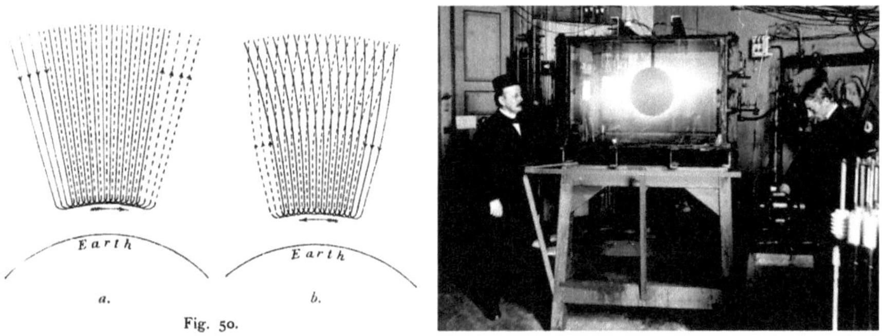

(a) (b)

Fig. 1. (a) The field aligned current system proposed by Kristian Birkeland [from pp. 105, ref. 4]. (b) Kristian Birkeland (left) and Olav Devik stand next to the terrella experiment in the plasma laboratory in Oslo [photo Norsk Teknisk Museum archive, NTM C 25395a].

2.1 Norwegian Infrastructure for Space Weather Research

In the following, a brief overview of the Norwegian research infrastructure is presented. This overview is intended to give a general picture of the space weather research landscape in Norway, but it is not exhaustive or detailed list of instruments.

In the high Arctic, the Svalbard archipelago is in a unique geographic position for studying the geospace. While its main island Spitsbergen is located between ca. 76–80 degrees north, it is easily accessible during the whole year. During daytime, around noon, it is located directly under the magnetospheric cusp, while at night it is mostly in the polar cap. Here, several months of the polar night between November to the end of February allow for optical studies of daytime and cusp aurora, and hence several countries run ionospheric observatories on Sptisbergen. Main locations for Norwegian instruments include Ny-Ålesund (NYA), which is a research town hosting several stations, and Longyearbyen (LYR), which is the main settlement on Svalbard.

In NYA, Norwegian research institutions operate auroral all-sky-imagers, GNSS receivers for monitoring the total electron content (TEC) and ionospheric scintillations (GISTM), ionosonde, riometers, and magnetometer. NYA has also launch facilities for the ionospheric sounding rockets operated by Andøya Space. LYR has been central for ionospheric research on Svalbard since 1978, when the first auroral station was established. Currently, the main site for space weather research in LYR is on the Breinosa mountain, further away from town. The Kjell Henriksen Observatory, run by the University Centre in Svalbard (UNIS), is a large auroral observatory hosting various instruments from several countries and operating there since 2008 [6]. The nearby incoherent EISCAT Svalbard Radar (ESR) has two dish antennas with operating frequency 500 MHz: a 32-m steerable antenna and 42-m antenna aligned with the magnetic field lines of the Earth. ESR has been in operation for about 30 years, and nowadays there are efforts to deploy a new generation incoherent radar on the site [7]. The recently established HF radar, also located at the Breinosa site, is now operational. It is a part of the SuperDARN network, and by covering the region east of Svalbard it improves the global picture of ionospheric plasma convection at high latitudes [8]. LYR is hosting several other instruments, such as HF receivers, MST radar or magnetometers. Other important stations with Norwegian instrumentation in the high Arctic include Hornsund, Hopen, Bjørnøya, and Jan Mayen. All of them have GISTMs and magnetometers, while in Hornsund there is also an all-sky-camera for auroral studies.

Northern mainland Norway is directly under the auroral oval. Here majority of instrumentation is at three sites: Andøya, Skibotn and Tromsø/Ramfjordmoen. Andøya is the launching site for ionospheric sounding rockets, and as such it also hosts several radars and observatories, including ALOMAR. Skibotn hosts all-sky imagers, riometers, and GISTM, and it is also the main site for EISCAT 3D, a new generation incoherent radar consisting of three sites with all together ca. 20000 cross-dipole antennas, shown in Fig. 2(a), which will allow for volumetric studies of ionospheric plasma [9, 10]. The radar should start its operations in late 2024. Note that EISCAT is an international consortium with several stakeholders. Other EISCAT sites are in Sweden and Finland. Ramfjordmoen is the previous main site of the EISCAT radars, which also hosts other radars, including ionosonde, heating system, MF radar, and all-sky-cameras.

Across the whole country there is a dense network of TEC and GISTM receivers operated by the Norwegian Mapping Authority (Kartverket) and a line of magnetometers maintained by Tromsø Geophysical Observatory (TGO). Smaller networks of GISTMs are maintained in the high Arctic by the University of Bergen (UiB) and University of Oslo (UiO).

Norwegian space weather instrumentation is also deployed in Antarctica to allow for interhemispheric studies of space weather. UiO has established ionospheric observatory at the Norwegian research station Troll in Droning Maud Land [11]. Currently, Troll Ionospheric Observatory includes all-sky-imagers, GISTM and ionosonde (to start operations in 2025), and it is a part of Troll Observing Network (TON-e) [12]. A magnetometer will also be deployed at Troll as a part of TON-e. UiO also operates GISTM receiver at the Polish Antarctic station Arctowski on King George Island in collaboration with the Institute of Biochemistry and Biophysics of the Polish Academy of Sciences [13].

Norway is also active in developing spaceborne instrumentation for space weather studies and monitoring. UiO has developed the multi-Needle Langmuir Probe (mNLP) system, which can sample the plasma density at kHz rates, and thus allows accessing the small scales in the ionospheric plasma [14]. mNLP has successfully flown on several sounding rockets, see Fig. 2(b). In collaboration with industry, mNLP has flown on satellites (e.g., Norsat-1, Brick-2), and is currently operational on the International Space Station (Bartolomeo platform). Another space weather instrument on the International Space Station is ESA's Atmosphere Space Interactions Monitor (ASIM), where UiB/Birkeland Centre for Space Science delivered Modular X-and Gamma-ray Sensor for observing transient effects related to thunderstorms, such as sprites, elves and blue jets [15].

(a) (b)

Fig. 2. (a) EISCAT 3D antennas on the main site in Skibotn. (b) Artist visualization of the launch of UiO's Investigation of Cusp Irregularities – 3 (ICI-3) sounding rocket [credits: Trond Abrahamsen, Andøya Space].

2.2 Main Space Weather Research Topics

In the context of space weather, main research activities in Norway concentrate on processes and phenomena related to magnetosphere-ionosphere-thermosphere coupling at high latitudes. Topics of research include for example magnetospheric reconnection and related processes, cusp aurora [16], ion outflow [17, 18], ionospheric plasma turbulence and irregularities [19], coupling between neutral and ionized parts of the atmosphere [20], energetic particle precipitation and its role in atmospheric processes at lower altitudes [21, 22], or the impact of ionospheric plasma irregularities on the propagation of trans-ionospheric radio waves and accuracy of navigational systems [23, 24]. Combining available ground-based instrumentation and in-situ measurements by sounding rockets and satellites allows for studying these problems at different spatiotemporal scales [25].

In addition to the high latitude ionosphere, fundamental processes in the solar wind as well as the solar atmosphere are studied extensively. Solar research is centered around the Rosseland Centre for Solar Physics, UiO [26].

3 Space Weather Community

The space weather community in Norway forms a basis for the development of operational space weather services in Norway. It includes universities, observatories, governmental agencies, and industry partners.

University of Oslo (UiO), University of Bergen (UiB), UiT The Arctic University of Norway (UiT), University Centre in Svalbard (UNIS), and the Norwegian University of Science and Technology (NTNU) are the main actors when it comes to space weather research. For NTNU and UiO space weather is also important in the context of satellite operations in the low Earth orbit.

Tromsø Geophysical Observatory (TGO) is a unit under UiT, which focuses on maintaining long-time series of geophysical measurements in northern Norway. TGO operates a network of 20 magnetometers, of which 17 in Norway, integrated in international networks. It also operates ionosondes in Tromsø and on Svalbard, MF-radar, riometers, meteor radars, all-sky-cameras, and MST radar on Svalbard. The ionosonde data from the Tromsø area are available from year 1935.

In 2014, the Norwegian Space Agency supported establishing the Norwegian Center for Space Weather (NOSWE). NOSWE is a part of TGO, and its role is to provide the status of space weather conditions in the Arctic, the so-called Nordic Situation Screen. Since 2019, NOSWE has been a regional warning center of the International Space Environmental Service.

Other important actors within the space weather in Norway are the Norwegian Space Agency, an agency under the Ministry of Trade and Fisheries of Norway, which coordinates the Norwegian space policy and industry to develop the space sector on the national level. It also coordinates collaboration with the European Space Agency. Norwegian Mapping Authority (Kartverket) is a national agency that operates a network of base stations for GNSS and prepares maps of TEC over Norway. It also focuses on precise positioning and related space weather effects. Meteorological Institute (MET) is a governmental administrative body focusing on analysis and forecasting of weather conditions and warnings. As a part of its activity, MET can also disseminate space weather conditions. Andøya Space operates launch facilities from Andøya and Svalbard, supporting space weather research by facilitating both ground-based and in-situ measurements by sounding rockets. There are also several industry partners that closely collaborate in research and development within space weather.

3.1 User Groups

The user groups for space weather services in Norway are many, and they are mostly concerned with communication and positioning services. Selected relevant end-users and aspects of their activity that can be impacted by space weather are given in Table 1. With the view of operational space weather in Norway, the space weather services should meet these user needs.

Table 1. Example of end-users for space weather services in Norway.

Industry/activity	Operations that depend on space weather
Oil and gas companies	Navigation, Precise positioning, Resource exploration
Aviation	Satellite-based navigation, HF communication
Maritime sector	Satellite-based navigation, HF communication
Power grid	GNSS-based time accuracy, Induced currents in the grid
Satellite operators	Radiation and communication related risks
Satellite ground stations	Downlink of data (interruptions, data errors)
Construction industry	GNSS-based precise positioning
Tourism	Satellite-based navigation, Aurora forecast
Automotive transport	Satellite based navigation, Communication
Search and rescue (SAR)	Precise positioning, Satellite-based navigation

4 Operational Space Weather

A fully operational and integrated space weather service in Norway is currently under development. It is expected that a continuous National Space Weather Forecasting service will be operational already in year 2026. This short time for the development is due to existing broad expertise and several products that have already been operational. While the full continuous, uninterrupted "24/7" support within space weather for all end-users requires significant resources, some end-users can receive such support already now. Selected space weather products are provided below.

4.1 Space Weather Products

Nordic Situation Screen. NOSWE provides current picture of space weather in Fennoscandia and in the high Arctic on its website: https://www.spaceweather.no. The real-time analysis of conditions and one hour forecast of space weather impact due to geomagnetic disturbances are provided for the sub-auroral region, auroral zone, and the polar cap area. It also provides a simple color-coded matrix that characterizes disturbances in respective regions. In addition, probability of radio blackouts is provided, as well as prediction of aurora, see Fig. 3(a). NOSWE also provides access to near-real time data for different elements of the space weather system: Sun and solar wind conditions, global geomagnetic conditions, ionospheric conditions, as well as aurora forecast. NOSWE and TGO also deliver services tailored for specific users, including geological surveys, undersea drilling, and offshore industry. TGO provides open-access magnetometer data in near real time [27]. These datasets have typically more than 12 000 users and 600 000 magnetograms downloaded per week. They are also used in many international space weather products also those coordinated by the European Space Agency.

seSolstom is a service (https://sesolstorm.kartverket.no/) provided by the Norwegian Mapping Authority, which gives an overview of ionospheric plasma conditions and their disturbances over Fennoscandia based on the network of 160 base stations that are operated by the authority. This service is run in collaboration with Sweden and Finland and is open to all users. Norwegian Mapping Authority monitors the quality of the GNSS signals due to space weather, as it is a supplier of the CPOS (Centimeter POSitioning) RTK service for professional GNSS users who require cm precision. CPOS is used in mapping and surveying, construction machinery, agriculture and nautical mapping.

værio.no is a specialized service developed by the Metrological Institute for advanced users. Under this service, users can collect the space weather data on their dashboards, to have an overview of space weather conditions. While MET does not collect space weather data, it facilitates data stream to the service. These include data from the Norwegian data providers as well as from international sources. A dedicated space weather analysis is also included in værio.no.

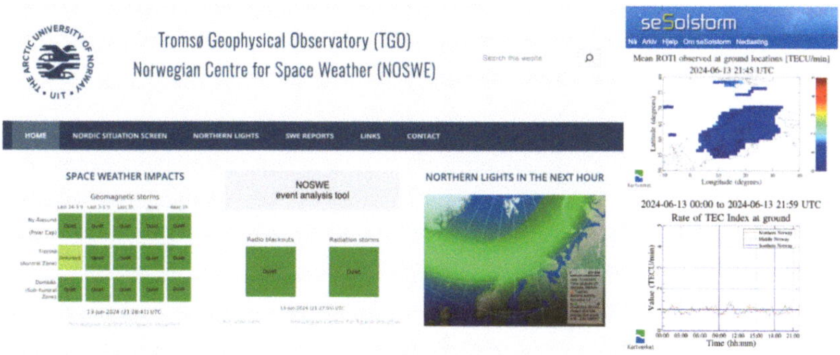

(a) (b)

Fig. 3. (a) The summary view of the NOSWE space weather service [28]. (b) Part of the seSolstorm service provided by the Norwegian Mapping Authority [29].

Aurora Forecast 3D is an auroral forecast developed by prof. Fred Sigernes at UNIS/KHO [30]. The service is now used in several services and is freely available on different platforms. Auroral tourism is growing in Norway, and this science-based application can facilitate operations of the aurora tourist industry and radio-amateur community.

Contributions to ESA's Space Weather Service Network. European Space Agency's Space Weather Service Network is the European platform for space weather products. Norway contributes to several of them. Norwegian Mapping Authority provides ionospheric monitoring and ionospheric conditions, TGO feeds magnetometer data to several other products, UiB contributes to Geomagnetic Expert Service Centre, while UiO together with Kartverket and GMV, is preparing to deploy ISPA service for nowcasting and forecasting ionospheric conditions related to scintillations [31]. Several other prototypes for space weather services and data products have been developed in international collaboration and with support from ESA. These include prototypes of forecasts of ionospheric irregularities, Swan and FORSWAR, as well as data products and models that can be used for the development of space weather services based on satellite data: Swarm-IPIR L2 dataset and Swarm-VIP-models [19, 32], and global models for ionospheric currents: Average Magnetic Field and Polar Current System (AMPS) [33].

5 Conclusions

The space weather research in Norway has a long tradition, and due to geographic location and accessibility it has focused on space weather phenomena in the polar regions, primarily on understanding phenomena in the auroral zone and the polar cap. Nowadays in Norway there is a wide network of space weather instruments, which facilitates both basic research and monitoring of geomagnetic conditions. Increased human activity in the polar regions and technology development require good knowledge on space weather conditions for safety and reliability of operations. In recent years, Norwegian institutions have been working toward developing space weather services to address the needs of end-users. Currently, several such services exist, and they are mostly related to reliability of positioning, navigation and communication. The space weather community in Norway is consolidating towards space weather forecasting service on the national level, which is expected to provide continuous, uninterrupted monitoring and broad spectrum of services for increasing number of end-users.

References

1. Bothmer, V., Daglis, I.A.: Space Weather - Physics and Effects. Springer, Berlin, Heidelberg (2006)
2. Kintner, P.M., Ledvina, B.M., de Paula, E.R.: GPS and ionospheric scintillations. Space Weather **5**(9), S09003 (2007)
3. Moen, J.I., et al.: Space weather challenges of the polar ionosphere. J. Space Weather Space Clim. **3**, A02 (2013)
4. Birkeland, K.: The Norwegian Aurora Polaris Expedition 1902–1903. H. Aschehoug & Co., Christiania (now Oslo) and New York (1908)
5. Cummings, W.D., Dessler, A.J.: Field-aligned currents in the magnetosphere. J. Geophys. Res. **72**(3), 1007–1013 (1967)
6. KHO homepage. http://kho.unis.no/index.html. last accessed 11 June 2024
7. Baddeley, L., et al.: Space and atmospheric physics on Svalbard: a case for continued incoherent scatter radar measurements under the cusp and in the polar cap boundary region. Prog Earth Planet Sci **10**, 53 (2023)
8. Chisham, G., et al.: A decade of the super dual auroral radar network (superDARN): scientific achievements, new techniques and future directions. Surv. Geophys. **28**(1), 33–109 (2007)
9. McCrea, I., et al.: The science case for the EISCAT_3D radar. Prog Earth Planet Sci **2**(1), 1–63 (2015)
10. EISCAT 3D homepage. https://eiscat.se/eiscat3d-information/. last accessed 11 June 2024
11. Skjæveland, A.H., Kotova, D.S., Miloch, W.: J: Case studies of ionospheric plasma irregularities over Queen Maud Land, Antarctica. J. Geophys. Res. **126**, e2021JA029963 (2021)

12. Pedersen C.A., et al.: The Troll Observing Network (TONe): a contribution to improving observations in the data-sparse region of Dronning Maud Land, Antarctica. Polar Res. **43** (2024)
13. Arctowski Polish Antarctic Station. https://arctowski.aq/. last accessed 11 June 2024
14. Hoang, H., et al.: The multi-needle Langmuir probe system on board NorSat-1. Space Sci. Rev. **214**, 75 (2018)
15. Neubert, T., et al.: A terrestrial gamma-ray flash and ionospheric ultraviolet emissions powered by lightning. Science **367**, 183–186 (2020)
16. Jin, Y., et al.: On the collocation of the cusp aurora and the GPS phase scintillation: a statistical study. J. Geophys. Res. **120**(10), 9176–9191 (2015)
17. Bjoland, L.M., et al.: Electron density depletion region observed in the polar cap ionosphere. J. Geophys. Res. **126**(1), 28432 (2021)
18. Lorentzen, D., et al.: Pulsating dayside aurora in relation to ion upflow events during a northward interplanetary magnetic field (IMF) dominated by a strongly negative IMF By. J. Geophys. Res. **112**(A3), 66 (2007)
19. Jin, Y., et al.: Ionospheric plasma irregularities characterized by the swarm satellites: statistics at high latitudes. J. Geophys. Res. **124**(2), 1262–1282 (2019)
20. Stamm, J., et al.: A technique for volumetric incoherent scatter radar analysis. Ann. Geophys. **41**, 55–67 (2023)
21. Nesse, H., et al.: Energetic electron precipitation during slot region filling events. J. Geophys. Res. **128**, e2023JA031606 (2023)
22. Jia, J., et al.: Energetic particle precipitation influences global secondary ozone distribution. Commun. Earth Environ. **5**, 270 (2024)
23. Oksavik, K., van der Meeren, C., et al.: Scintillation and loss of signal lock from poleward moving auroral forms in the cusp ionosphere. J. Geophys. Res. **120**, 9161–9175 (2015)
24. Enengl, F., Spogli, L., et al.: Investigation of ionospheric small-scale plasma structures associated with particle precipitation. Space Weather **22**, e2023SW003605 (2024)
25. Buschmann, L.M.: The role of particle precipitation on plasma structuring at different altitudes by in-situ measurements. J. Space Weather Space Clim. **13** (2023)
26. RoCS – Rosseland Centre for Solar Physics website, https://www.mn.uio.no/rocs/english/, last accessed 2024/06/11
27. TGO homepage. https://geo.phys.uit.no. last accessed 11 June 2024
28. NOSWE homepage. https://www.spaceweather.no. last accessed 11 June 2024
29. seSoltorm homepage. https://sesolstorm.kartverket.no/. last accessed 11 June 2024
30. Sigernes, F., et al.: Two methods to forecast auroral displays. J. Space Weather Space Clim. **1**, A03 (2011)
31. Jin, Y., Miloch, W.J., et al.: Modeling TEC irregularities in the Arctic ionosphere using empirical orthogonal function method. Space Weather **21**, e2023SW003531 (2023)
32. Wood, A.G., et al.: Statistical models of the variability of plasma in the topside ionosphere: 1. Development and optimisation, J. Space Weather Space Clim. **14** (2024)
33. Laundal, K.M., et al.: Solar wind and seasonal influence on ionospheric currents from swarm and CHAMP measurements. J. Geophys. Res. **123**, 4402–4429 (2018)

Open Access This chapter is licensed under the terms of the Creative Commons Attribution 4.0 International License (http://creativecommons.org/licenses/by/4.0/), which permits use, sharing, adaptation, distribution and reproduction in any medium or format, as long as you give appropriate credit to the original author(s) and the source, provide a link to the Creative Commons license and indicate if changes were made.

The images or other third party material in this chapter are included in the chapter's Creative Commons license, unless indicated otherwise in a credit line to the material. If material is not included in the chapter's Creative Commons license and your intended use is not permitted by statutory regulation or exceeds the permitted use, you will need to obtain permission directly from the copyright holder.

Ionosphere Monitoring and Prediction Center (IMPC)

Martin Kriegel(✉), Paul David, Youssef Tagargouste, Dmytro Vasylyev, David Wenzel, Fredrik Nistor, and Jens Berdermann

Deutsches Zentrum für Luft- und Raumfahrt e.V. (German Aerospace Center – DLR), Kalkhorstweg 53, 17235 Neustrelitz, Germany
martin.kriegel@dlr.de

Abstract. Reaching the maximum of solar cycle 25 the influence of space weather is becoming increasingly critical, with phenomena like solar flares and geomagnetic storms occurring more frequently, affecting satellite technology, aerospace operations, telecommunications, and navigation systems more severely. The DLR Institute for Solar-Terrestrial Physics in Neustrelitz (DLR-SO) studies these impacts on critical infrastructures. Central to these activities is the Ionosphere Monitoring and Prediction Center (IMPC) which provides near real-time monitoring and prediction capabilities for space weather impacts, enhancing system resilience and reliability. It leverages advanced instrumentation and data processing technologies, including GNSS-based remote sensing and ionospheric modeling. IMPC instruments like the Global Ionospheric Flare Detection System (GIFDS) and eCALLISTO offer comprehensive space weather data assets. Integrated into international networks, the IMPC contributes to a global effort to understand and mitigate space weather impacts to protect and improve our technology and its operational performance during space weather events.

Keywords: Space Weather · Realtime Monitoring · Remote Sensing

1 Introduction

In an era of high-precision and increasingly autonomous Global Navigation Satellite Systems (GNSS) applications, the influence of space weather on their system performance has become an area of growing concern and active research. Space weather phenomena, such as solar flares, geomagnetic storms, and ionospheric disturbances, can significantly affect satellite technology, aerospace operations, telecommunications, navigation systems, and power grids. The DLR Institute for Solar-Terrestrial Physics (DLR-SO) in Neustrelitz is a leading institution investigating these effects on critical systems and infrastructures on the ground- and in space. Central to this research is the Ionosphere Monitoring and Prediction Center (IMPC), developed and operated by the Pre-operational Services working group in collaboration with the German Remote Sensing Data Center of DLR. As illustrated in Fig. 1, a significant strength of the IMPC is its utilization of state-of-the-art data processing technologies, enabling rapid and precise

© The Author(s) 2026
N. Gopalswamy et al. (Eds.): ISWI 2024, SPPHY 431, pp. 295–302, 2026.
https://doi.org/10.1007/978-981-95-1121-1_33

analysis of large ground and space based data quantities. This facilitates near real-time monitoring and response to space weather events. The IMPC's capabilities extend beyond monitoring to include a range of products and services designed to mitigate space weather impacts on contemporary technologies, crucial for sectors reliant on GNSS and RF communications. The IMPC acts as a bridge between scientific expertise and the diverse requirements of users from national and international public and private sectors, academia, and industry. Providing near real-time monitoring and predictive capabilities for space weather impacts, it is enhancing the resilience and reliability of modern technological systems.

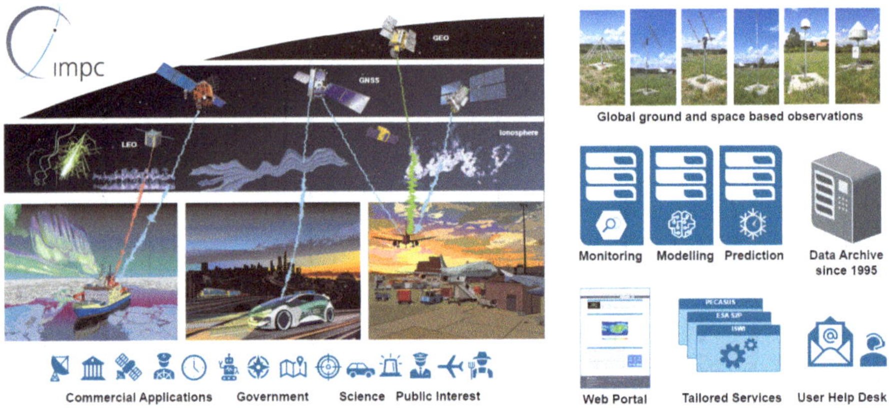

Fig. 1. Schematic overview of the core components of the IMPC to address the impact of space weather on the various end-user sectors.

2 Capabilities

2.1 IMPC Framework

The IMPC provides a scalable and robust platform for the automated processing, archiving and dissemination of ionospheric products on a 24/7 near real-time basis [1]. Each data product is currently processed, archived and delivered in approximately one minute. The developed system is highly efficient and generates thousands of products daily. The platform integrates data from real-time data streams and files and follows a three-step processing environment that transitions from development through a test zone to operational deployment. The framework is highly containerized and being developed towards open data standards, e.g. driven by the Open Geospatial Consortium (OGC), to enhance system interoperability and user-friendliness. To ensure high-quality service management in development and maintenance, the FitSM approach, a lightweight framework for effective IT service management, is applied. The IMPC User Help Desk (impc-uhd@dlr.de) is available to address technical, functional, or thematic issues and requests from registered users, project partners, and anyone interested in space weather. Additionally, the platform's services are continuously monitored by an automated system to ensure reliability and performance required by safety-critical applications.

2.2 Observational Networks, Products and Services

In order to provide services of interest to the very complex end user domain it is essential to assess the space weather situation in real time. This section provides a brief overview of selected space weather observation techniques utilized by the IMPC, showcasing its unique observational capabilities throughout different space weather domains and characteristics.

Multi-instrument Ionospheric Radio observation Array (MIRA) DLR-SO established MIRA, an antenna measurement and test field for various passive reception measurement techniques ranging from 10 kHz to 2 GHz. This easily accessible and configurable, yet undisturbed site serves two purposes: we can check antenna and measurement equipment before shipping to international partners, and it is used for long-term measurement of ionospheric and solar disturbances. MIRA also facilitates the development of various radio reception techniques and hosts multiple receivers for GNSS, Beacon, and VLF signals, as well as solar spectrometers, supporting various DLR-SO projects. The field's size allows for the installation and testing of additional receivers upon request from cooperation partners.

GNSS Realtime Data Processing System Well-established ground and space based GNSS measurements offer a unique opportunity to continuously monitor the electron density and structure of the ionosphere-plasmasphere system. IMPC processes hundreds of real-time GNSS stations in real time, providing detailed measurements for GPS, Galileo, and GLONASS and other GNSS across multiple frequencies with 1 Hz cadence. These data are pre-processed continuously to mitigate outliers, to correct cycle slips and to reconstruct the geometry of each observed GNSS connection. Afterwards the total electron content (TEC) is estimated and distributed to other near real-time processing applications, supporting various scientific tools and algorithms. Leveraging long-term expertise in ionospheric modeling [2], IMPC develops near real-time products like Total Electron Content (TEC) maps, Rate of Change of TEC index (ROTI), and normalized information on ionospheric disturbances [3] as shown in Fig. 2. Delivered with high cadence and low latency, these products significantly support internal and global research activities in space weather and related applications.

Monitoring and Modelling of Ionospheric Scintillations Especially at very high or low latitudes, small-scale ionospheric disturbances can cause severe radio scintillations of signals transmitted from global navigation satellite systems (GNSS). These plasma irregularities can significantly degrade GNSS performance, posing challenges for safety-critical applications. IMPC's capabilities to model and forecast these disturbances are important, as simply upgrading GNSS equipment is insufficient to mitigate scintillation effects. IMPC operates the Experimentation and Verification Network (EVNet) of high-rate, multi-frequency GNSS receivers across polar to equatorial regions, collecting data to analyze space weather events and study ionospheric scintillations [4, 5]. By applying data assimilation, these datasets significantly enhance IMPC's modeling capabilities based on the Global Ionospheric Scintillation Model (GISM), which includes numeric simulations with multiple phase screens (1D and 2D), utilization of the NeQuick or

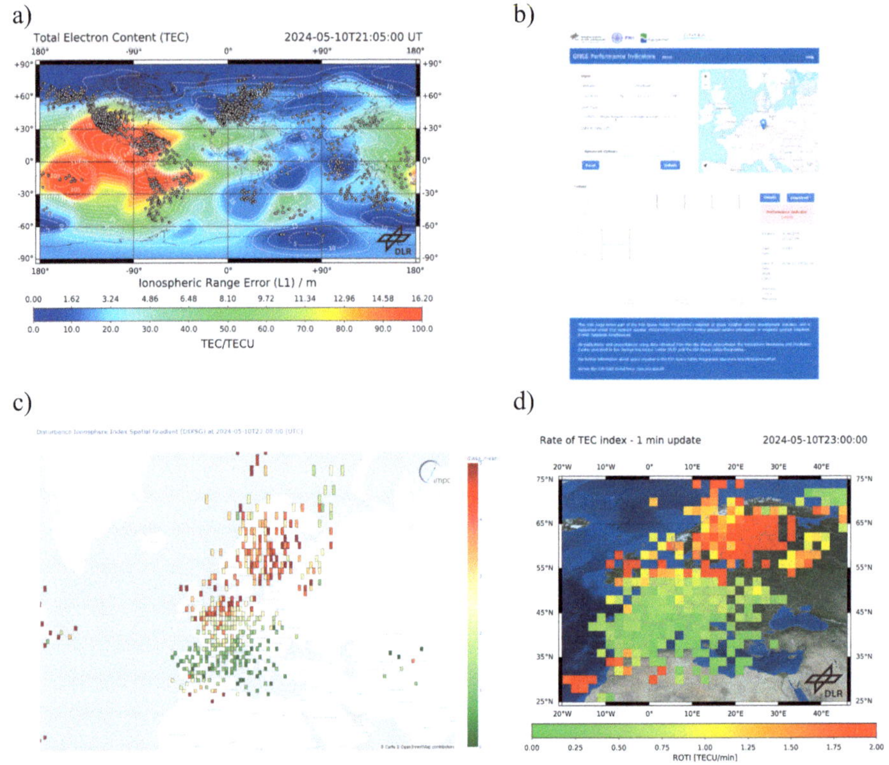

Fig. 2. Response on the severe space weather activity during 10–12 March 2024 observed by several low latency IMPC products provided by IMPC in near-real-time to the end users: TEC maps based on 1 Hz real-time GNSS observations and the Neustrelitz TEC model (NTCM) (a), the response of the GNSS performance indicators application utilized to provide simple indicators over Germany (b), real-time map of the disturbance index (DIX-SG) (c) and maps of the rate of TEC index (ROTI) with an update rate of one minute to real time (d).

Neustrelitz Electron Density Model (NEDM) models of ambient ionospheres, refractional ray bending, and synthetic scintillation time series simulation and extraction [6, 7].

CALLISTO for Solar Radio Burst Monitoring Sudden eruptions of high-energy particles and electromagnetic radiation from the Sun may disrupt GNSS and high-frequency communication systems. The global e-Callisto network, with around 70 stations worldwide, monitors these bursts as part of the ISWI activities. DLR contributes to this network and is developing and operating three types of receivers across different spectral ranges. Each CALLISTO receiver (Compound Astronomical Low frequency Low cost Instrument for Spectroscopy and Transportable Observatory) includes an antenna, preamplifier, heterodyne receiver, and computer, recording frequencies from 45 to 870 MHz, adjustable to other ranges. DLR's dual-band receivers can cover 10 to 1,600 MHz, with antennas following the Sun's position for best measurements. We are currently preparing the automatic data integration into the IMPC for further processing and interactive visualization.

Global Ionosphere Flare Detection System (GIFDS) The ISWI instrument GIFDS has been developed by DLR-SO to monitor sudden ionospheric disturbances (SIDs) in the D-layer ionosphere caused by solar flares [8]. As severe solar flares can have a strong impact on communications and navigation signals, users of such systems can be informed of potential performance degradation or malfunctioning. The major advantage is that the system is ground-based, simplifying hardware maintenance compared to x-ray detectors on satellites. Moreover, the actual influence on Earth is measured instead of the source strength in space, which suits better to ground technology. The integration into IMPC for user-friendly data provision and search are in active development, as is extending the models behind.

Reception and Processing of Solar Wind Parameters Solar wind information is crucial for evaluating and predicting solar storms and their impacts on Earth, making it extremely important for the IMPC services. DLR is the only European member of the Real Time Solar Wind (RTSW) observation network and plays a key role in the data transfer and analysis of information from NASA's Advanced Composition Explorer (ACE) and the Deep Space Climate Observatory (DSCOVR). Within the IMPC framework, the received solar wind data is analyzed in real-time to generate space weather warnings and forecasts.

GNSS Performance Indicators (GPI) The GPI application utilizes products available within the SWE Service Network (TEC, ROTI, scintillation indices, Kp index), subsequently processing them to provide the end user with an indication of positioning uncertainty caused by space weather impact at their location. The application interface was designed to allow the user to indicate a given geographic position in order to receive an estimate of the performance in positioning services they can expect based on current space weather conditions and GNSS satellites in view. These results are retrievable via a web interface designed by be easy to interpret by end users and through an API enabling machine to machine communication [9, 10].

2.3 Contribution to International Networks and Outreach

International Space Weather Initiative (ISWI) There are strong connections between ISWI and DLR-SO, and we are coordinating their actions in Germany. With CALLISTO and GIFDS, DLR-SO is constructing and further developing two acknowledged ISWI instruments. Both systems are expanding in terms of the number of stations as well as their functionality and presentation.

ESA Space Weather Service Network Since several years DLR is coordinating the Expert Service Center Ionospheric Weather (I-ESC) within the Space Weather Service Network developed in the ESA Space Safety Programme (S2P). The I-ESC provides, implements and supports the Ionospheric and upper Atmosphere Weather products and capabilities of the network, including the observation, monitoring, interpretation, modelling and forecasting of ionospheric and upper Atmosphere weather conditions. Within this activity, the IMPC plays an important role by continuously providing products of ionospheric key parameters e.g. nowcast and forecasts of TEC and ROTI maps, scintillation indices accessible via a Web Feature Service (WFS), MUF alarms for skip distances of 750 km, slab thickness, TEC gradients via TechTIDE as well as developing and operating interactive web services e.g. to provide GNSS performance indicators [11].

PECASUS – Global SWX Center for ICAO The PECASUS consortium is one of the four global centers providing space weather advisories according to ICAO regulations. These advisories are sent to airliners using the existing aeronautical fixed network for international aviation (AFS). The three other services providing advisories are the Space Weather Prediction Center of NOAA, the ACFJ consortium formed by Australia, Canada, France and Japan and the CRC consortium formed by China and Russia. DLR leads the GNSS domain (DLR, INGV, SRC, STCE, FMI) and continuously delivers near real time products e.g. maps of the total electron content and rate of change of TEC (ROTI) as well as amplitude and phase scintillation indices (S4 and σ_φ) in near real-time with five minutes cadence [12, 13].

International Space Environment Service (ISES) In April 2023, the DLR Institute for Solar-Terrestrial Physics joined ISES as a Collaborative Expert Center (CEC) [14]. This partnership aims to provide real-time forecasting and monitoring of space weather to mitigate risks to technology, critical infrastructure, and human activities. Within this frame, IMPC will facilitate international communication and coordination during periods of increased space weather activity and extreme events. Additionally, IMPC will enhance its space weather services, promoting understanding among users, researchers, the media, and the public.

PITHIA-NRF – Building European Research Infrastructure PITHIA-NRF aims to establish a European distributed network integrating research infrastructures focused on the ionosphere, thermosphere, and plasmasphere, providing access to facilities like EISCAT, LOFAR, and Ionosondes. IMPC is involved in developing the eScience Center, providing access to high-rate GNSS data and a web service for NEDM runs. The project advances upper atmosphere research by integrating satellite data and prediction models and supports technology development and standardization. Through the Trans-National Access programme, PITHIA-NRF facilitates access for external research teams to its nodes such as the IMPC activities provided through the DLR-SO node [15, 16].

3 Conclusion

In conclusion, the IMPC plays a central role in addressing the challenges posed by space weather to modern technologies. Its advanced monitoring and prediction capabilities, supported by integration into global networks, ensure stakeholders across various sectors receive accurate and timely space weather information. This enables the mitigation of space weather impacts, safeguarding critical systems and enhancing the reliability of services that underpin modern society. The joint efforts of the IMPC framework represents a significant advance in applied space weather research and its practical applications and emphasize the importance of further investment in this area.

Acknowledgements. The cooperation with the German Federal Agency for Cartography and Geodesy (BKG), International GNSS Service (IGS), European Permanent Reference Network (EUREF), Agenzia Spaciale Italiana (ASI), South African's TrigNet, UNAVCO, NOAA Space Weather Prediction Center (SWPC), World Data Center for Geomagnetism (WDC-2) Kyoto and the consortium of the EU FP7 project AFFECTS (grant agreement no. 263506, www.affects-fp7.eu) is gratefully acknowledged. This material is based on services provided by the NSF GAGE Facility, operated by EarthScope Consortium, with support from the National Science Foundation, the National Aeronautics and Space Administration, and the U.S. Geological Survey under NSF Cooperative Agreement EAR-1724794. EVNet is operated by DLR KN-NAV and DLR SO. We thank all institutions that host EVNet equipment: Bahir Dar University, JAXA, Swedish Institute of Space Physics in Kiruna, Stanford University, University of Tromso, Universidad de La Laguna, Hartebeesthoek Radio Astronomy Observatory, University of Toulouse, EISCAT Ramfjordbotn, University Fortaleza (UFC).

References

1. IMPC Homepage. https://impc.dlr.de. last accessed 01 July 2024
2. EUSPA Homepage. https://www.gsc-europa.eu/news/good-performance-less-computation-a-new-ionospheric-model-for-the-galileo-open-service. last accessed 01 July 2024
3. Wilken, V., Kriegel, M., Jakowski, N., Berdermann, J.: An ionospheric index suitable for estimating the degree of ionospheric perturbations. J. Space Weather Space Clim. (2018). https://doi.org/10.1051/swsc/2018008
4. Kriegel, M. et al.: An experimentation and verification network to analyze small scale ionospheric disturbances. In: VII Brazilian Symposium on Space Geophysics and Aeronomy, Santa Maria (2018)
5. Kriegel, M., Jakowski, N., Berdermann, J., Sato, H., Wassaie Merhsa, M.: Scintillation measurements at Bahir Dar during the high solar activity phase of solar cycle 24. Ann. Geophys. **35**, 97–106 (2017). https://doi.org/10.5194/angeo-35-97-2017
6. Beniguel, Y.: Ionospheric scintillations: indices and modeling. Radio Sci. **54**, 618–632 (2017)
7. Vasylyev, D., Beniguel, Y., Wilken, V., Kriegel, M., Berdermann, J.: Modeling of ionospheric scintillation. J. Space Weather Space Clim. **12**, 22 (2022)
8. Wenzel, D., Jakowski, N., Berdermann, J., Mayer, C., Valladares, C., Heber, B.: Global ionospheric flare detection system (GIFDS). J. Atmos. Sol. Terr. Phys. **138**, 233–242 (2016). https://doi.org/10.1016/j.jastp.2015.12.011
9. David, P., et al.: Performance indicator development addressing mitigation of the space weather impacts on GNSS. J. Space Saf. Eng. **10**(3), 324–330 (2023). https://doi.org/10.1016/j.jsse.2023.07.004
10. Kriegel, M., et al.: A performance indicator application to rate the space weather impact on GNSS positioning. Eur. J. Navig. **23**(1), 24–33 (2023)
11. ESA I-ESC Homepage. https://swe.ssa.esa.int/de/ionospheric-weather. last accessed 01 July 2024
12. PECASUS Homepage. https://pecasus.eu/. last accessed 01 July 2024
13. Kauristie, K., et al.: Space weather services for civil aviation-challenges and solutions. Remote Sens. **13**(18), 3685 (2021). https://doi.org/10.3390/rs13183685
14. ISES Homepage. http://www.spaceweather.org/ISES/rwc/rwc_de.html. last accessed 01 July 2024
15. PITHIA-NRF Homepage. https://pithia-nrf.eu/pithia-nrf-users/tna/nodes/dlr-so-node-1. last accessed 01 July 2024

16. PITHIA-NRF Success Stories ASPIS. https://pithia-nrf.eu/pdf/pithia-nrf_tna-success-story_aspis.pdf. last accessed 01 July 2024

Open Access This chapter is licensed under the terms of the Creative Commons Attribution 4.0 International License (http://creativecommons.org/licenses/by/4.0/), which permits use, sharing, adaptation, distribution and reproduction in any medium or format, as long as you give appropriate credit to the original author(s) and the source, provide a link to the Creative Commons license and indicate if changes were made.

The images or other third party material in this chapter are included in the chapter's Creative Commons license, unless indicated otherwise in a credit line to the material. If material is not included in the chapter's Creative Commons license and your intended use is not permitted by statutory regulation or exceeds the permitted use, you will need to obtain permission directly from the copyright holder.

Author Index

© The Editor(s) (if applicable) and The Author(s) 2026
N. Gopalswamy et al. (Eds.): ISWI 2024, SPPHY 431, pp. 303–304, 2026.
https://doi.org/10.1007/978-981-95-1121-1

MIX
Papier aus verantwortungsvollen Quellen
Paper from responsible sources
FSC® C105338

If you have any concerns about our products,
you can contact us on
ProductSafety@springernature.com

In case Publisher is established outside the EU,
the EU authorized representative is:
Springer Nature Customer Service Center GmbH
Europaplatz 3, 69115 Heidelberg, Germany

Printed by Libri Plureos GmbH
in Hamburg, Germany